Heat Treatment of Steels

Heat Treatment of Steels

Editors

Claudio Testani
Andrea Di Schino

MDPI • Basel • Beijing • Wuhan • Barcelona • Belgrade • Manchester • Tokyo • Cluj • Tianjin

Editors
Claudio Testani
CALEF
CALEF C/O ENEA Casaccia
Rome
Italy

Andrea Di Schino
Engineering
University of Perugia
Perugia
Italy

Editorial Office
MDPI
St. Alban-Anlage 66
4052 Basel, Switzerland

This is a reprint of articles from the Special Issue published online in the open access journal *Metals* (ISSN 2075-4701) (available at: www.mdpi.com/journal/metals/special_issues/heat_treatment_steels).

For citation purposes, cite each article independently as indicated on the article page online and as indicated below:

LastName, A.A.; LastName, B.B.; LastName, C.C. Article Title. *Journal Name* **Year**, *Volume Number*, Page Range.

ISBN 978-3-0365-2389-7 (Hbk)
ISBN 978-3-0365-2388-0 (PDF)

© 2021 by the authors. Articles in this book are Open Access and distributed under the Creative Commons Attribution (CC BY) license, which allows users to download, copy and build upon published articles, as long as the author and publisher are properly credited, which ensures maximum dissemination and a wider impact of our publications.

The book as a whole is distributed by MDPI under the terms and conditions of the Creative Commons license CC BY-NC-ND.

Contents

About the Editors vii

Preface to "Heat Treatment of Steels" ix

Andrea Di Schino and Claudio Testani
Heat Treatment of Steels
Reprinted from: *Metals* **2021**, *11*, 1168, doi:10.3390/met11081168 1

Marcel Carpio, Jessica Calvo, Omar García, Juan Pablo Pedraza and José María Cabrera
Heat Treatment Design for a QP Steel: Effect of Partitioning Temperature
Reprinted from: *Metals* **2021**, *11*, 1136, doi:10.3390/met11071136 5

Han Jiang, Yanlin He, Li Lin, Rendong Liu, Yu Zhang, Weisen Zheng and Lin Li
Microstructures and Properties of Auto-Tempering Ultra-High Strength Automotive Steel under Different Thermal-Processing Conditions
Reprinted from: *Metals* **2021**, *11*, 1121, doi:10.3390/met11071121 21

Eliseo Hernandez-Duran, Luca Corallo, Tanya Ros-Yanez, Felipe Castro-Cerda and Roumen H. Petrov
The Effect of Different Annealing Strategies on the Microstructure Development and Mechanical Response of Austempered Steels
Reprinted from: *Metals* **2021**, *11*, 1041, doi:10.3390/met11071041 39

Manuel de J. Barrena-Rodríguez, Francisco A. Acosta-González and María M. Téllez-Rosas
A Review of the Boiling Curve with Reference to Steel Quenching
Reprinted from: *Metals* **2021**, *11*, 974, doi:10.3390/met11060974 59

Huasong Liu, Yannan Dong, Hongguang Zheng, Xiangchun Liu, Peng Lan, Haiyan Tang and Jiaquan Zhang
Precipitation Criterion for Inhibiting Austenite Grain Coarsening during Carburization of Al-Containing 20Cr Gear Steels
Reprinted from: *Metals* **2021**, *11*, 504, doi:10.3390/met11030504 77

Michio Shimotomai
Heuristic Design of Advanced Martensitic Steels That Are Highly Resistant to Hydrogen Embrittlement by -Carbide
Reprinted from: *Metals* **2021**, *11*, 370, doi:10.3390/met11020370 95

Idurre Kaltzakorta, Teresa Gutierrez, Roberto Elvira, Pello Jimbert and Teresa Guraya
Evolution of Microstructure during Isothermal Treatments of a Duplex-Austenitic 0.66C11.4Mn.9.9Al Low-Density Forging Steel and Effect on the Mechanical Properties
Reprinted from: *Metals* **2021**, *11*, 214, doi:10.3390/met11020214 103

Jin Young Jung, Kang Suk An, Pyeong Yeol Park and Won Jong Nam
Correlation between Microstructures and Ductility Parameters of Cold Drawn Hyper-Eutectoid Steel Wires with Different Drawing Strains and Post-Deformation Annealing Conditions
Reprinted from: *Metals* **2021**, *11*, 178, doi:10.3390/met11020178 117

Mattia Franceschi, Luca Pezzato, Claudio Gennari, Alberto Fabrizi, Marina Polyakova, Dmitry Konstantinov, Katya Brunelli and Manuele Dabalà
Effect of Intercritical Annealing and Austempering on the Microstructure and Mechanical Properties of a High Silicon Manganese Steel
Reprinted from: *Metals* **2020**, *10*, 1448, doi:10.3390/met10111448 . . . . **133**

Andrea Di Schino, Matteo Gaggiotti and Claudio Testani
Heat Treatment Effect on Microstructure Evolution in a 7
Reprinted from: *Metals* **2020**, *10*, 808, doi:10.3390/met10060808 . . . . **153**

Mengjia Hu, Kejian Li, Shanlin Li, Zhipeng Cai and Jiluan Pan
Analytical Model to Compare and Select Creep Constitutive Equation for Stress Relief Investigation during Heat Treatment in Ferritic Welded Structure
Reprinted from: *Metals* **2020**, *10*, 688, doi:10.3390/met10050688 . . . . **167**

Chiara Soffritti, Annalisa Fortini, Ramona Sola, Elettra Fabbri, Mattia Merlin and Gian Luca Garagnani
Influence of Vacuum Heat Treatments on Microstructure and Mechanical Properties of M35 High Speed Steel
Reprinted from: *Metals* **2020**, *10*, 643, doi:10.3390/met10050643 . . . . **185**

Andrea Di Schino and Claudio Testani
Corrosion Behavior and Mechanical Properties of AISI 316 Stainless Steel Clad Q235 Plate
Reprinted from: *Metals* **2020**, *10*, 552, doi:10.3390/met10040552 . . . . **199**

Stefanie Hoja, Matthias Steinbacher and Hans-Werner Zoch
Compound Layer Design for Deep Nitrided Gearings
Reprinted from: *Metals* **2020**, *10*, 455, doi:10.3390/met10040455 . . . . **213**

About the Editors

Claudio Testani

Dr. Claudio Testani graduated in Aerospace Structural Engineering in 1987, University of Rome, La Sapienza; then, in 2006, he defended his PhD thesis in Material Engineering, Università di Roma "Tor-Vergata". He has long-standing experience in teaching and is a Member of the Teaching Board of the PhD school of TorVergata University Rome, Italy.

He is a qualified engineer and has worked for about 30 years in research centres for advanced structural materials development, applied metallurgy, industrial research and management with eight patents, producing more than 100 scientific publications. He attained a habilitation to become an associate professor and has worked as an R&D manager for an italian industry before coming back to research activities. He is also the technical director of CALEF, Italian R&D research Consortium of Universities and Industries.

He has received awards for his achievements, including: Top Peer Reviewer 2019 for "Material Science" and Top Peer Reviewer 2019 for "Cross-Field" on the Publons global reviewer database.

Andrea Di Schino

Prof. Andrea Di Schino completed his degree in Physics at the University of Pisa and PhD in Materials Engineering at the University of Napoli. He then joined Centro Sviluppo Materiali SpA (CSM), a european leading centre for materials research. After spending more than 10 years in CSM, he joined the University of Perugia as a Professor of Metallurgy. His main interests lie in combining fundamental theoretical research and experiments to explain mechanisms and discovering new insights in the field of steel metallurgy. He is the author of more than 200 papers relating to physical metallurgy and product development and is member of the Editorial Board of several journals. As of 2021, he is listed in the Top Italian Scientists (TIS) List.

Preface to "Heat Treatment of Steels"

This book is a compilation of recent articles presenting developments in the processing, innovative heat treatments, corrosion and mechanical properties characterisation of steels, including ultra-high strength steel families.

The subjects are multidisciplinary and embrace interdisciplinary work covering physical metallurgy and processes, reporting on experimental and theoretical progress concerning microstructural evolution during the heat treatment of steels, including a review concerning heat-transfer phenomena when quenching metals at high temperatures.

This book can provide a source for data and presents information on the emerging processes and thermal treatments of steels that have been proposed and discussed, with interesting results on complex heat treatments and precipitation phenomena.

Some of the results can be a base for an effective application in the industrial application of automotive and high-stress structures, but obviously there are still many challenges to overcome.

As Guest Editors of this Special Issue, we really hope that all scientific results in this Special Issue can contribute to the advancement of the future research of steels and find application in real industrial environments and markets.

We would like to warmly thank the authors of the research articles in this Special Issue for their contributions and all of the reviewers for their efforts in ensuring high-quality publications.

Finally, thanks to the editors of *Metals* and to Kinsee Guo, the Assistant Editors and assistants for their valuable and continuous support during the preparation of this volume.

Claudio Testani, Andrea Di Schino
Editors

metals

Editorial

Heat Treatment of Steels

Andrea Di Schino [1,*] and Claudio Testani [2]

1 Dipartimento di Ingegneria, Università degli Studi di Perugia, Via G. Duranti 93, 06125 Perugia, Italy
2 CALEF-ENEA CR Casaccia, Via Anguillarese 301, Santa Maria di Galeria, 00123 Rome, Italy; claudio.testani@consorziocalef.it
* Correspondence: andrea.dischino@unipg.it

1. Introduction and Scope

Steels represent an interesting family of materials, both from a scientific and commercial point of view, considering the many innovative applications they can be used for [1]. It is therefore essential to understand the relations between properties and microstructure and how to drive them via a specific process. Despite their wide use as a consolidated material, many research fields remain active in finding new applications for steels. Particularly within this framework, the role of heat treatments in obtaining even complex microstructures is still a relatively open matter, thanks to the design of these innovative heat treatments [2–6].

The Special Issue scope embraces interdisciplinary work covering physical metallurgy and processes, reporting on experimental and theoretical progress concerning microstructural evolution during the heat treatments of steels.

2. Contributions

The volume collects contributions from academic and industrial researchers with stimulating new ideas and original results. The present volume consists of 14 research papers and two review papers. Hoja et al. [7] report on deep nitriding as a tool to obtain a nitriding hardness depth beyond 0.6 mm. They show how long-nitriding processes, which are necessary to reach high-nitriding hardness depths, mostly have a negative influence on the hardness and strength of the nitrided layer of steel, as well as on the bulk material. It is shown that thick and compact compound layers have the potential for a high-flank load capacity of gears. The investigations focus on the simultaneous formation of a high-nitriding depth and a thick and compact compound layer. Beside the preservation of its strength, a challenge is to control the porosity of the compound layer, which should be as low as possible. The investigations were carried out using the common nitriding and heat-treatable mild steel, 31CrMoV9, which is often used for gear applications. The article gives an insight on the development of multistage nitriding processes, studied by short- and long-term experiments, which aim for a specific compound-layer build-up with low porosity and the high strength of the nitride layer and core material.

Di Schino et al. [8] report on the corrosion behavior and the mechanical properties of a Q235 plate clad by AISI 316 stainless steel. The reported cladding process is performed by the submerged-arc welding overlay process. Due to element diffusion (Fe, Cr, Ni, and Mn), a 1.5 mm wide diffusion layer is formed between the stainless steel and carbon steel interface of the cladded plate affecting corrosion resistance. Pitting resistance and intergranular corrosion are both evaluated.

Soffritti et al. [9] report on the influence of vacuum heat treatments on the microstructure and mechanical properties of M35 high-speed steel. Their study investigates the influence of vacuum heat treatments, at different pressures of quenching gas, on the microstructure and mechanical properties of taps made of M35 high-speed steel.

Hu et al. [10] report on an analytical model for comparing and selecting a creep constitutive equation for a stress-relief investigation into ferritic welded structure during heat treatment. A one-dimensional analytical model is proposed to help select the creep constitutive equation and predict the temperature of heat treatment in a ferritic welded structure, while neglecting the impact of structural constraint and deformation compatibility. The analytical solutions were compared with simulation results, which were validated with experimental measurements in a ferritic welded rotor. The residual stresses from welding and post-welding heat treatment on the inner and outer cylindrical surfaces were measured with the hole-drilling method for validation.

Di Schino et al. [11] describe the heat treatment effect on microstructure evolution in 7% Cr steel for forging. In their paper, different heat treatments are considered and their effect on a 7% Cr steel for forging is reported. Results show that, following the high-intrinsic steel hardenability, significant differences were not found in comparison to the cooling-step treatment, although prior austenite grain size was significantly different. Moreover, retained austenite content was lower in double-tempered specimens after heat treatments at higher temperatures.

Franceschi et al. [12] describe the effect of inter-critical annealing and austempering on the microstructure and mechanical properties of high-silicon manganese steel. In their study, they report on the achievement of a multiphase ferritic-martensitic microstructure characterized by a hardness of 426 HV and a tensile strength of 1650 MPa, together with an elongation of 4.5%. They compare their results with those values obtained with annealing and Q&T treatments.

Jung et al. [13] report on the relationship between microstructures and ductility parameters, including the reduction in area, elongation to failure, occurrence of delamination, and the number of turns until failure in torsion in hypereutectoid pearlitic steel wires.

Kaltzakorta et al. [14] describe the results obtained for a 0.66C11.4Mn9.9Al duplex austenitic low-density steel after applying a set of isothermal treatments at different combinations of time and temperature, aiming to promote kappa-carbide precipitation and improve the mechanical properties obtained with a water-quenching treatment. The effects of the different isothermal treatments on the microstructure and mechanical properties of the steel were analyzed.

Shimotomai [15] reports on advanced, tempered, martensitic steels. He shows how their mechanical strength is characterized by fine sub-grain structures with a high density of free dislocations and metallic carbides and/or nitrides. This paper is a preliminary experimental survey of the hydrogen absorption and hydrogen embrittlement of a tempered martensitic steel with ε-carbide precipitates. Shimotomai suggests that the proper use of carbides in steels can promote a high resistance to hydrogen embrittlement.

Liu et al. [16] report on the precipitation criterion for inhibiting austenite grain coarsening during carburization of 20Cr gear steels containing Al. In their paper, the quantitative influence of Al and N on grain size after carburization is studied through experiments based on 20Cr steel. According to the grain-structure features and kinetic theory, the abnormal grain growth is demonstrated as the mode of austenite grain coarsening in carburization.

Hernandez-Duran et al. [17] report on the effect of different annealing strategies on the microstructure development and the mechanical response of bainitic steels. Their study focuses on the effect of non-conventional annealing strategies on the microstructure and mechanical properties of austempered steels. Multistep thermo-cycling and ultrafast heating annealing were carried out and compared with the outcome obtained from a conventionally annealed 0.3C-2Mn-1.5Si steel.

Acosta-González et al. [18] contribute to the volume with a review concerning heat-transfer phenomena when quenching metals from high temperatures. The purpose of their work is to acquaint the field engineer, the graduate student, or any research professional with the current knowledge on heat-transfer phenomena during metal quenching and the methods used for their interpretation and application.

Jiang et al. [19] report on automotive steels with ultra-high strength and low alloy content under different heating and cooling processes. It was shown that those processes exhibited great influence on the performance of the investigated steels due to the different auto-tempering effects.

Carpio et al. [20] report on the effect of Q&P heat treatment on a new family of steels for the automotive industry.

Acknowledgments: As Guest Editors, we would like to especially thank Kinsee Guo, Assistant Editor, for his support and his active role in the publication. We are also grateful to the entire staff of Metals Editorial Office for this precious collaboration. Last but not least, we express our gratitude to all the contributing authors and reviewers: without your excellent work it would not have been possible to accomplish this Special Issue that we hope will be a piece of interesting reading and reference literature.

Conflicts of Interest: The authors declare no conflict of interest.

References

1. Di Schino, A. Manufacturing and application of stainless steels. *Metals* **2020**, *10*, 327–329. [CrossRef]
2. Di Schino, A.; Alleva, L.; Guagnelli, M. Microstructure evolution during quenching and tempering of martensite in a medium C steel. *Mater. Sci. Forum* **2012**, *715–716*, 860–865. [CrossRef]
3. Di Schino, A.; Di Nunzio, P.E. Metallurgical aspects related to contact fatigue phenomena in steels for back up rolling. *Acta Metall. Slovaca* **2017**, *23*, 62–71. [CrossRef]
4. Di Schino, A. Analysis of phase transformation in high strength low alloyed steels. *Metalurgija* **2017**, *56*, 349–352.
5. Di Schino, A.; Di Nunzio, P.E.; Turconi, G.L. Microstructure evolution during tempering of martensite in a medium C steel. *Mater. Sci. Forum* **2007**, *558–559*, 1435–1441. [CrossRef]
6. Mancini, S.; Langellotto, L.; Di Nunzio, P.E.; Zitelli, C.; Di Schino, A. Defect reduction and quality optimization by modeling plastic deformation and metallurgical evolution in ferritic stainless steels. *Metals* **2020**, *10*, 186. [CrossRef]
7. Hoja, S.; Steinbacher, M.; Zoch, H.W. Compound layer design for deep nitride gearings. *Metals* **2020**, *10*, 455. [CrossRef]
8. Di Schino, A.; Testani, C. Corrosion behavior and mechanical properties of AISI 316 stainless steel clad Q235 plate. *Metals* **2020**, *10*, 552. [CrossRef]
9. Soffritti, C.; Fortini, A.; Sola, R.; Fabbri, E.; Merlin, M.; Garagnani, G.L. Influence of vacuum heat treatments on microstructure and mechanical properties of M35 high speed steel. *Metals* **2020**, *10*, 643. [CrossRef]
10. Hu, M.; Li, K.; Li, S.; Cai, Z.; Pan, J. Analytical model to compare and select creep constitutive equation for stress relief investigation during heat treatment in ferritic welded structure. *Metals* **2020**, *10*, 688. [CrossRef]
11. Di Schino, A.; Gaggiotti, M.; Testani, C. Heat treatment effect on microstructure evolution in 7% Cr steel for forging. *Metals* **2020**, *10*, 808. [CrossRef]
12. Franceschi, M.; Pezzato, L.; Gennari, C.; Fabrizi, A.; Polyakova, M.; Konstantinov, D.; Brunelli, K.; Dabalà, M. Effect of intercritical annealing and austempering on the microstructure and mechanical properties of high silicon manganese steel. *Metals* **2020**, *10*, 1448. [CrossRef]
13. Jung, J.Y.; An, K.S.; Park, P.Y.; Nam, W.Y. Correlation between Microstructures and Ductility Parameters of Cold Drawn Hyper-Eutectoid Steel wires with different drawing strains and post-deformation annealing conditions. *Metals* **2021**, *11*, 178. [CrossRef]
14. Kaltzakorta, I.; Gutierrez, T.; Elvira, R.; Jimbert, P.; Guraya, T. Evolution of Microstructure during Isothermal Treatments of a Duplex-Austenitic 0.66C11.4Mn.9.9Al Low-Density Forging Steel and Effect on the Mechanical Properties. *Metals* **2021**, *11*, 214. [CrossRef]
15. Shimotomai, M. Heuristic Design of advanced martensitic steels that are highly resistant to hydrogen embrittlement by ε-carbide. *Metals* **2021**, *11*, 370. [CrossRef]
16. Liu, H.; Dong, Y.; Zheng, H.; Liu, X.; Lan, P.; Tang, H.; Zhang, J. Precipitation criterion for inhibiting austenite grain coarsening during carburization of Al-containing 20Cr gear steels. *Metals* **2021**, *11*, 504. [CrossRef]
17. Hernandez-Duran, E.; Corallo, L.; Ros-Yanez, T.; Castro-Cerda, F.; Petrov, R.H. The effect of different annealing strategies on the microstructure development and the mechanical response of bainitic steels. *Metals* **2021**, *11*, 1041. [CrossRef]
18. Acosta-González, F.A.; Barrena-Rodriguez, M.; Tellez-Rosas, M.M. A review of the boiling curve with reference to steel quenching. *Metals* **2021**, *11*, 974.
19. Jiang, H.; He, Y.; Lin, L.; Liu, R.; Zhang, Y.; Zheng, W.; Li, L. Microstructures and properties of auto-tempering ultra-high strength automotive steels under different thermo-processing. *Metals* **2021**, *11*, 1121. [CrossRef]
20. Carpio, M.; Calvo, J.; Garcia, O.; Pedraza, J.P.; Cabrera, J.M. Heat Treatment Design for a QP Steel: Effect of Partitioning Temperature. *Metals* **2021**, *11*, 1136. [CrossRef]

Article

Heat Treatment Design for a QP Steel: Effect of Partitioning Temperature

Marcel Carpio [1,*], Jessica Calvo [1], Omar García [2], Juan Pablo Pedraza [2] and José María Cabrera [1]

1. Departament de Ciència i Enginyeria de Materials (CEM), EEBE, Campus Diagonal Besòs, Univesitat Politècnica de Catalunya, Av. Eduard Maristany 10-14, 08019 Barcelona, Spain; jessica.calvo@upc.edu (J.C.); jose.maria.cabrera@upc.edu (J.M.C.)
2. TERNIUM México, Av. Universidad 992 Nte., Col. Cuauhtémoc, San Nicolás de los Garza 66450, Mexico; ogarciar@ternium.com.mx (O.G.); jpedraza@ternium.com.mx (J.P.P.)
* Correspondence: marcel.francisco.carpio@upc.edu; Tel.: +34-934037218

Abstract: Designing a new family of advanced high-strength steels (AHSSs) to develop automotive parts that cover early industry needs is the aim of many investigations. One of the candidates in the 3rd family of AHSS are the quenching and partitioning (QP) steels. These steels display an excellent relationship between strength and formability, making them able to fulfill the requirements of safety, while reducing automobile weight to enhance the performance during service. The main attribute of QP steels is the TRIP effect that retained austenite possesses, which allows a significant energy absorption during deformation. The present study is focused on evaluating some process parameters, especially the partitioning temperature, in the microstructures and mechanical properties attained during a QP process. An experimental steel (0.2C-3.5Mn-1.5Si (wt%)) was selected and heated according to the theoretical optimum quenching temperature. For this purpose, heat treatments in a quenching dilatometry and further microstructural and mechanical characterization were carried out by SEM, XRD, EBSD, and hardness and tensile tests, respectively. The samples showed a significant increment in the retained austenite at an increasing partitioning temperature, but with strong penalization on the final ductility due to the large amount of fresh martensite obtained as well.

Keywords: QP; retained austenite; low carbon steel

Citation: Carpio, M.; Calvo, J.; García, O.; Pedraza, J.P.; Cabrera, J.M. Heat Treatment Design for a QP Steel: Effect of Partitioning Temperature. *Metals* **2021**, *11*, 1136. https://doi.org/10.3390/met11071136

Academic Editor: Ulrich Prahl

Received: 3 June 2021
Accepted: 13 July 2021
Published: 19 July 2021

Publisher's Note: MDPI stays neutral with regard to jurisdictional claims in published maps and institutional affiliations.

Copyright: © 2021 by the authors. Licensee MDPI, Basel, Switzerland. This article is an open access article distributed under the terms and conditions of the Creative Commons Attribution (CC BY) license (https://creativecommons.org/licenses/by/4.0/).

1. Introduction

In order to improve automobile performance, the automotive industry is continuously developing new steels aiming at reducing the automobile's weight while enhancing the passenger's safety. Accordingly, new families of advanced high-strength steels (AHSSs) are constantly being proposed and analyzed. Very promising candidates are the so-called quenching and partitioning (QP) steels because they exhibit a good combination of strength and formability [1]. These steels are based on the QP process which was first proposed by Speer et al. [2,3]. They were developed to create thin steel sheets with different fractions of martensite and retained austenite. Basically, the process implies a partial or full austenitization treatment followed by a quenching step between the martensite start (M_s) and martensite finish (M_f) temperatures to control the fractions of untransformed austenite and martensite. The martensitic transformation is diffusionless, so the martensite has the same composition as its parent austenite. Subsequently, a sort of annealing (usually called partitioning) treatment is applied where the carbon in the supersaturated martensite starts to diffuse into the untransformed austenite. This step enhances carbon enrichment in the austenite which, in turn, might be stabilized at room temperature. In the subsequent quenching, fresh martensite and untransformed austenite can be obtained. Throughout the entire QP process (see Figure 1), and particularly during the partitioning treatment, the selection of an appropriate chemical composition is essential to avoid competing and undesirable reactions such as the formation of pearlite, bainitic structures, and carbide precipitation. On the other hand, the lack of carbon enrichment in the gamma phase may result in an unstable austenite that may promote the formation of too much fresh martensite during the final quenching step at the end of the process [4–8].

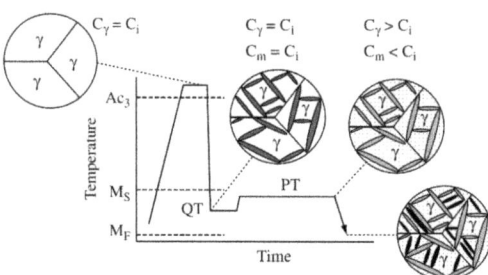

Figure 1. Thermal process to develop QP steels. QT and PT correspond to quenching and partitioning temperatures [5]. Copyright obtained from Elsevier. C_i, C_γ, and C_m are the carbon contents of the initial alloy, austenite, and martensite, respectively. Reproduced with permission of the editor.

The final aim of the QP process is to obtain an appropriate combination of mechanical properties as a result of a controlled microstructure formed by tempered and fresh martensite and a carbon-enriched retained austenite, avoiding or minimizing the decomposition of austenite into other secondary microstructures during the final quenching step. In addition, ferrite can also be present in the QP process by partial austenization (austenitization at intercritical temperatures). Accordingly, plenty of microconstituents (ferrite, bainite, retained austenite, fresh martensite, tempered martensite) with their respective mechanical properties can be achieved at the end of this novel heat treatment. Particularly, the retained austenite in QP steels plays a major role as it provides a transformation-induced plasticity (TRIP) effect, associated with the transformation of the austenite into martensite as a result of local deformation, thus contributing to enhancing the formability and energy absorption of the steel [9–13].

Normally, different alloying elements are considered in the design of QP steels, although the most usual grades are based on the C-Si-Mn and C-Si-Mn-Al systems, with carbon being the main element [6]. Moreover, the addition of other alloying elements such as Si, Al, and P plays an important role by delaying the carbide formation [2,9,12]. On the other hand, several studies have reported the immobility of substitutional elements during the partitioning step as a result of typical low partitioning temperatures, in the range of 350–450 °C. Therefore, the diffusion of these elements through martensite and austenite can be considered negligible [14,15]. However, their role as alloying elements is more pronounced in the initial austenitization temperatures. It is well known, for example, that Mn additions may enhance the austenite stabilization while decreasing the M_s temperature [4,7,12,15–22].

The present work aims at designing an optimal QP process. For this purpose, a 0.2%C–1.5%Si steel was used with 3.5 wt% manganese. In order to design an optimal thermal cycle, the critical temperatures and continuous cooling transformation diagram of the steel were determined using a quenching dilatometry, which was also used to apply the designed QP cycles in the laboratory. The final microstructure for each condition was characterized by a scanning electron microscope (SEM), X-ray diffraction (XRD), and electron backscatter diffraction (EBSD). This was correlated with the mechanical properties in terms of hardness and tensile tests.

2. Materials and Methods

The chemical composition of the QP steel selected in this study is given in Table 1. The amount of Mn is expected to enhance the stabilization of austenite while retarding the formation of secondary microstructures at the quenching stage. The addition of Si is commonly used to avoid carbide precipitation during the partitioning treatment. A laboratory ingot was cast after vacuum induction melting. The ingot was subsequently homogenized and hot rolled to obtain an approximately 7 mm thick sheet, followed by air cooling to room temperature.

Table 1. Chemical composition (wt%) of the present steel.

C	Mn	Si
0.2	3.5	1.5

The austenite transformation temperatures, as well as the temperatures corresponding to the martensitic transformation, were obtained by dilatometry. For this purpose, cylindrical samples of 10 mm in length and 4 mm in diameter were machined parallel along the rolling direction of the steel plate and heat treated in a DIL 805A/D (TA Instruments, New Castle, DE, USA) quenching dilatometer. Samples were austenitized up to 1100 °C at a heating rate of 10 °C/s and held for 1 min to ensure a full homogenization. After austenitization, a direct quenching was applied at a cooling rate of 50 °C/s to room temperature.

Continuous cooling transformation (CCT) diagrams were also obtained by dilatometry using the previously mentioned cylindrical samples according to the following set of experiments: austenization at 920 °C at a heating rate of 10 °C/s and held for 1 min, followed by cooling down to room temperature at different cooling rates, namely: 100, 50, 20, 10, 5, 1, and 0.1 °C/s.

In order to design an appropriate QP process, the methodology developed by Speer et al. [2] was used to find the optimal quenching temperature, after which the maximum retained austenite could be attained. The theoretical model was based on the constrained carbon equilibrium, considering a full partitioning of carbon between martensite and austenite, ignoring the partitioning kinetics, and avoiding carbide precipitation or bainite formation. It was also assumed that substitutional atoms cannot diffuse at the partitioning temperatures [2,5,7,9,13,23–26].

The model predicts the fraction of martensite and untransformed austenite at the quenching temperature (QT) during an undercooling below M_s, based on the Koistinen–Marburger (K-M) [2,5,9,23,27] relationship:

$$F_m = 1 - e^{\alpha(M_s - QT)}, \quad (1)$$

where F_m corresponds to the austenite fraction which transforms into martensite during a quenching treatment at QT, below the M_s temperature, and α corresponds to a rate parameter. For the present research, α and M_s (as-quenched martensite) were obtained by following the methodology reported in [28], with α being equal to -1.1×10^{-2}. A typical representative plot of the estimated variation in microstructure according to the theoretical model is shown in Figure 2. The bold solid line represents the maximum fraction of austenite as a function of QT and the optimal quenching temperature can be depicted as the QT for which the maximum austenite fraction can be obtained. The solid lines show the martensite (M) and austenite (γ) fraction during the first quenching temperature condition, as the dash-dotted line represents the fraction of martensite (as-quenched) that forms during the final quenching step, and the dashed line the estimated carbon content of the austenite [5].

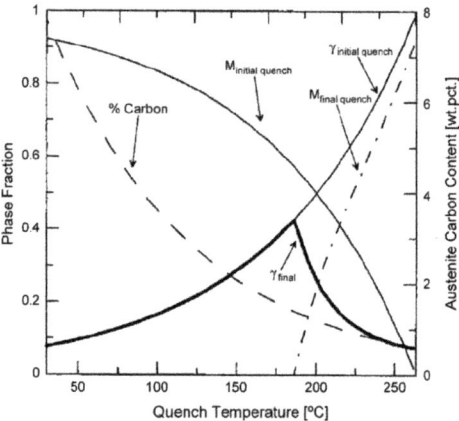

Figure 2. Predicted phase fractions of the microstructure after the QP process [5]. Copyright obtained from Elsevier. The solid bold line corresponds to the final austenite fraction at room temperature. Solid lines illustrate the austenite (γ) and martensite (M) fraction at the first quenching temperature, and the dash-dotted line represents the as-quenched or fresh martensite that forms during the final quench. Reproduced with permission of the editor.

In consequence, once the experimental critical temperatures are known, and the corresponding diagram of Figure 2 is obtained for a given steel, a physical simulation of the QP industrial process can be performed. The QP route designed here is illustrated in Figure 3. The heat treatments of this physical simulation were carried out by dilatometry in the cylindrical samples previously mentioned. First a full austenitization at 920 °C for 60 s was carried out and followed by quenching at 50 °C/s to the optimal temperature (as will be shown later, this temperature was 261 °C). Then, a partitioning treatment was applied at TP1 = M_s, TP2 = TP1 + 50 °C, and TP3 = TP2 + 50 °C for a partitioning time (tp) of 100 s, prior to a final quench to room temperature. The value of tp was selected as one closer to real industrial facilities. In consequence, the current study verified the effect of the partitioning temperature. Samples from this thermal cycle were named QP-x, where x stands for the TP involved.

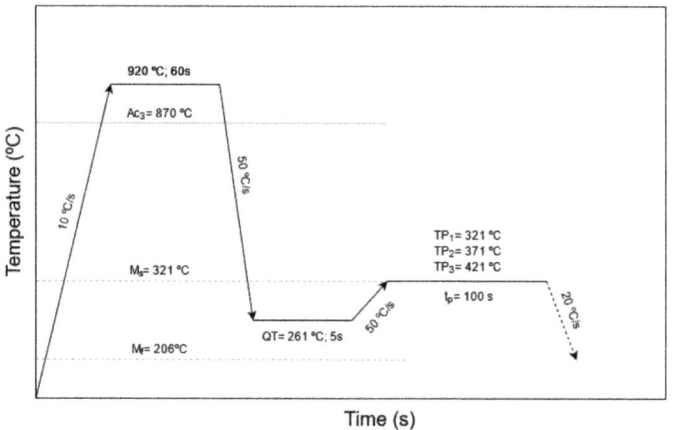

Figure 3. Schematic of the heat treatments applied to the present steel.

The as-received samples, and those from the CTT experiments and physical simulation, were analyzed by SEM, (JEOL USA Inc., MA, USA), EBSD (Oxford instruments, Abingdon, UK), and XRD (Malvern Panalytical Ltf, Malvern, UK). EBSD (Oxford instruments, Abingdon, UK) and XRD were used in order to identify and measure the amount of retained austenite at the different Q&P conditions. For observation purposes, samples were mechanically ground with abrasive papers followed by a final precision polishing step until 1 µm. EBSD samples were further polished with a 0.03 µm colloidal silica suspension. Microstructure characterization by JEOL JSM-70001F scanning electron microscope (JEOL USA Inc., MA, USA) was carried out after nital (2% nitric acid and ethanol) etching. EBSD scans on the physical simulation samples were performed using a 30 nm step size and 20 kV in a JEOL JSM-70001F scanning electron microscope (JEOL USA Inc., MA, USA) and EBSD detector Oxford Instruments HKL Nordlys (Oxford instruments, Abingdon, UK) using HKL Channel 5 software (A/S 2007, Oxford Instruments HKL, Hobro, Denmark) for data processing. It has been reported that the size of retained austenite found between martensite blocks is around 20–100 nm, therefore the EBSD analysis cannot detect such small retained austenite particles, due to the technique's limitations (spatial resolution of 0.08 µm). Instead, an alternative measurement by XRD was also performed [29,30]. XRD analyses were executed in a PANalytical X'Pert PRO MPD diffractometer (Malvern Panalytical Ltf, Malvern, UK) with CuKα, using a secondary graphite flat crystal monochromator operated at 45 kV and 40 mA. The 2 θ range was 30° to 125° with a step size of 0.017° and a measuring time of 125 s per step. The fraction of retained austenite in each condition was quantified using MAUD software and Rietveld analysis [31]. Finally, using the XRD spectra, the carbon content in austenite was calculated using the following expression [32]:

$$a_\gamma = 3.555 + 0.044 x_c, \qquad (2)$$

where a_γ is the lattice parameter of austenite in Angstroms (Å) and x_c is the average carbon amount in weight percentage. The average lattice parameter was obtained from the (220) and (311) austenite peaks of the XRD diagrams.

In order to measure the mechanical properties, tensile tests were performed in each QP condition. Tensile flat samples of 10 mm in gauge length and 5 mm in gauge width were subjected to each QP treatment in the quenching dilatometer. The tensile samples were used according to the standard. In addition, the fracture surface of each condition was analyzed by SEM after the tensile test. The tensile tests were carried out in an Instron 100kN 4507 universal testing machine frame (Instron, MA, USA) with hydraulic grips at room temperature with a cross-head speed of 2 mm/min. Deformation was recorded by a Basler Ace 5 MegaPixels camera and data were registered by Vic-Gauge 2D Digital Image Correlation software (V6, Correlated Solutions, SC, USA). Hardness measurements were taken in an Akashi MVK-HO Vickers (Mitutoyo, Kanagawa, JP) hardness tester with a 1 Kg applied load for 15 s. In doing so, 9 measurements were taken for each sample to obtain the average hardness.

3. Results and Discussion

3.1. Dilatometric Study

The experimental critical transformation temperatures Ac_1, Ac_3, M_s, and M_f obtained by dilatometry are listed in Table 2. As expected, the addition of Mn significantly reduces the critical transformation temperatures, validating the stabilizing effect of this element on the austenitic phase [33]. According to the results, the experimental CCT diagram was constructed as illustrated in Figure 4. As shown in Figure 4a, a full martensitic microstructure is reached even at cooling rates as low as 5 °C/s due to the good hardenability, as a result of the addition of Mn that suppresses the formation of allotriomorphic ferrite and perlite [34]. However, the slowest cooling rate (0.1 °C/s) showed the presence of a secondary microstructure (e.g., bainite) at 450 °C [35]. On the other hand, the hardness measurements in the CCT conditions (Figure 4b) show that for cooling rates lower than

0.1 °C/s, a hardness of 429 HV can be achieved, while increasing the cooling rate enhanced the hardness and it reached a value of about 509 HV.

Table 2. Critical transformation temperatures measured by dilatometry.

Ac_1	Ac_3	M_s	M_f
729 °C	869 °C	321 °C	206 °C

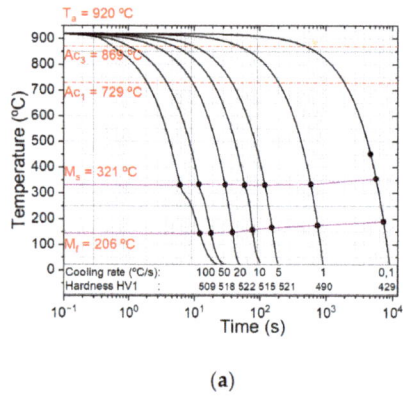

(a)

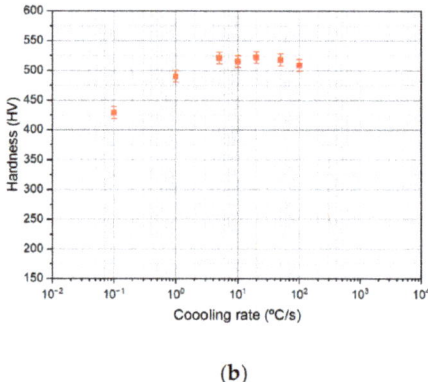

(b)

Figure 4. (a) Continuous cooling transformation (CCT) diagram of the present steel, (b) hardness values in each CCT condition.

3.2. QP Model Simulation

The Speer model to predict the maximum retained austenite and the corresponding optimal quenching temperature after a QP process was applied to the present steel. The result when assuming a full austenitization condition is illustrated in Figure 5. Accordingly, the optimal quenching temperature is 261 °C. The predicted maximum amount of retained austenite (γ_F) at this quenching temperature was 29%, assuming full partitioning of carbon from martensite to austenite.

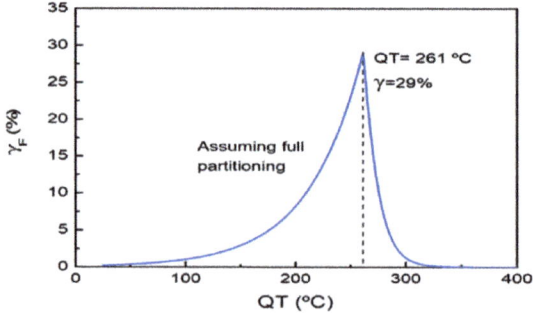

Figure 5. Estimated austenite fraction after the QP process in a full austenitization condition for the current steel. Derivation of the optimal quenching temperature.

3.3. Microstructural Characterization of the As-Received Condition

The as-received microstructure of the current steel is shown in Figure 6. The microstructure corresponds to an almost fully martensitic steel. However, some noticeable features can be seen. For instance, some carbide precipitation (Figure 6b) within coarse martensite regions (zone A) can be seen, suggesting a sort of self-tempering; this phenomenon may occur directly during cooling in steels with high M_s [36,37]. Moreover, zones of as-quenched martensite (B) were found throughout the microstructure. Nevertheless, a different morphology zone (C) was found in the microstructure. In Figure 4a, at the slowest cooling rate (0.1 °C/s), the presence of a secondary microstructure is shown, which can be suggested to be bainite. The formation of this microstructure under a slow cooling condition after steel processing has already been suggested. For instance, Navarro-López et al. [38] reported similar morphologies during an isothermal holding of 1 h at 300 °C or higher temperature in a 0.2C-3.51Mn-1.52Si-0.25Mo-0.04Al (wt%) steel.

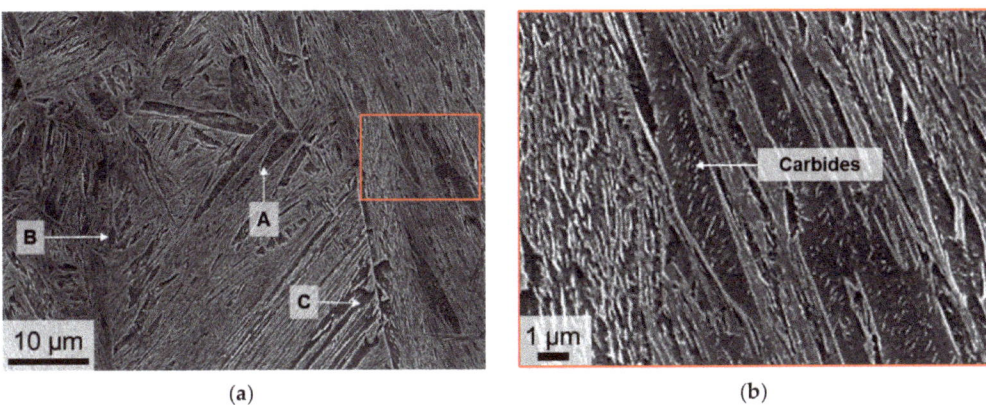

Figure 6. SEM micrograph of the as-received microstructure of the present steel. (**a**) General view, (**b**) magnification of red area in (**a**).

3.4. Microstructural Characterization after the QP Process

Figure 7 shows the resulting microstructure observed with SEM after the QP process where (a) QP-321, (b) QP-371, and (c) QP-421 refer to the partitioning treatment temperatures. Two types of martensitic regions can be observed. The heavily etched one consists in tempered martensite formed during the first quenching (M1) plus the partitioning, and the rest of the microstructure corresponds to as-quenched or fresh martensite (M2) with thin lath morphology surrounded by retained austenite (RA) forming M2/RA islands after the final quench [7,39]. According to Figure 7, M1 displays representative carbides inside the martensite blocks. On the other hand, M2 martensite can be identified as less etched zones in comparison with M1 as a result of carbon enrichment [12].

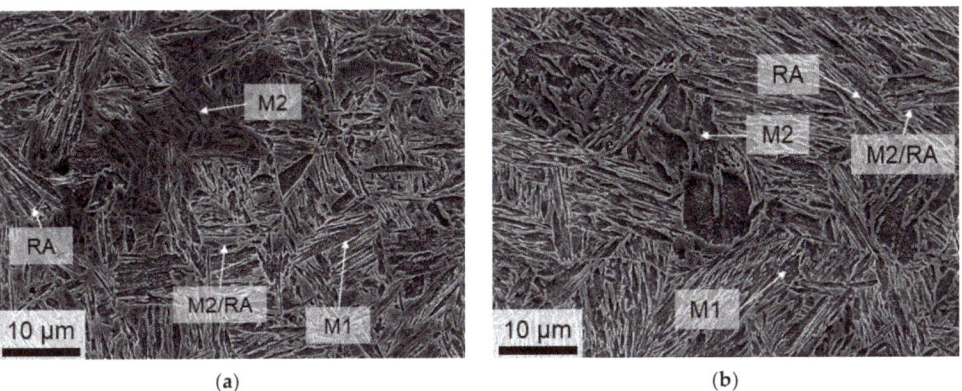

Figure 7. *Cont.*

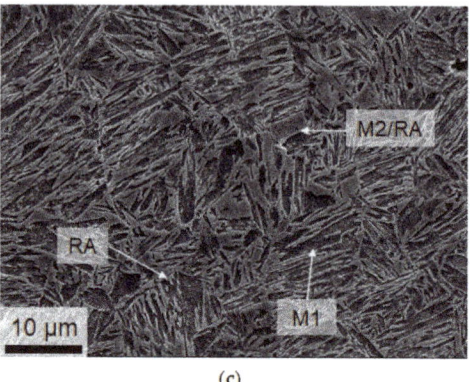

(c)

Figure 7. SEM micrographs after the QP process at (**a**) QP-321, (**b**) QP-371, (**c**) QP-421 s. M1 stands for tempered martensite, M2 for fresh martensite, and RA for retained austenite.

A first remark is the relatively homogeneous microstructure obtained at the QT partitioning temperatures, although the one corresponding to QP-371 has some isolated M2 areas. Primary or tempered martensite M1 displays a coarse blocky morphology and is defined by the presence of transitional needle-type carbides within the block structure [12,40]. This kind of martensite is observed in all samples, although in different amounts. Fresh martensite M2 presents a very thick morphology, and is more evident in treatments at a high partitioning temperature, i.e., in specimen QT-421, as shown in Figure 8.

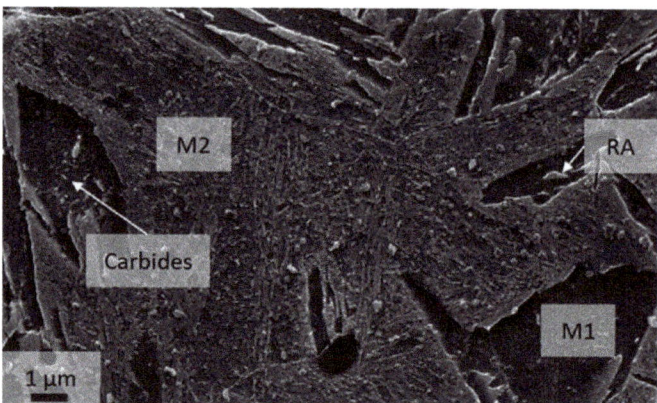

Figure 8. SEM micrograph of sample QP-421. M1 stands for tempered martensite, M2 for fresh martensite, RA for retained austenite and carbides of M1.

On the other hand, retained austenite can be noticed as having two morphologies: a thin and film-like morphology that is found between laths of martensite M1, and a coarser block-like morphology that is observed next to the prior austenite grain boundaries or martensite packet boundaries [7,12]. Moreover, regions formed by M2/RA islands were observed in all samples (Figure 7a–c). These regions are developed by fresh martensite M2 and large grains of retained austenite RA with a ring-like shape [12]. It has been reported [41] that these islands appear in the microstructure when carbon is not homogeneously dispersed along the austenite grain during partitioning treatment. The number of these M2/RA islands also depends on the quenching temperature [39]. One possible source of these heterogenous microstructures may come from Mn segregation. Hidalgo et al. [12] have reported that Mn-rich and Mn-poor regions in 0.3C-4.5Mn-1.5Si steel can cause the fraction of tempered martensite M1 and fresh martensite M2 to change from Mn-rich zones to Mn-poor areas. This indicates the key role that Mn segregation can play in the heterogeneity of the microstructures.

The EBSD image quality (IQ) images overlapped with the phase maps are shown in Figure 9. Here, the RA fraction discriminated by the EBSD analysis is highlighted in blue. In correspondence with the SEM images, the microstructure is mainly formed by a mixture of tempered M1 and fresh M2 martensite. De Diego-Calderón et al. [29] reported that martensite M2 can be detected in the IQ and phase maps as a darker region next to RA grains which can help to discern the presence of these regions in all conditions, as shown in Figure 9. In any case, as already pointed out, the amount of retained austenite was estimated by (i) EBSD using Channel 5 software and by (ii) XRD. Results are listed in Table 3 and will be discussed later. Three RA morphologies were observed in the microstructure after the QP process, namely a film-like RA, blocky RA, and lamellar RA, in agreement with the literature [29]. According to the EBSD images, retained austenite ($\leq$1%) is practically absent in the condition QP-321 (Figure 9a) in comparison with the conditions QP-371 (Figure 9b) and QP-421 (Figure 9c) that display 4.6% and 8.4%, respectively. This result shows the effect of increasing the partitioning temperature on increasing retained austenite. These values are also in agreement with literature reports indicating the relatively large amounts of RA evaluated by EBSD obtained at a partitioning temperature of 400 °C [29].

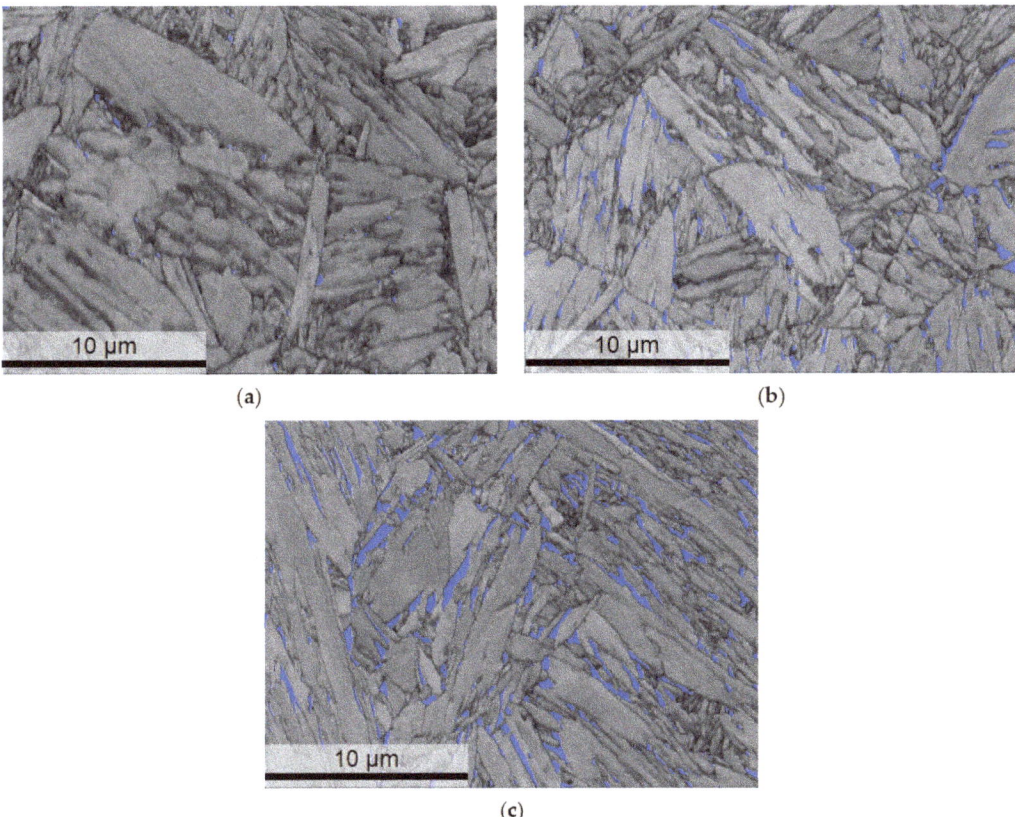

Figure 9. EBSD image quality (IQ) and phase maps after the current QP process. (**a**) QP-321, (**b**) QP-371, (**c**) QP-421. Blue corresponds to areas identified as retained austenite (RA).

Table 3. Volume fraction of retained austenite measured by EBSD and XRD.

Sample	Retained Austenite by EBSD	Retained Austenite by XRD
QP-321	1.1%	8.8%
QP-371	4.6%	12.9%
QP-421	8.4%	18.3%

The amount of RA evaluated by XRD, listed in Table 3, is also shown in Figure 10. The same trends as those of the EBSD analysis can be seen, but with clearly larger amounts of RA found in all conditions. It is worth noticing the lower fraction of retained austenite measured by XRD than those predicted by the model proposed by Speer. The difference in measurements of retained austenite between both techniques is due to the limitation of EBSD to detect the smallest austenite laths next to martensite blocks. The size range of lath austenite is around 20–100 nm, which can be detectable by XRD [29]. This can be explained by an incomplete partitioning, formation of secondary phases, or carbide precipitation during the entire QP process [23,42]. The present amounts of RA agree with some authors. For instance, Zhao et al. [43] reported 19.3% retained austenite by XRD measurements in a nearly similar chemical composition of 0.28C-1.42Si-4.08Mn. On the other hand, De Diego-Calderon et al. [29] reported a proportion of 20.2% of RA in 0.25C–1.5Si–3Mn–0.023Al.

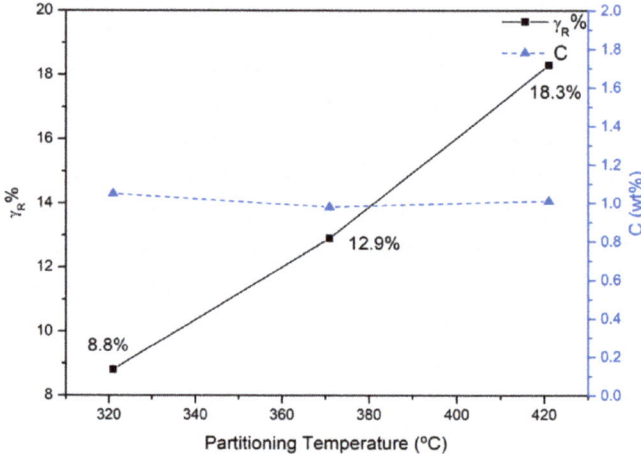

Figure 10. Retained austenite volume fraction and carbon amount in wt% obtained by XRD.

The estimated amount of carbon in the RA measured by XRD is shown in Figure 10. The evolution of the amount of carbon in the RA during the QP process is roughly similar in the different partitioning conditions. A larger amount was found at QP-321 with 1.05 wt%, whereas an increment in partitioning temperature showed a slight reduction in the amount of carbon of 0.98 wt% at QP-371 and another small increment at the highest partitioning temperature of 1.01 wt%. This slight change in carbon concentration can be explained by the presence of different RA morphologies. It has been reported that the amount of carbon in lath RA reduces the carbon mean value while the block RA increases this value [44]. However, a high carbon concentration (1.05%C) and low fraction of RA measured in the QP-321 condition can be explained as a result of a large amount of M1, in which the carbon enrichment from M1 to austenite exceeds that required to stabilize the untransformed austenite [41]. The large amount of carbon in the RA in comparison with the initial carbon concentration proved that enough carbon diffused to austenite during the partitioning step, enhancing the stability at room temperature [45]. As already mentioned, the martensitic transformation is diffusionless, so the carbon content of this RA should be the same (or similar) to the one of the fresh martensite. It is well known that the morphology of the martensite changes from a lath-like shape to a plate-like shape depending on carbon contents [46]. This transition is around 0.8–1%C. In consequence, the fresh martensites displayed in Figure 7 are in agreement with the carbon content evaluated by XRD.

The mechanical properties of the present steel are listed in Table 4 and plotted in Figure 11. The QP-321 condition shows the highest hardness value of 448 HV in comparison with QP-371 and QP-421 specimens that have values of 408 HV and 421 HV, respectively. It can be discerned that the high hardness of the QP-321 sample may be explained by the low fraction of retained austenite present in the microstructure. It is also remarkable that, except sample QP-321, the other samples present nearly the same hardness.

Table 4. Results of mechanical properties of studied steel.

Sample	Hardness (HV)	Ys (MPa)	UTS (MPa)	El (%)	(n)	Ag (MPa)
QP-321	448	1190	1556	11.1	0.12	9.68
QP-371	408	1140	1583	8.3	0.13	7.15
QP-421	404	832	1305	3	0.3	2.20

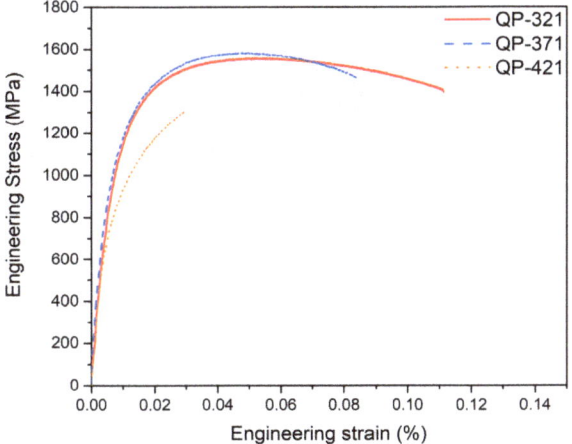

Figure 11. Tensile properties of the studied steel QP-x.

Figure 11 shows the engineering stress–strain graphs corresponding to the two thermal cycles selected in this work. The results from mechanical testing are also listed in Table 4. The as-received condition exhibits a YS of 1201 MPa, 1671 Mpa of UTS, total elongation (EL%) of 10.8% and n-value of 0.10. On the other hand, the uniform elongation (Ag) presents the same trend as El% in which the QP-421 condition exhibits a low work-hardening behavior. Interestingly, one can notice that the total elongation is, in general, below 5%, except in the QP-321 and QP-371 conditions. Indeed, the flow curves with the largest hardening exponents correspond, surprisingly, to the lower elongation values. The large work hardening exponents agree with the larger amount of retained austenite displayed in these conditions, as expected. However, the lack of ductility in terms of elongation must be associated with the heterogeneous microstructure, so although the RA fraction is larger, the big islands of fresh (brittle) martensite are heterogeneously dispersed, and, consequently, they can promote early necking and even failure, i.e., local brittleness. The low elongation found in the QP-421 condition which displays 18.3% of RA and 1.01%C proves that RA does not fully help during the work-hardening step. Hidalgo et al. [12] reported the same behavior in 0.3C–4.5Mn–1.5Si (wt%). They attributed the low elongation due to low stability of RA in regions with a small amount of carbon during deformation.

The fast transformation into a harder martensite inhibits the work-hardening behavior of the soft untransformed RA. Accordingly, although the n-value can be initially large, the total final elongation can be low.

The surface fracture analysis of different QP conditions is shown in Figure 12. The fracture surface of the QP-321 sample (Figure 12a) displays a large density of dimples and elongated dimple zones representative of ductile fracture. A mixture of dimple zones and intergranular fracture was observed in the QP-371 condition (Figure 12b) that, in turn, decreases the elongation percentage in comparison with sample QP-321. On the other hand, small regions of microdimples and dominant regions of intergranular fracture were found along the fracture surface in the QP-421 condition (Figure 12c) as a result of a brittle fracture [47]. According to Figure 11, the QP-321 condition reached 11.1% of elongation while QP-371 and QP-421 conditions reached 8.3% and 3%, respectively. This behavior shows that the increment in partitioning temperature decreases the elongation percentage. The low elongation found in the QP-421 sample can be explained as a result of brittle fracture. The amount of retained austenite does not enhance the work hardening in this condition, and it instead transforms into hard and brittle martensite, promoting an overall brittle behavior [12].

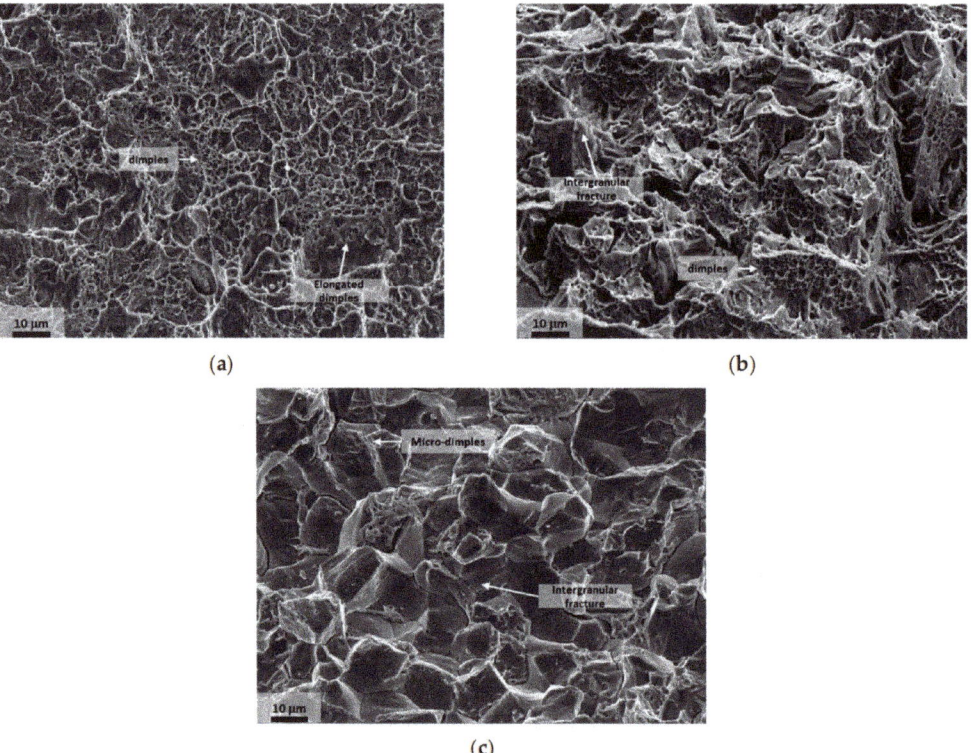

Figure 12. Fracture surface of tensile samples. (**a**) QP-321, (**b**) QP-371, (**c**) QP-421.

On the other hand, the increment in the n-exponent found in QP-421 can be explained by the austenite's transformation into martensite during the deformation. This transformation behavior depletes part of the strain energy and a softening phase transformation is reached. However, new martensite blocks the dislocation movement and affects the work hardening [48]. In addition, the same results were reported by Kozłowska and Grajcar for a hot-rolled Si-Al-alloyed 0.24C–1.5Mn-0.87Si in which there was found an increment of work-hardening exponent with the gradual transformation of retained austenite into martensite [49].

4. Conclusions

The present study showed the designing process of a QP heat treatment route of a given steel in an optimal condition. The following conclusions can be drawn:

- Low fractions of retained austenite were measured at QP in all conditions in comparison with the one estimated by the theoretical model. This difference is explained in terms of the formation of secondary phases and carbide precipitation.
- QP-421 displays the highest fraction of retained austenite of 18.3%. These results are still below the estimated RA by the Speers model (29%).
- The increment in partitioning temperature shows a significant increment in RA, from 8.8 to 18.3%, in the QP process measured by XRD.
- High mechanical properties were observed due to the presence of tempered and fresh martensite. The work-hardening exponent increased in conditions with a higher RA fraction. However, improvement in ductility in terms of elongation was not evidenced at increasing RA fractions. In other words, although RA increments are in principle beneficial for formability, elongation is penalized with a large amount of fresh martensite. According to the results, the optimum QP treatment would correspond to the QP-321 condition that shows a more homogeneous microstructure, with a total amount of finely distributed RA close to 8.8% (after XRD measurements).

Author Contributions: Conceptualization, J.M.C. and O.G.; methodology, M.C.; validation, M.C., O.G., J.P.P. and J.C.; formal analysis, M.C. and J.M.C.; investigation, M.C.; resources, O.G., J.P.P. and J.M.C.; writing—original draft preparation, M.C.; writing—review and editing, J.M.C., O.G., J.C. and J.P.P.; visualization, M.C.; supervision, J.M.C. and J.C.; project administration, J.M.C. and O.G.; funding acquisition, O.G. and J.P.P. All authors have read and agreed to the published version of the manuscript.

Funding: This research received no external funding.

Institutional Review Board Statement: Not applicable.

Informed Consent Statement: Not applicable.

Data Availability Statement: Data can be provided upon request.

Acknowledgments: M.C. acknowledges his scholarship funded by CONACYT México.

Conflicts of Interest: The authors declare no conflict of interest.

References

1. Fonstein, N. Advanced high strength sheet steels. In *Advanced High Strength Sheet Steels*; Springer: Berlin/Heidelberg, Germany, 2015; pp. 12–16.
2. Speer, J.G.; Streicher, A.M.; Matlock, D.K.; Rizzo, F.; Krauss, G. Quenching and partitioning: A fundamentally new process to create high strength trip sheet microstructures. In Proceedings of the A Symposium on the Thermodynamics, Kinetics, Characterization and Modeling of Austenite Formation and Decomposition Held at the Materials Science & Technology, Chicago, IL, USA, 9–12 November 2003; pp. 505–522.
3. Speer, J.G.; De Moor, E.; Findley, K.; Matlock, D.K.; De Cooman, B.C.; Edmonds, D.V. Analysis of Microstructure Evolution in Quenching and Partitioning Automotive Sheet Steel. *Met. Mater. Trans. A* **2011**, *42*, 3591–3601. [CrossRef]
4. Ayenampudi, S.; Celada-Casero, C.; Sietsma, J.; Santofimia, M. Microstructure evolution during high-temperature partitioning of a medium-Mn quenching and partitioning steel. *Materials* **2019**, *8*, 100492. [CrossRef]
5. Edmonds, D.; He, K.; Rizzo, F.; De Cooman, B.; Matlock, D.; Speer, J. Quenching and partitioning martensite—A novel steel heat treatment. *Mater. Sci. Eng. A* **2006**, *438–440*, 25–34. [CrossRef]
6. Wang, L.; Speer, J.G. Quenching and Partitioning Steel Heat Treatment. *Met. Microstruct. Anal.* **2013**, *2*, 268–281. [CrossRef]
7. Santofimia, M.; Zhao, L.; Petrov, R.; Kwakernaak, C.; Sloof, W.; Sietsma, J. Microstructural development during the quenching and partitioning process in a newly designed low-carbon steel. *Acta Mater.* **2011**, *59*, 6059–6068. [CrossRef]
8. Jiang, H.-T.; Zhuang, B.-T.; Duan, X.-G.; Wu, Y.-X.; Cai, Z.-X. Element distribution and diffusion behavior in Q&P steel during partitioning. *Int. J. Miner. Met. Mater.* **2013**, *20*, 1050–1059. [CrossRef]

9. Speer, J.G.; Assunção, F.C.R.; Matlock, D.K.; Edmonds, D.V. The "quenching and partitioning" process: Background and recent progress. *Mater. Res.* **2005**, *8*, 417–423. [CrossRef]
10. De Knijf, D.; Petrov, R.; Föjer, C.; Kestens, L.A. Effect of fresh martensite on the stability of retained austenite in quenching and partitioning steel. *Mater. Sci. Eng. A* **2014**, *615*, 107–115. [CrossRef]
11. Tsuchiyama, T.; Tobata, J.; Tao, T.; Nakada, N.; Takaki, S. Quenching and partitioning treatment of a low-carbon martensitic stainless steel. *Mater. Sci. Eng. A* **2012**, *532*, 585–592. [CrossRef]
12. Hidalgo, J.; Celada-Casero, C.; Santofimia, M. Fracture mechanisms and microstructure in a medium Mn quenching and partitioning steel exhibiting macrosegregation. *Mater. Sci. Eng. A* **2019**, *754*, 766–777. [CrossRef]
13. De Avillez, R.R.; da Costa e Silva, A.L.V.; Martins, A.R.F.A.; Assunção, F.C.R. The effect of alloying elements on constrained carbon equilibrium due to a quench and partition process. *Int. J. Mater. Res.* **2008**, *99*, 1280–1284. [CrossRef]
14. Seo, E.J.; Cho, L.; De Cooman, B.C. Kinetics of the partitioning of carbon and substitutional alloying elements during quenching and partitioning (Q&P) processing of medium Mn steel. *Acta Mater.* **2016**, *107*, 354–365. [CrossRef]
15. Speer, J.; Matlock, D.K.; De Cooman, B.C.; Schroth, J.G. Carbon partitioning into austenite after martensite transformation. *Acta Mater.* **2003**, *51*, 2611–2622. [CrossRef]
16. Hou, Z.R.; Zhao, X.M.; Zhang, W.; Liu, H.L.; Yi, H.L. A medium manganese steel designed for water quenching and partitioning. *Mater. Sci. Technol.* **2018**, *34*, 1168–1175. [CrossRef]
17. Lee, S.; Lee, S.-J.; De Cooman, B.C. Austenite stability of ultrafine-grained transformation-induced plasticity steel with Mn partitioning. *Scr. Mater.* **2011**, *65*, 225–228. [CrossRef]
18. De Moor, E.; Matlock, D.K.; Speer, J.G.; Merwin, M.J. Austenite stabilization through manganese enrichment. *Scr. Mater.* **2011**, *64*, 185–188. [CrossRef]
19. De Moor, E.; Kang, S.; Speer, J.G.; Matlock, D.K. Manganese diffusion in third generation advanced high strength steels. In Proceedings of the International Conference on Mining, Materials and Metallurgical Engineering, Prague, Czech Republic, 11–12 August 2011; pp. 1–7.
20. Lee, S.-J.; Lee, S.; De Cooman, B.C. Mn partitioning during the intercritical annealing of ultrafine-grained 6% Mn transformation-induced plasticity steel. *Scr. Mater.* **2011**, *64*, 649–652. [CrossRef]
21. Lee, S.; De Cooman, B.C. On the Selection of the Optimal Intercritical Annealing Temperature for Medium Mn TRIP Steel. *Met. Mater. Trans. A* **2013**, *44*, 5018–5024. [CrossRef]
22. Jirková, H.; Kučerová, L.; Mašek, B. Effect of Quenching and Partitioning Temperatures in the Q-P Process on the Properties of AHSS with Various Amounts of Manganese and Silicon. *Mater. Sci. Forum* **2012**, *706–709*, 2734–2739. [CrossRef]
23. Sun, J.; Yu, H. Microstructure development and mechanical properties of quenching and partitioning (Q&P) steel and an incorporation of hot-dipping galvanization during Q&P process. *Mater. Sci. Eng. A* **2013**, *586*, 100–107. [CrossRef]
24. Santofimia, M.; Zhao, L.; Sietsma, J. Model for the interaction between interface migration and carbon diffusion during annealing of martensite–austenite microstructures in steels. *Scr. Mater.* **2008**, *59*, 159–162. [CrossRef]
25. Clarke, A.; Speer, J.; Matlock, D.; Rizzo, F.; Edmonds, D.; Santofimia, M. Influence of carbon partitioning kinetics on final austenite fraction during quenching and partitioning. *Scr. Mater.* **2009**, *61*, 149–152. [CrossRef]
26. Clarke, A.; Speer, J.; Miller, M.; Hackenberg, R.; Edmonds, D.; Matlock, D.; Rizzo, F.; Clarke, K.; De Moor, E. Carbon partitioning to austenite from martensite or bainite during the quench and partition (Q&P) process: A critical assessment. *Acta Mater.* **2008**, *56*, 16–22. [CrossRef]
27. Koistinen, D.; Marburger, R. A general equation prescribing the extent of the austenite-martensite transformation in pure iron-carbon alloys and plain carbon steels. *Acta Met.* **1959**, *7*, 59–60. [CrossRef]
28. Van Bohemen, S.M.C. Bainite and martensite start temperature calculated with exponential carbon dependence. *Mater. Sci. Technol.* **2012**, *28*, 487–495. [CrossRef]
29. De Diego-Calderón, I.; De Knijf, D.; Molina-Aldareguia, J.M.; Sabirov, I.; Föjer, C.; Petrov, R. Effect of Q&P parameters on microstructure development and mechanical behaviour of Q&P steels. *Rev. Met.* **2015**, *51*, e035. [CrossRef]
30. Santofimia, M.; Zhao, L.; Petrov, R.; Sietsma, J. Characterization of the microstructure obtained by the quenching and partitioning process in a low-carbon steel. *Mater. Charact.* **2008**, *59*, 1758–1764. [CrossRef]
31. Lutterotti, L.; Gialanella, S. X-ray diffraction characterization of heavily deformed metallic specimens. *Acta Mater.* **1998**, *46*, 101–110. [CrossRef]
32. Cullity, B.D. *Elements of X-ray Diffraction*; Adison–Wesley Publishing: Boston, MA, USA, 1967.
33. Lee, S.; De Cooman, B.C. Tensile Behavior of Intercritically Annealed 10 pct Mn Multi-phase Steel. *Met. Mater. Trans. A* **2013**, *45*, 709–716. [CrossRef]
34. Fadel, A.; Glišić, D.; Radović, N.; Drobnjak, D. Influence of Cr, Mn and Mo Addition on Structure and Properties of V Microalloyed Medium Carbon Steels. *J. Mater. Sci. Technol.* **2012**, *28*, 1053–1058. [CrossRef]
35. Caballero, F.; Santofimia, M.J.; Garcia-Mateo, C.; De Andrés, C.G. Time-Temperature-Transformation Diagram within the Bainitic Temperature Range in a Medium Carbon Steel. *Mater. Trans.* **2004**, *45*, 3272–3281. [CrossRef]
36. Matsuda, H.; Mizuno, R.; Funakawa, Y.; Seto, K.; Matsuoka, S.; Tanaka, Y. Effects of auto-tempering behaviour of martensite on mechanical properties of ultra high strength steel sheets. *J. Alloys Compd.* **2013**, *577*, S661–S667. [CrossRef]

37. Ramesh Babu, S.; Nyyssönen, T.; Jaskari, M.; Järvenpää, A.; Davis, T.P.; Pallaspuro, S.; Kömi, J.; Porter, D. Observations on the Relationship between Crystal Orientation and the Level of Auto-Tempering in an As-Quenched Martensitic Steel. *Metals* **2019**, *9*, 1255. [CrossRef]
38. Navarro-López, A.; Hidalgo, J.; Sietsma, J.; Santofimia, M.J. Characterization of bainitic/martensitic structures formed in isothermal treatments below the Ms temperature. *Mater. Charact.* **2017**, *128*, 248–256. [CrossRef]
39. Huyghe, P.; Malet, L.; Caruso, M.; Georges, C.; Godet, S. On the relationship between the multiphase microstructure and the mechanical properties of a 0.2C quenched and partitioned steel. *Mater. Sci. Eng. A* **2017**, *701*, 254–263. [CrossRef]
40. HajyAkbary, F.; Sietsma, J.; Miyamoto, G.; Furuhara, T.; Santofimia, M.J. Interaction of carbon partitioning, carbide precipitation and bainite formation during the Q&P process in a low C steel. *Acta Mater.* **2016**, *104*, 72–83. [CrossRef]
41. Celada-Casero, C.; Kwakernaak, C.; Sietsma, J.; Santofimia, M.J. The influence of the austenite grain size on the microstructural development during quenching and partitioning processing of a low-carbon steel. *Mater. Des.* **2019**, *178*, 107847. [CrossRef]
42. HajyAkbary, F.; Sietsma, J.; Petrov, R.H.; Miyamoto, G.; Furuhara, T.; Santofimia, M.J. A quantitative investigation of the effect of Mn segregation on microstructural properties of quenching and partitioning steels. *Scr. Mater.* **2017**, *137*, 27–30. [CrossRef]
43. Zhao, Z.Z.; Liang, J.H.; Zhao, A.M.; Liang, J.T.; Tang, D.; Gao, Y.P. Effects of the austenitizing temperature on the mechanical properties of cold-rolled medium-Mn steel system. *J. Alloys Compd.* **2017**, *691*, 51–59. [CrossRef]
44. Li, Y.; Kang, J.; Zhang, W.; Liu, D.; Wang, X.; Yuan, G.; Misra, R.; Wang, G. A novel phase transition behavior during dynamic partitioning and analysis of retained austenite in quenched and partitioned steels. *Mater. Sci. Eng. A* **2018**, *710*, 181–191. [CrossRef]
45. Liu, L.; He, B.B.; Cheng, G.J.; Yen, H.W.; Huang, M.X. Optimum properties of quenching and partitioning steels achieved by balancing fraction and stability of retained austenite. *Scr. Mater* **2018**, *150*, 1–6. [CrossRef]
46. Krauss, G. *Steels: Processing, Structure, and Performance*, 2nd ed.; ASM International, Metals Park: Russell Township, OH, USA, 2015.
47. Dong, B.; Hou, T.; Zhou, W.; Zhang, G.; Wu, K. The Role of Retained Austenite and Its Carbon Concentration on Elongation of Low Temperature Bainitic Steels at Different Austenitising Temperature. *Metals* **2018**, *8*, 931. [CrossRef]
48. Zhang, L.; Wang, C.-Y.; Lu, H.-C.; Cao, W.-Q.; Wang, C.; Dong, H.; Chen, L. Austenite transformation and work hardening of medium manganese steel. *J. Iron Steel Res. Int.* **2018**, *25*, 1265–1269. [CrossRef]
49. Kozłowska, A.; Grajcar, A. Effect of Elevated Deformation Temperatures on Microstructural and Tensile Behavior of Si-Al Alloyed TRIP-Aided Steel. *Materials* **2020**, *13*, 5284. [CrossRef] [PubMed]

Article

Microstructures and Properties of Auto-Tempering Ultra-High Strength Automotive Steel under Different Thermal-Processing Conditions

Han Jiang [1], Yanlin He [1,*], Li Lin [2], Rendong Liu [2,3], Yu Zhang [3], Weisen Zheng [1] and Lin Li [1]

1. School of Materials Science and Engineering, Shanghai University, Shanghai 200044, China; 18362208055@163.com (H.J.); wszheng@shu.edu.cn (W.Z.); liling@shu.edu.cn (L.L.)
2. Ansteel Group Co., Anshan 114021, China; ag_linli@126.com (L.L.); ag_lrd@126.com (R.L.)
3. Ansteel Group Beijing Research Institute, Beijing 102211, China; 15104124606@163.com
* Correspondence: ylhe@t.shu.edu.cn

Abstract: Automotive steels with ultra-high strength and low alloy content under different heating and cooling processes were investigated. It was shown that those processes exhibited a great influence on the performance of the investigated steels due to the different auto-tempering effects. Compared with the steels under water quenching, there was approximately a 70% increase in the strength and elongation of steels under air cooling, in which the martensite was well-tempered. Although the elongation of the steel with a microstructure composed of ferrite, well-tempered martensite and less-tempered martensite could exceed 15%, the hole expansion ratio was still lower because of the undesirable hardness distribution between the hard phases and the soft phases. It followed from the calculation results based on SEM, TEM and XRD analyses, that for the steel under air cooling, the strengthening mechanism was dominated by the solid solution strengthening and the elongation was determined by the auto-tempering of martensite. Experiments and analyses aimed to explore the strengthening and plasticity mechanisms of auto-tempering steels under the special process of flash heating.

Keywords: ultra-high strength steel; auto-tempering; martensite; hole expansion ratio; flash heating

1. Introduction

With the increasing demand for energy saving and environmental protections, the strength level of the automotive steel sheet has been raised and many works have contributed to the development of steels that perform better [1–3]. These steels are referred to as Advanced High Strength Steel (AHSS), among which, twinning-induced plasticity (TWIP) steels are characterized by an austenite phase with extremely large uniform elongation and high ultimate tensile strength [4,5]. However, the high Mn content (15–30 wt%) in the steels leads to high costs as well as problems in the production line which limits its further application in car industries. In addition, quenching-partition (Q&P) steel is designed to be quenched into a certain temperature range between M_S and M_f, then carbon partition from the martensite to the retained austenite starts, which results in a certain amount of austenite that is stabilized at room temperature to ensure the excellent strength and plasticity of the steel [1]. Though there are no expensive elements included, its complex process route does not fit the present continuous annealing line (CAL) and supplemental investment is required. In recent years, in view of the urgent demand for green manufacturing in the iron and steel industries and for advanced manufacturing processes to produce steel with a high performance, the near net shape process represented by compact strip production (CSP) to produce automotive steel has been widely studied [6,7]. Obviously, steel with a high alloy content and a complex process is not an ideal candidate for this special process.

As is well known, auto-tempering is a phenomenon in which the first-formed martensite near the martensitic transformation start temperature (M_S) is tempered during the following up process of quenching [8]. When auto-tempering happens, martensite in steel with a lower carbon content can be decomposed and the mechanical performance is improved due to the formation of tempered martensite and metastable carbides. Thus, the steel exhibits better strength and plasticity. Recently, some works have developed auto-tempering steel with a strength above 1200 MPa and a good elongation above 10% [9–11], but the strengthening and plasticity mechanisms of the steel under different thermal-processing conditions are still obscure. In the present work, the effect of auto-tempering on microstructures and the properties of two low alloy steels with different carbon contents are investigated to elucidate the intrinsic mechanism.

2. Experimental Procedure

The chemical compositions of the steel used in the present work are listed in Table 1. Ingots were prepared by pure raw materials and vacuum induction melting at 10^{-1} Pa vacuum value. Slabs with a 35 mm thickness were hot rolled after reheating at 1200 °C to produce a 3.5 mm thick sheet. The hot-rolled sheets were pickled and cold rolled to a 60 pct reduction. The samples with a size of $\varphi 4 \times 10$ mm heated to 880 °C at a rate of 2 k/s, were prepared to measure the phase transformation temperature with a DIL805 thermal expansion analyzer, and liquid nitrogen was used in the quenching process. The dilatometric curves of the samples are shown in Figure 1 and the experimental results are listed in Table 1.

Table 1. Chemical compositions (wt%) and the phase transformation temperature (°C) of the investigated steels.

Steel	C	Mn	Si	Cr	Ti	A_{C1}	A_{r3}	M_S
A	0.13	2.10	1.35	0.98	0.010	728	819	395
B	0.18	2.13	1.40	1.00	0.012	715	833	371

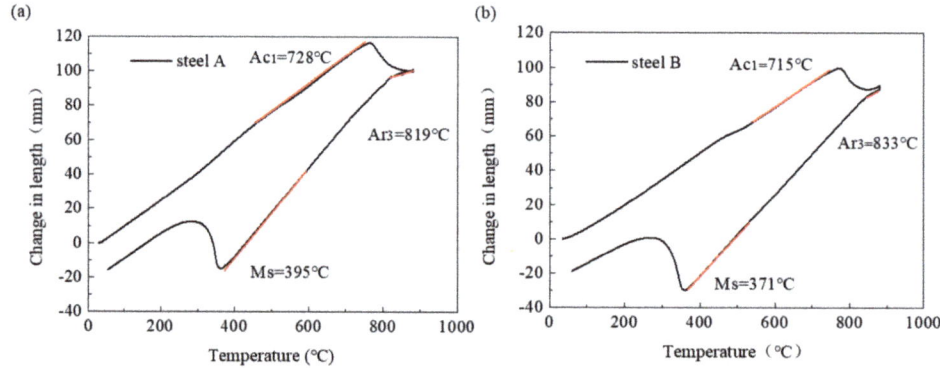

Figure 1. Dilatometric curves and phase transformation temperature of the experimental steels: (a) steel A and (b) steel B.

Steels A and B with different compositions were designed to compare the effect of carbon on the auto-tempering behavior. The heat-treatment cycle is schematically shown in Figure 2, and was operated in salt bath furnace. The samples were austenized at 880 °C and intercritical annealed at 800 °C for 3 min and then air cooled to room temperature to compare the auto-temper behavior of the full martensite microstructure and the ferrite–martensite microstructure. Three cooling processes, such as air cooling (denoted by the red line), water quenching (denoted by the blue line) and quenching to 200 °C in a salt bath for 3 and 30 min and then air cooled to room temperature (denoted by the green line) were applied to study the effect of different cooling procedures on auto-temper behavior. In addition, by using Gleeble3500 (DSI, Saint Paul, MN, USA), flash-heating with a heating rate 300 °C/S was adopted to study the effect of heating processes on auto-temper behavior.

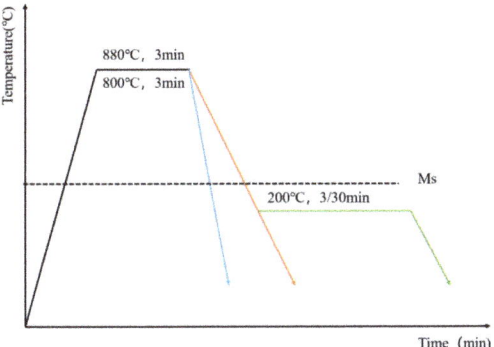

Figure 2. Schematic diagram of heat treatments.

The tensile properties of the samples were measured by the uniaxial tensile test with the standard specimen (according to the GB/T228.1-2010 standard, gauge length: 50 mm, width: 12.5 mm, thickness: 1.2 mm), the tensile direction of which was paralleled to the rolling direction. The tests were performed by a MTS C45.305E electronmechanical universal testing machine with a strain rate of about $1 \times 10^{-3}\,\text{S}^{-1}$ at room temperature, and an extensometer was used to measure the strain of the gauge length of the samples. The experimental results were determined by the average value of three tensible samples. The microstructures of the samples were etched with picric acid and 4% of nital was observed by means of OM and SEM analysis, respectively. Prior austenite grain size was measured by the linear intercept method (according to GB/T 3488.2 standard). The samples for transmission electron microscopy (TEM) were sliced from bulk specimens and mechanically polished to thick discs of about 50 μm with a diameter of 3 mm. Electrolytic polishing was conducted using 10 vol.% perchloric acid in ethanol at −35 °C in a twin-jet electrolytic polisher. The samples prepared were detected in a JEM-2010F microscope (JEOL Ltd, Tokyo, Japan) with an accelerated voltage of 200 kV. To calculate the dislocation density, specimens were measured in a 18KW D/MAX2500 X-ray diffractometer (Rigaku, Tokyo, Japan) with Cu-Kα radiation. Scanning was carried out with a 0.02° step and a 3 s stay for each step over a 2θ range from 40° to 100°.

The hole expansion test (HET) was carried out according to the ISO16630-2009 standard using ITC-SP225 equipment with a sample size of 90 mm × 90 mm × 1.2 mm and the diameter of the initial hole D_0 was 10 mm in the sample center. The test speed was 3 mm/min and the sample blank holder force was 50 KN. The hole expansion ratio (HER) was calculated by the following equation [8]:

$$\text{HER}(\%) = (D_f - D_0)/D_0 \times 100\% \qquad (1)$$

where D_0 and D_f represent the initial hole diameter (mm) and the ultimate hole diameter (mm) when a crack is initiated, respectively.

3. Results and Discussion

3.1. Effects of Various Cooling Processes on the Microstructures and the Mechanical Properties of the Experimental Steels

Figure 3 shows the microstructures of steel A and steel B heated at 880 °C for 3 min after being treated by water-quenching, and after being air-cooled and soaked at 200 °C for 3 min during the air-cooling. Combined with Figure 4, it can be seen that the microstructures of the experimental steels are mainly composed of lath martensite and film retained austenite. Compared with the structures after water-quenching, the lath martensite in the steels following air-cooling looks broad and the boundary is blurry, and tempered martensite (TM) is formed. In addition, the tempering of the martensite in steel A with the lower carbon content is more obvious than that in steel B. As shown in Figure 5, nano-scale TiC carbide can be observed in the experimental steel.

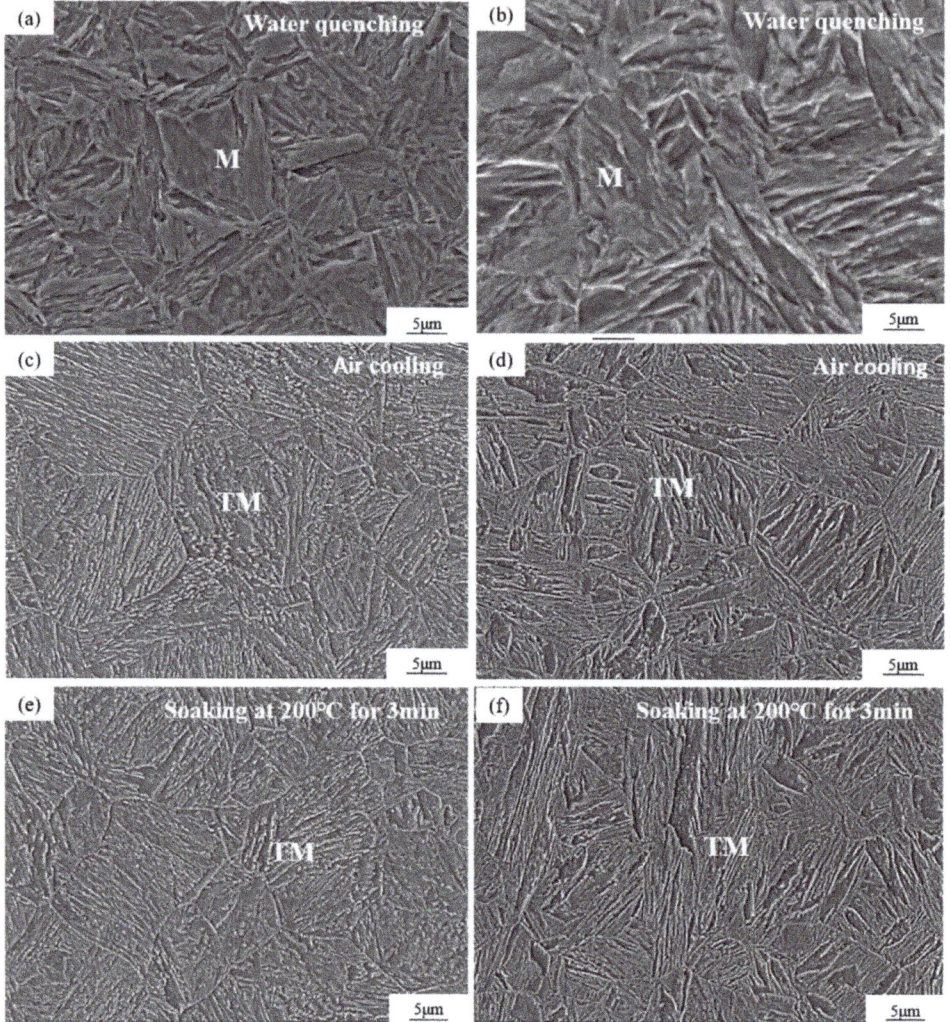

Figure 3. Microstructures of steel A (**a,c,e**) and steel B (**b,d,f**) heated at 880 °C for 3 min then treated with different cooling processes.

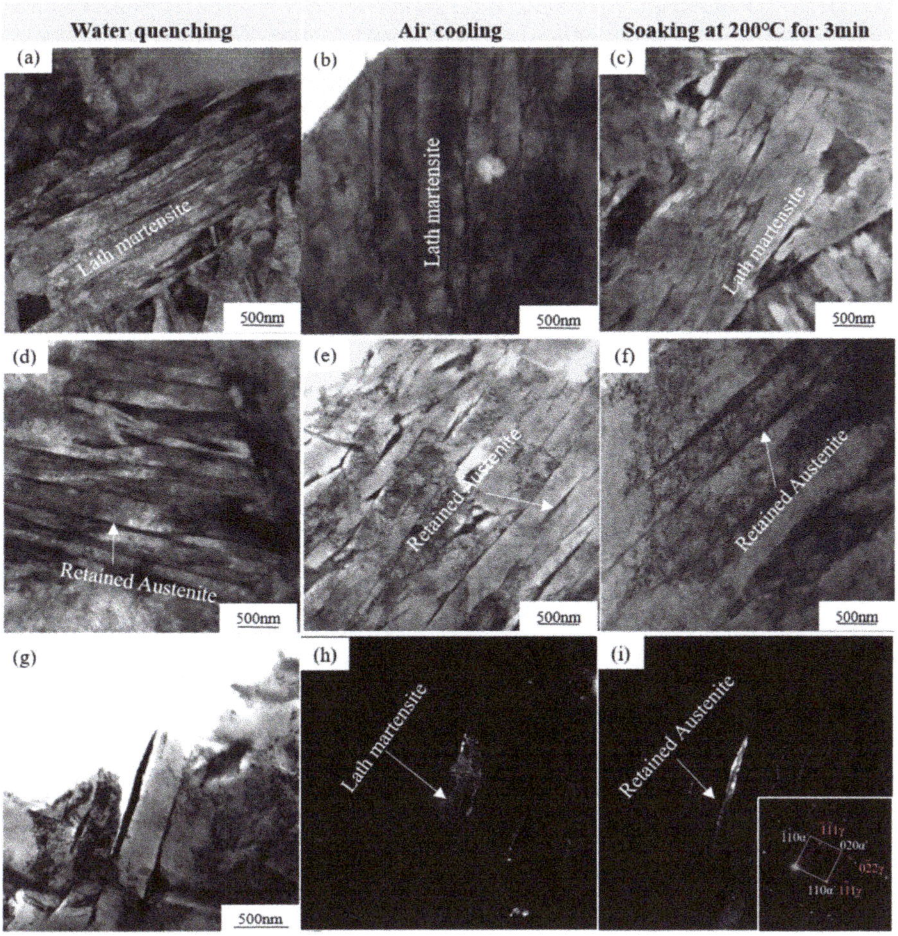

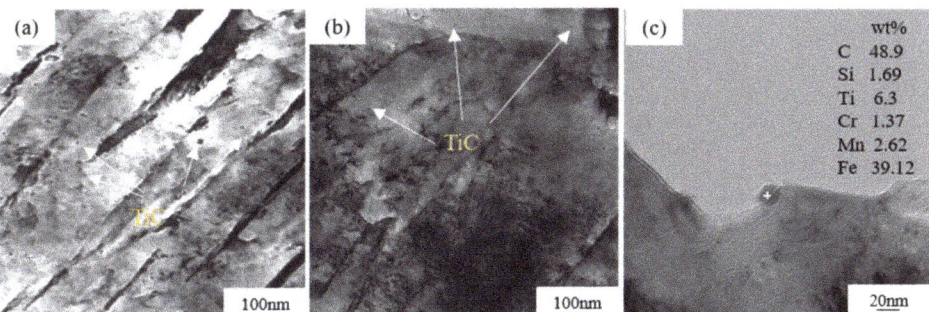

Figure 4. TEM analysis of lath martensite and retained austenite in steel A (**a**–**c**) and steel B (**d**–**f**) heated at 880 °C for 3 min then treated by different cooling processes, (**g**–**i**) show the bright field image, the dark field image and the corresponding selected area electron diffraction (SAED) pattern.

Figure 5. TEM analysis of nano-scale TiC carbides precipitated in experimental steels heated at 880 °C for 3 min then air cooling: (**a**) steel A, (**b**) steel B and (**c**) chemical composition of TiC.

Figure 6 shows the microstructures of steel A and steel B soaked at 200 °C for 30 min during air cooling. Compared with those soaked at 200 °C for 3 min, the martensite lath of the experimental steels is obviously coarsened. In steel B, the precipitation of ε-carbides can also be found, as shown in Figure 7. So, the yield strength is further improved by the interactions between nano-scale carbides and dislocations.

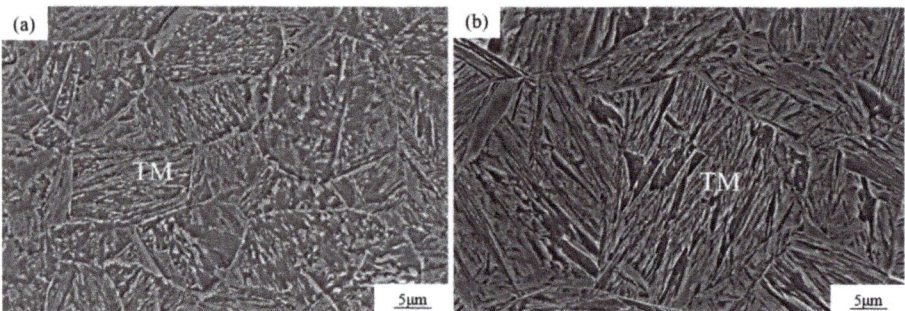

Figure 6. Microstructures of steel A (**a**) and steel B (**b**) heated at 880 °C for 3min then soaked at 200 °C for 30 min during air cooling.

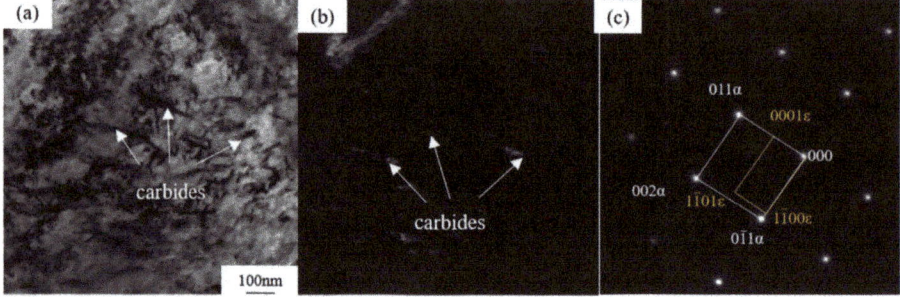

Figure 7. TEM analysis of steel B heated at 880 °C for 3 min, soaked at 200 °C for 30 min then air-cooled. (**a**) bright field TEM image and (**b**) dark field TEM image and (**c**) corresponding SAED pattern of lath martensite and ε-carbides.

The tensile properties at room temperature of the experimental steels under different cooling processes are shown in Table 2. The tensile strength of all the steels exceeded 1000 MPa. The steels which underwent water quenching had the highest tensile strength and the lowest elongation. For steel A, the product of strength and elongation (PSE) under the air-cooling condition was the highest, while for steel B, the highest PSE was obtained under soaking at 200 °C for 3 min during air cooling.

Table 2. Mechanical properties of steels heated at 880 °C for 3 min then treated by different cooling processes.

Cooling Process	TS (MPa)		YS (MPa)		TE (%)		PSE (MPa·%)	
	A	B	A	B	A	B	A	B
Water quenching	1384 ± 30	1646 ± 39	1184 ± 33	1242 ± 37	5.9 ± 0.2	6.6 ± 0.3	8166	10,864
Air cooling	1258 ± 35	1466 ± 23	843 ± 29	916 ± 26	10.9 ± 0.4	8.9 ± 0.2	13,712	13,047
Soaking at 200 °C for 3 min	1195 ± 13	1410 ± 24	752 ± 21	826 ± 17	11.4 ± 0.7	10.1 ± 0.5	13,623	14,241
Soaking at 200 °C for 30 min	1176 ± 22	1422 ± 31	743 ± 14	1038 ± 13	8.1 ± 0.3	9.0 ± 0.1	9526	12,798

As shown in Table 2, compared with the samples under water quenching, there was approximately a 70% increase in the PSE of steel A under air cooling, while there was a 20% increase for steel B under the same cooling conditions. As indicated in Table 1, the martensite transformation temperature M_S for steel A and steel B were 395 °C and 371 °C, respectively. When martensite is formed below M_S, it may have the opportunity of tempering during the remainder of the cooling. This phenomenon, which is referred to as auto-tempering, is more likely to occur in steels with a higher M_S, when the temperature is about or above 300 °C. This is because at this temperature, carbon possesses a diffusion activation energy between 60 and 80 kJ mol^{-1}, which is favorable to the diffusion in martensite, i.e., the interstitial carbon atoms in the tetragonal martensite lattice can easily diffuse from the octahedral interstices to the position of defects such as dislocations and/or the martensite boundary [12,13]. Then the solid solution strengthening effect of martensite is weakened, and the elongation increases as the tensile strength decreases. Compared with steel B, the M_S of steel A with a lower carbon level is higher, so auto-tempering is more likely to take place.

Detected by an infrared thermometer, the average cooling rates during the martensite transformation (between 400 and 20 °C) for the samples were about 300 °C/S and 3 °C/S, respectively. The mean diffusion distance of C atoms d_c in martensite can be integrated by Equations (2) and (3) [14–16]:

$$d_c = \sqrt{Dt} \tag{2}$$

$$D = 2 \times 10^{-6} \times \exp(-1.092 \times 10^5 / 8.314T) \tag{3}$$

where t is the time (s), D is the diffusion coefficient of carbon and T is the temperature (K). The diffusion distances of the C atoms corresponding to water quenching and air cooling are plotted in Figure 8. Carbon migrated 3.8 μm in the first martensite laths of steel A under air cooling to room temperature while under water quenching, the value was only 0.95 μm. Similarly, for steel B, the diffusion distances were 2.5 μm and 0.73 μm, respectively. According to the calculated results, it can be inferred that for the steel under water quenching, the probability of the occurrence of auto-tempering is less because of the limited diffusion capacity of the carbon atom.

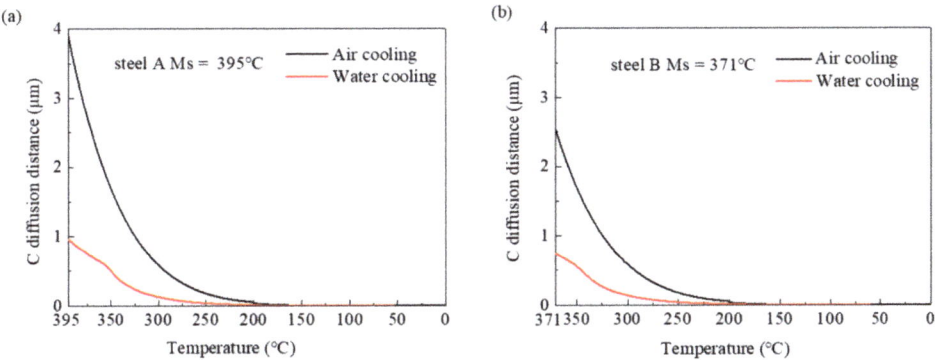

Figure 8. C diffusion distance during martensite transformation for steel A (**a**) and steel B (**b**).

In order to have a clear understanding of the strengthening mechanisms of the two steels under air cooling, all strengthening factors which play a role in σ_Y were carefully considered. According to the previous study [17,18], σ_Y of the experimental steel is attributed to multiple strengthening mechanisms, which can be expressed using the following formula:

$$\sigma_Y = \sigma_0 + \sigma_d + \sigma_g + \sigma_s + \sigma_p \tag{4}$$

where σ_0 is the internal frictional stress of body centered cubic (BCC) iron 54 MPa [19], σ_d is the dislocation strengthening in martensite; σ_s is the solid solution strengthening, σ_g

is the effective grain boundary strengthening, and σ_p is the precipitation strengthening contribution from nano-scale precipitates.

The dislocation density can be estimated by the modified Williamson–Hall (MWH) method based on an XRD analysis [20,21]. The MWH equation is written as follows:

$$\frac{2\delta \cos\theta}{\lambda} = \frac{0.9}{D} + Mb\sqrt{\frac{\pi\rho}{2}}\frac{2\sin\theta}{\lambda}C^{1/2} \quad (5)$$

where, δ, θ, and λ represent the XRD full width at half maximum (FWHM) of the diffraction peak, the diffraction angle, and the wavelength of the X-ray, respectively. For Cu radiation, the value of λ is 0.154 nm. D, ρ, and b are the average grain size, dislocation density and the Burgers vector of 0.248 nm, respectively. M is a constant of 3, and it depended on the effective out cut-off radius of dislocation density. C is the contrast factor of the dislocations and it can be expressed as follows [22]:

$$C = 0.285\left[1 - q\frac{h^2k^2 + k^2l^2 + l^2h^2}{(h^2 + k^2 + l^2)^2}\right] \quad (6)$$

where h, k, and l are the Miller's indices of each peak of martensite.

In the present work, the XRD diffraction peaks used for this estimation were the (110), (200) and (211) peaks, as shown in Figure 9a. Using Equations (5) and (6) combined, the value of $0.9/D$ was obtained as the intercept in the Figure 9b, imposing a linear relationship between $2\sin\theta/\lambda$ and $2\delta\cos\theta/\lambda$ by Origin data analysis software (OriginLab, Northamptom, MA, USA). The parameter q can be derived from the ratio between absolute value of slope and the intercept in Figure 9c, according to the linear relationship between $(2\delta\cos\theta/\lambda - 0.9/D)^2/(2\sin\theta/\lambda)^2$ and $(h^2k^2 + k^2l^2 + l^2h^2)/(h^2 + k^2 + l^2)^2$. Then, based on the value of the slope, as shown in Figure 9d, the dislocation density was calculated as 1.51×10^{14} and 2.32×10^{14} m^{-2} for steel A and steel B by the fitted curves of $2\sin\theta \cdot C^{1/2}/\lambda$ and $2\delta\cos\theta/\lambda$, as shown in Figure 9d.

The increased yield stress resulting from the dislocation strengthening can be estimated by the Baile–Hirsch relationship [23]:

$$\sigma_d = MG\alpha b\rho^{1/2} \quad (7)$$

where G is the shear modulus of 82 GPa, and α and ρ are constants with the value of about 0.24 and 3 [19]. ρ is the total dislocation density. Based on Equation (7), the σ_d of steel A and B was calculated as 179.9 MPa and 223.0 MPa.

The σ_s contribution is expressed using the following empirical equation [24]:

$$\sigma_s = 4570Xc + 84XSi + 32XMn - 30XCr + 80XTi \quad (8)$$

where Xc, XMn, XSi, XCr, and XTi are the weight percentages of C, Mn, Si, Cr, and Ti dissolved in the matrix, respectively. The average content of Xc, XMn, XSi, XCr, and XTi were obtained via SEM and EDS analyses, which were performed using at least five-spot analyses per condition, as shown in Table 3. The carbon content is difficult to accurately detect by EDS. It can be seen from Equation (8) that a small amount of carbon would make a great contribution to the solution strengthening. Therefore, the solution strengthening effect evaluated would be lower than the actual contribution value. The calculated σ_s is shown in Table 3.

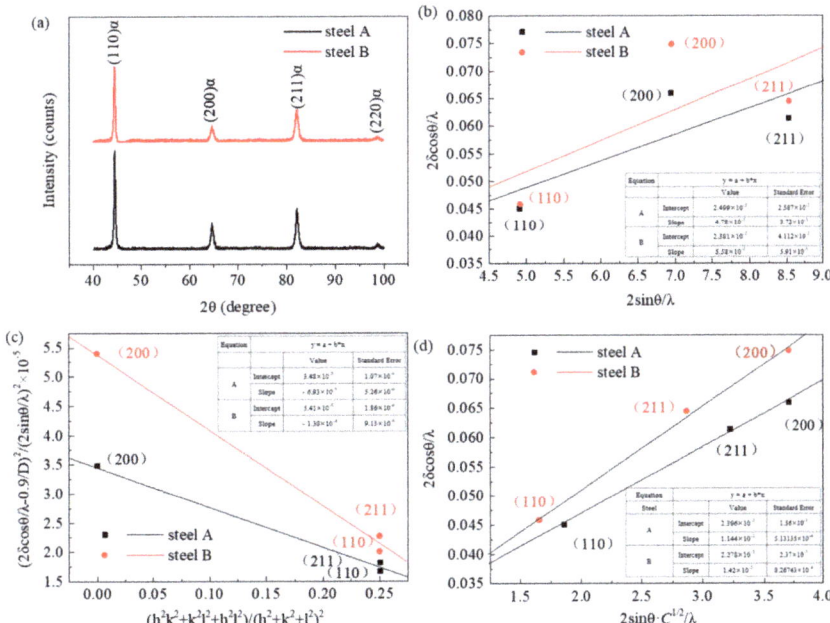

Figure 9. (a) The measured XRD profiles, and the estimation of dislocation density for steel A and B heated to 880 °C then air cooling according to a linear relationship between $2\sin\theta/\lambda$ and $2\delta\cos\theta/\lambda$ (b), $(h^2k^2 + k^2l^2 + h^2l^2)/(h^2 + k^2 + l^2)^2$ and $(2\delta\cos\theta/\lambda - 0.9/D)^2/(2\sin\theta/\lambda)^2$ (c), $2\sin\theta \cdot C^{1/2}/\lambda$ and $2\delta\cos\theta/\lambda$ (d).

Table 3. Chemical composition of the matrix (wt%) of steel A, B and the calculated σ_s.

Samples	Mn	Si	Cr	Ti	σ_s/MPa
A	2.30	1.32	1.03	0.010	154.4
B	2.36	1.45	1.04	0.013	167.2

The σ_g is calculated using the Hall–Petch principle [25]:

$$\sigma_g = k \times d^{-1/2} \quad (9)$$

where k is the Hall–Petch slope 120 MPa/·μm$^{1/2}$, and d is the average width of the martensite lath in μm. According to the analysis of the microstructure in Figure 4, the average widths of the lath martensite of steel A and B are 0.367 μm and 0.332 μm, respectively. Therefore, the calculated results of σ_g are 198.1 and 208.3 MPa for steel A and B, respectively.

The precipitation strengthening caused by carbide can be calculated quantitatively by the Ashby–Orowan equation [26] under the assumption of particle by-passing, as follows:

$$\sigma_p = \left(\frac{0.538Gb\sqrt{V_f}}{X}\right) \ln\left(\frac{X}{2b}\right) \quad (10)$$

where X and V_f are the mean diameter of the precipitates and the volume fraction of the precipitates, respectively.

Combined with the particle size observed in the TEM analysis, as shown in Figure 5, and the volume fraction of these particles calculated by Thermo-Calc software with the TCFE10 database, the resulting strength increase was calculated using Equation (10) and is shown in Table 4.

Table 4. Calculated volume fraction, average size of the carbide, and the calculated σ_p of the experimental steel.

Samples	Volume Fraction	Average Size/nm	σ_p/MPa
A	1.937×10^{-4}	9.676	47
B	2.376×10^{-4}	10.231	50

Based on the above calculations, the σ_Y estimated by Equation (4) is about 633.4 and 702.5 MPa for steel A and B. It is far below the measured yield strength of 843 and 916 MPa.

According to Equation (8), the σ_s will obviously increase if minor carbon is considered. So, the solid solution strengthening is the dominating strengthening mechanism for the experimental steels, and steel B with the higher carbon content in the matrix shows a higher yield strength.

3.2. Effects of Different Heating Temperaturse on Microstructures and Mechanical Properties

As shown in Figure 10, when the heating temperature decreases from 880 °C to 800 °C, the size of the martensite lath becomes shorter due to the refinement of the grain, and the amount of well-tempered martensite decreases. In addition, the recrystallization of ferrite was found in the samples that were soaked at 200 °C for 3 min then air-cooled. According to Table 5, the PSE of the steels under air cooling is better than that under water quenching, the value of which is similar to that obtained for the steels heated at 880 °C.

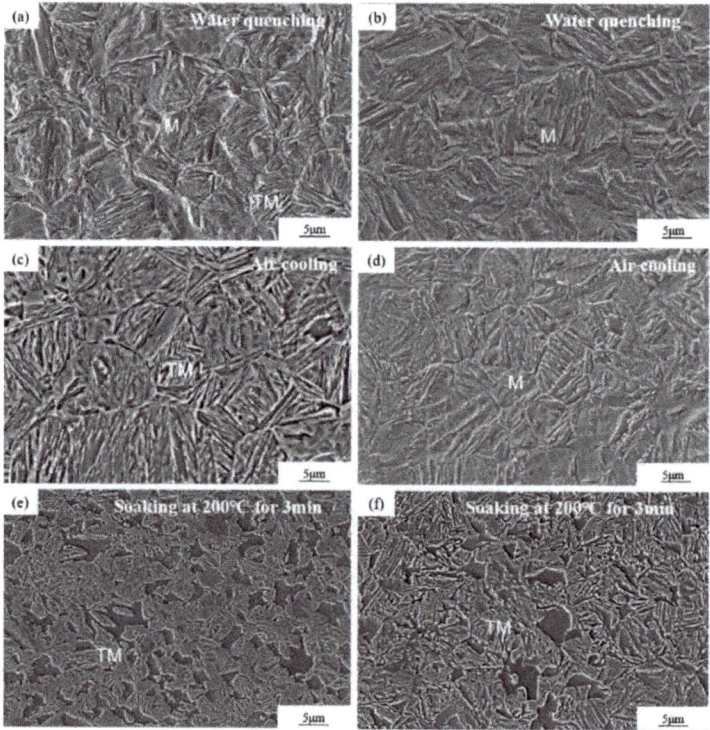

Figure 10. Microstructures of steel A (**a,c,e**) and steel B (**b,d,f**) heated to 800 °C for 3 min then treated by different cooling processes.

Table 5. Mechanical properties of steels heated to 800 °C for 3 min then under different cooling processes.

Cooling Process	TS (MPa)		YS (MPa)		TE (%)		PSE (MPa·%)	
	A	B	A	B	A	B	A	B
Water quenching	1491 ± 22	1726 ± 43	1069 ± 33	1225 ± 19	7.6 ± 1.1	6.9 ± 0.3	11,332	11,909
Air cooling	1271 ± 18	1553 ± 11	828 ± 42	952 ± 17	9.0 ± 0.4	8.2 ± 0.2	11,439	12,735
Soaking at 200 °C for 3 min	1181 ± 27	1517 ± 15	640 ± 23	1056 ± 28	15.0 ± 0.7	10.2 ± 0.5	17,715	15,473

The equilibrium composition of austenite in steel A and steel B at 800 °C was calculated by Thermo-Calc software (Thermo-Calc Software, Stockholm, Sweden) with the TCFE10 database and is shown in Table 6, where it can be seen that the carbon content in austenite in steel A and B is 0.179 wt% and 0.210 wt%, respectively., and both values are higher than those in the matrix. The M_S of steel A and B can be calculated as 396 °C and 369 °C, respectively by MUCG83 [27], which are close to the experimental results listed in Table 1. However, according to the calculated composition of austenite listed in Table 6, the M_S of steel A and B heated at 800 °C can be calculated as 353 °C and 344 °C, respectively, so smaller amounts of auto tempered martensite were obtained due to the decrease in the M_S. It is obvious that the PSE of steels heated at 880 °C under air cooling is much higher than those at 800 °C. The PSE of steel A heated at 800 °C under air cooling is 11,439 MPa·%, a bit higher than that under water quenching, which is 11,332 MPa·%. However, the PSE of steel A after soaking at 200 °C for 3 min is 17,715 MPa·%, much higher than that after water quenching; this is because the interstitial carbon atoms can diffuse more easily from the martensite lattice during soaking. which leads to an increased elongation through the auto-tempering of martensite.

Table 6. Calculated equilibrium composition of austenite and the volume fraction of ferrite and austenite in experimental steels at 800 °C.

Steel	Ferrite (%)	Austenite (%)	Elements in Austenite (wt%)			
			C	Mn	Si	Cr
Steel A	30.3	69.7	0.179	2.47	1.29	1.04
Steel B	16.2	83.8	0.210	2.31	1.36	1.03

In order to further investigate the effect of the auto-tempering behavior of martensite on the properties of experimental steels heated at different temperatures, the hole expansion ratio of steels is listed in Table 7 and the crack shapes in the hole-edge regions of steels after HET are shown in Figure 11. The main crack around the hole-edge region after HET seems to occur along RD. The HER is 25.9% and 16.6% in steel A, which are obviously higher values than those for steel B. Moreover, the HER of the steels heated at 880 °C is better than that at 800 °C.

As was mentioned above, compared with steel B, the higher martensitic transformation temperature of steel A promotes the auto-tempering of martensite, and the well-tempered martensite can be more severely deformed compared to the less-tempered martensite. The HER of the steel heated at 800 °C is lower, which can be mainly attributed to the hardening of the martensite that could accelerate crack initiation at the interface of martensite and ferrite because of the large difference in hardness between the two phases [28]. The HER of the steel A heated at 880 °C is much higher because of the softening of its well-tempered martensite.

Table 7. HER of steel A and B under different temperatures.

Heating Temperature	Hole Expansion Ratio (%)	
	Steel A	Steel B
800 °C	16.6	2.4
880 °C	25.9	8.9

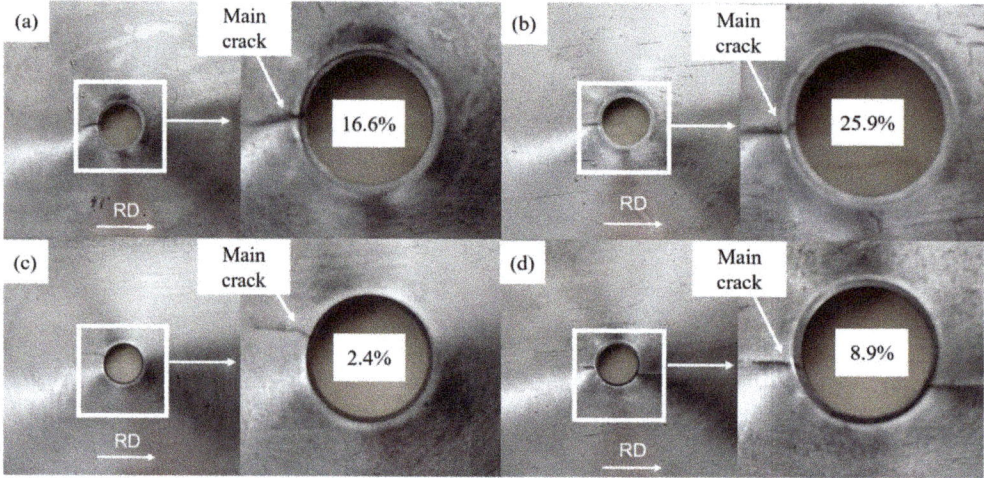

Figure 11. Crack shape in the hole-edge regions after HET of steels A (**a,b**) and steel B (**c,d**), heated at 800 °C and 880 °C for 3 min then air cooled.

3.3. Effects of Flash Heating on Microstructures and Mechanical Properties

Recently, flash heating technology with a heating rate of 100–300 °C/s has been proposed to study the effects of the mechanical properties of AHSS [29]. Flash heating is found to effectively refine the multiphase microstructures of DP steels and QP steels resulting from a retardation of recrystallization to a large extent and the induction of explosive nucleation of intercritical austenite [30–32].

The influence of conventional heating and flash heating on the structure and mechanical properties of the experimental steels is compared. The microstructures of steel A and steel B composed of martensite and ferrite are shown in Figure 12. Compared with the steels under conventional heating conditions, the refined martensite microstructure with equiaxed ferrite of the steels under flash heating was obtained and there was no obvious auto-tempering phenomenon because flash heating can result in the transformation of ferrite to austenite to be delayed and can raise the transit temperature above A_3 temperature; the M_S of austenite is decreased accordingly [33]. As shown in Figure 13, the size of the martensite lath of steel A and B under flash heating was about 0.349 µm and 0.273 µm, which are also smaller sizes than those obtained under conventional heating. The prior austenite grain morphology and the size distribution of steels under different heating conditions are shown in Figures 14 and 15, where it can be seen that the the average size of the austenite grain was 12.19 µm and 12.83 µm under conventional heating, and 6.28 and 6.39 µm under flash heating. According to Equation (9), the effect of the grain boundary strengthening can be calculated as 203.1 MPa and 229.6 MPa.

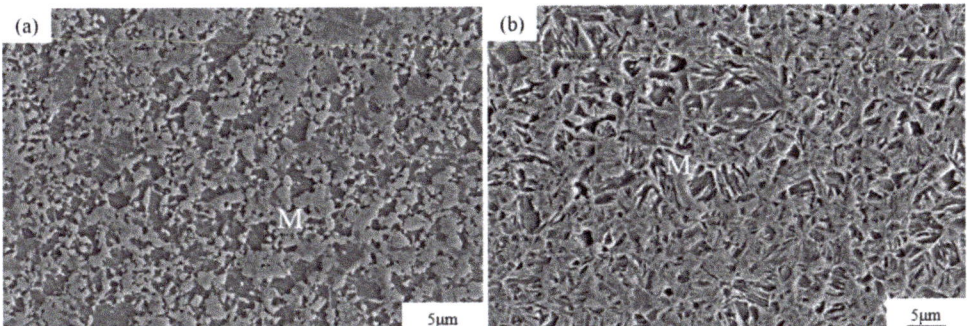

Figure 12. Microstructures of steel A (**a**) and B (**b**) under flash heating to 880 °C then air cooled.

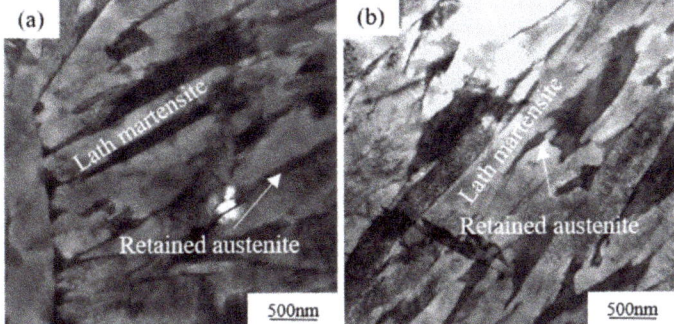

Figure 13. TEM analysis of lath martensite in steel A (**a**) and steel B (**b**) under flash heating to 880 °C then air cooled.

As shown in Figure 16b, the yield strength of the steels that were flash heated is obviously improved compared with that under conventional heating and air cooling. According to the XRD analysis in Figure 17a, the value of $0.9/D$ could be obtained as the intercept in the Figure 17b, the parameter q can be derived from the ratio between the absolute value of slope and the intercept in Figure 17c, and based on the value of the slope, the dislocation density can be calculated as 1.80×10^{14} m^{-2} and 2.65×10^{14} m^{-2} for steel A and steel B, as shown in Figure 17d. Then the increased yield stress that resulted from the dislocation strengthening can be calculated by Equation (7). The values of σ_d of steel A and B under flash heating are 196.4 MPa and 238.4 MPa, respectively. It can be seen that under flash heating, the effect of the grain boundary strengthening, and dislocation strengthening is not obvious for the experimental steel, while the yielding strength of the sample is about 100 MPa higher than that under conventional heating. As was stated in the above discussion, under flash heating, the carbon content in the martensite of steel is higher because of the existence of ferrite, so the contribution of solid solution strengthening is the main reason for the higher yield strength of the steels.

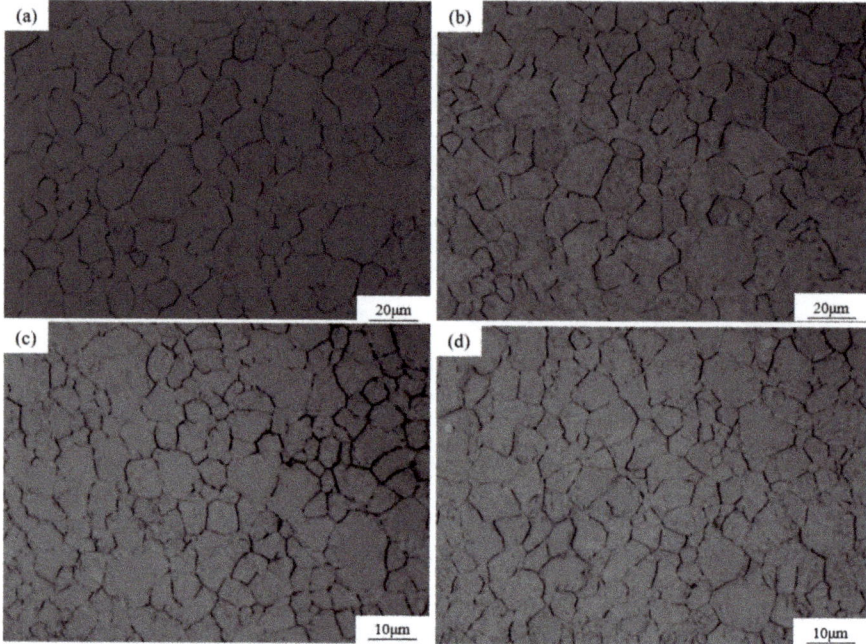

Figure 14. Prior austenite grain morphology of steels observed by OM: (**a**) steel A and (**b**) steel B under conventional heating at 880 °C then air cooled (CHA); (**c**) steel A and (**d**) steel B under flash heating to 880 °C then air cooled (FHA).

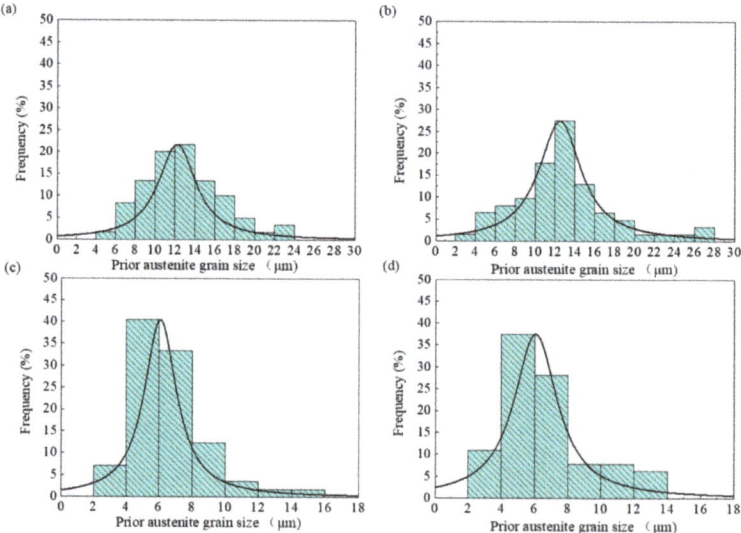

Figure 15. Prior Austenite grain size distribution of steels: (**a**) steel A and (**b**) steel B under conventional heating at 880 °C then air cooling (CHA); (**c**) steel A and (**d**) steel B under flash heating to 880 °C then air cooling (FHA).

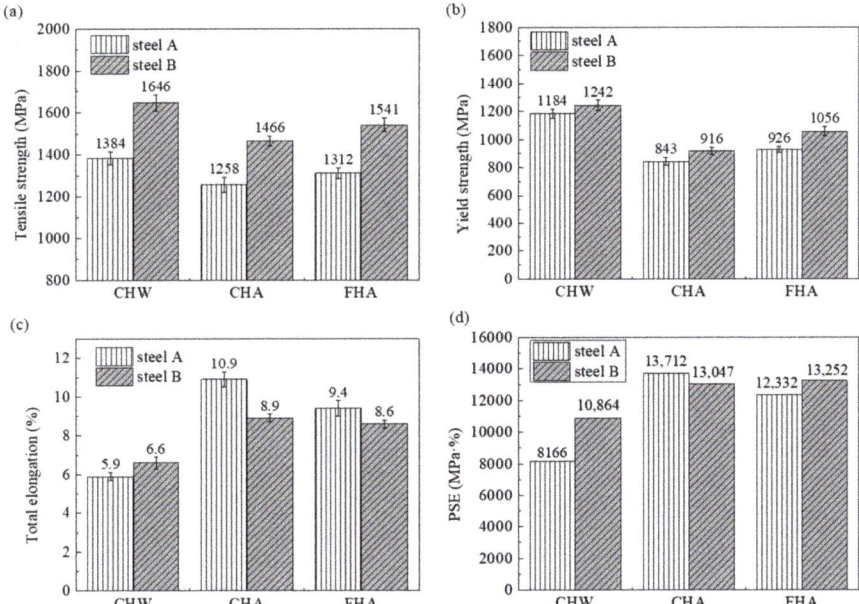

Figure 16. Mechanical properties of steels: (**a**) tensible strength, (**b**) yield strength, (**c**) total elongation and (**d**) PSE under conventional heating at 880 °C then water quenching (CHW), air cooling (CHA) and flash heating to 880 °C then air cooling (FHA).

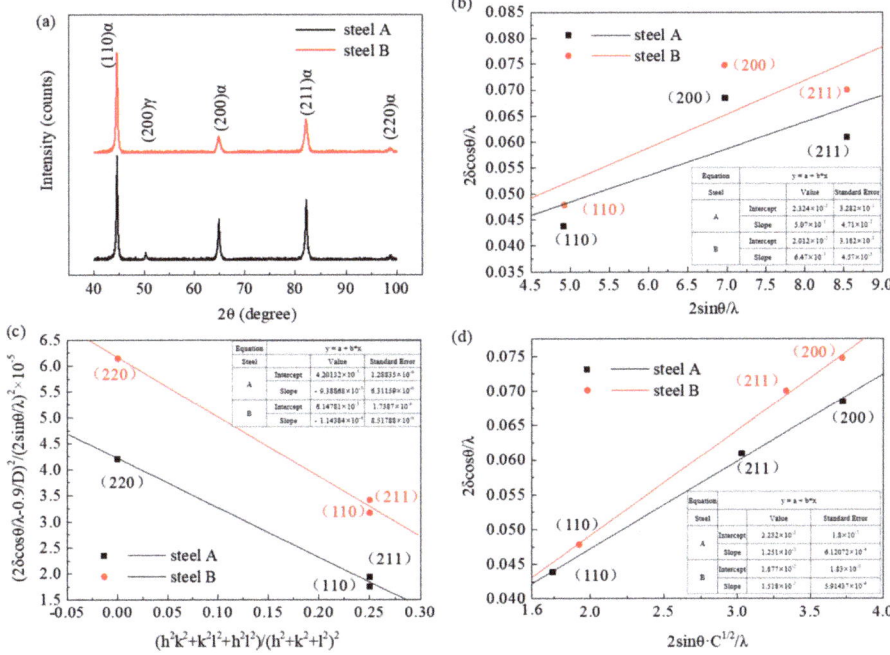

Figure 17. (**a**) The measured XRD profiles, the estimation of dislocation density for steel A and B under flash heating to 880 °C then air cooling according to a linear relationship between $2\sin\theta/\lambda$ and $2\delta\cos\theta/\lambda$ (**b**), $(h^2k^2+k^2l^2+h^2l^2)/(h^2+k^2+l^2)^2$ and $(2\delta\cos\theta/\lambda-0.9/D)^2/(2\sin\theta/\lambda)^2$ (**c**), $2\sin\theta\cdot C^{1/2}/\lambda$ and $2\delta\cos\theta/\lambda$ (**d**).

As is well known, the elongation of steel could be improved through uniform plastic deformation in a refined matrix. However, for the steels under flash heating, the elongation of steel A and B was lower than that under conventional heating as shown in Figure 16c. The grain refinement of the steel under flashing heating would enhance the thermal stability of austenite [34]. Moreover, although the heating temperature was 880 °C, far higher than its A$_3$ temperature, the ferrite phase still remained in the microstructure of the steel, which increased the carbon content in austenite. With the stability of the undercooling austenite increased, the M$_S$ of the experimental steel is so low that the martensite transformation is suppressed and hard to auto-temper. So, the retained austenite can be obviously observed in the microstructure of steel B, as shown in Figure 13. It is reported that Q&P steel under flash heating exhibited good elongation resulting from the increase in the content and stability of the retained austenite [34]. However, the elongation of the experimental steel was not improved. The relationship between the instantaneous strain hardening exponent (*n* value) and the true strain of the two steels was obtained according to the engineering stress–strain curve, as shown in Figure 18. It can be seen that the n value decreases constantly. According to the authors of [35], if transformation-induced plasticity happened in experimental steel, there would be a platform on the n value–true strain curve, resulting from transformation hardening and stress relaxation softening that coexist in the matrix during the gradual transformation of retained austenite to martensite. Obviously, there is no transformation-induced plasticity effect for the retained austenite in the steels during the tensile deformation. So, although the grain refinement can be found in the samples under flash heating, the microstructure composed of the ferrite soft phase and the less tempered martensite hard phase easily cracked under tensile stress. On the contrary, the samples under conventional heating had a better elongation because of their microstructure which is composed of well-tempered martensite.

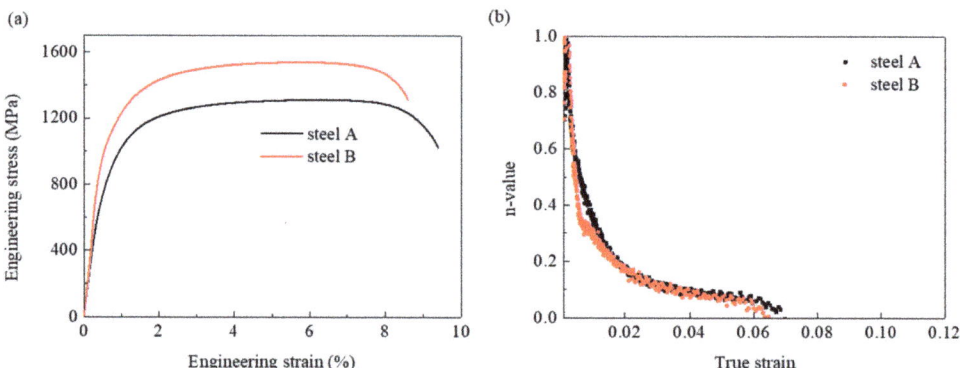

Figure 18. (a) Engineering stress versus strain and (b) *n*-value versus true strain of steel A and B under flash heating.

4. Conclusions

(1) Different cooling processes have a great influence on the performance of the investigated steels resulting from the different auto-tempering effect. Compared with the steels under water quenching, there was approximately a 70% increase of the PSE of steel A with well-tempered martensite under air cooling, which was caused by the fact that the probability of auto-tempering was less for the steel under water quenching according to the calculated average diffusion distance of the carbon atom in martensite for each steel. For steel B with a higher carbon content, its PSE was only 20% higher under air cooling than that of water quenching because its M$_S$ is lower and this limits the auto-tempering of martensite. A tensile strength of over 1400 MPa and a total elongation rate exceeding 10% can be obtained for steel B under soaking at 200 °C for 3 min.

(2) Steels heated at 800 °C then treated by either air cooling or water quenching both exhibited a low PSE since the M_S temperatures of the austenite in these steels were low. On the contrary, the PSE of the steels can be obviously improved under soaking at 200 °C for 3 min where the well-tempered martensite dominated. Although the steel with the microstructure composed of the ferrite, well-tempered martensite, and less-tempered martensite had an elongation that exceeded 15%, the hole expansion ratio was still lower because of the undesirable hardness distribution between the phases.

(3) Flash heating obviously refines the microstructure with equiaxed ferrite. The contribution of solution strengthening is the main cause of the higher yield strength for the steels under flash heating. Moreover, the improvement of elongation could not be obtained because cracks easily occurred due to the stress concentration at the interface of the soft phase ferrite and the hard phase martensite. It was difficult to obtain steel with ideal properties and less temped martensite under flash heating.

Author Contributions: Conceptualization, Y.H. and H.J.; methodology, Y.H.; software, H.J.; validation, H.J.; formal analysis, L.L. (Li Lin); investigation, R.L.; resources, Y.H.; data curation, W.Z.; writing—original draft preparation, H.J.; writing—review and editing, H.J.; visualization, Y.Z.; supervision, L.L. (Lin Li); project administration, Y.H.; funding acquisition, Y.H. All authors have read and agreed to the published version of the manuscript.

Funding: The present work was supported by the National Key R&D Program of China (Grant No. 2017YFB0304402) and National Natural Science Foundation of China (Grant No. 51971127).

Informed Consent Statement: Informed consent was obtained from all subjects involved in the study.

Conflicts of Interest: The authors declare no conflict of interest.

References

1. He, J.; Han, G.; Li, S.; Zhou, D. To correlate the phase transformation and mechanical behavior of QP steel sheets. *Int. J. Mech. Sci.* **2019**, *152*, 198–210. [CrossRef]
2. Navarro-López, A.; Sietsma, J.; Santofimia, M.J. Effect of prior athermal martensite on the isothermal transformation kinetics below Ms in a low-C high-Si steel. *Metall. Mater. Trans. A* **2016**, *47*, 1028–1039. [CrossRef]
3. Taylor, T.; Clough, A. Critical review of automotive hot-stamped sheet steel from an industrial perspective. *Mater. Sci. Technol.* **2018**, *34*, 809–861. [CrossRef]
4. Cooman, D.; Bruno, C.; Estrin, Y.; Kim, S.K. Twinning-induced plasticity (TWIP) steels. *Acta Mater.* **2018**, *142*, 283–362. [CrossRef]
5. Gibbs, P.; de Moor, E.; Merwin, M.; Clausen, B.; Speer, J.; Matlock, D. Austenite stability effects on tensile behavior of manganese-enriched-austenite transformation-induced plasticity steel. *Metall. Mater. Trans. A* **2011**, *42*, 3691–3702. [CrossRef]
6. Griffin, P.; Geoffrey, W.; Hammond, P. The prospects for 'green steel' making in a net-zero economy: A U5K perspective. *Glob. Transit.* **2021**, *3*, 72–86. [CrossRef]
7. Arribas, M.; López, B.; Rodriguez-Ibabe, J.M. Additional grain refinement in recrystallization controlled rolling of Ti-microalloyed steels processed by near-net-shape casting technology. *Mater. Sci. Eng. A* **2008**, *485*, 383–394. [CrossRef]
8. Matsuda, H.; Reiko, M.; Yoshimasa, F.; Kazuhiro, S.; Saiji, M.; Yasushi, T. Effects of auto-tempering behavior of martensite on mechanical properties of ultra high strength steel sheets. *J. Alloys Compd.* **2013**, *577*, 661–667. [CrossRef]
9. Li, C.; Guo, Y.; Ji, F.; Ren, D.; Wang, G. Effects of auto-tempering on microstructure and mechanical properties in hot rolled plain C-Mn dual phase steels. *Mater. Sci. Eng. A* **2016**, *665*, 98–107. [CrossRef]
10. Saastamoinen, A.; Antti, K.; Tun, T.; Pasi, S.; David, P.; Jukka, K. Direct-quenched and tempered low-C high-strength structural steel: The role of chemical composition on microstructure and mechanical properties. *Mater. Sci. Eng. A* **2019**, *760*, 346–358. [CrossRef]
11. Saastamoinen, A.; Antti, K.; David, P.; Pasi, S.; Yang, J.; Tsai, Y. The effect of finish rolling temperature and tempering on the microstructure, mechanical properties and dislocation density of direct-quenched steel. *Mater. Charact.* **2018**, *139*, 1–10. [CrossRef]
12. Bhadeshia, H.; Honeycombe, S.R. *The Tempering of Martensite—ScienceDirect*, 3rd ed.; Butterworth-Heinemann: Oxford, UK, 2006; pp. 183–208.
13. Morito, S.; Keiichiro, O.; Kazuhiro, H.; Takuya, O. Carbon Enrichment in Retained Austenite Films in Low Carbon Lath Martensite Steel. *ISIJ Int.* **2011**, *51*, 1200–1202. [CrossRef]
14. Hillert, M. The kinetics of the first stage of tempering. *Acta Metall.* **1959**, *7*, 653–658. [CrossRef]
15. Liu, Y. Internal friction associated with dislocation relaxations in virgin martensite—I. Experiments. *Acta Metall. Mater.* **1993**, *41*, 3277–3287. [CrossRef]

16. Chang, Z.Y.; Li, Y.J.; Wu, D. Enhanced ductility and toughness in 2000 MPa grade press hardening steels by auto-tempering. *Mater. Sci. Eng. A* **2020**, *784*, 139342. [CrossRef]
17. Morito, S.; Yoshida, H.; Maki, T.; Huang, X. Effect of block size on the strength of lath martensite in low carbon steels. *Mater. Sci. Eng. A* **2006**, *438–440*, 237–240. [CrossRef]
18. Zhao, Y.L.; Shi, J.; Cao, W.Q.; Wang, M.Q.; Xie, G. Effect of direct quenching on microstructure and mechanical properties of medium-carbon Nb-bearing steel. *J. Zhejiang Univ. Sci. A* **2010**, *11*, 776–781. [CrossRef]
19. Chen, W.; Gao, P.; Wang, S.; Zhao, X.; Zhao, Z. Strengthening mechanisms of Nb and V microalloying high strength hot-stamped steel. *Mater. Sci. Eng. A* **2020**, *797*, 140115. [CrossRef]
20. Ungár, T.; Borbély, A. The effect of dislocation contrast on x-ray line broadening: A new approach to line profile analysis. *Appl. Phys. Lett.* **1996**, *69*, 3173–3175. [CrossRef]
21. HajyAkbary, F.; Sitetsma, J.; Böttger, A.J.; Santofimia, M.J. An improved X-ray diffraction analysis method to characterize dislocation density in lath martensitic structures. *Mater. Sci. Eng. A* **2015**, *639*, 208–218. [CrossRef]
22. Takebayashi, S.; Kunieda, T.; Yoshinaga, N.; Ushioda, K.; Ogata, S. Comparison of the dislocation density in martensitic steels evaluated by some X-ray diffraction methods. *ISIJ Int.* **2010**, *50*, 875–882. [CrossRef]
23. Hu, B.; He, B.; Cheng, G.; Yen, H.; Huang, M.; Luo, H. Super-high-strength and formable medium Mn steel manufactured by warm rolling process. *Acta Mater.* **2019**, *174*, 131–141. [CrossRef]
24. Yong, Q.L. *Secondary Phase in the Steel*; Metallurgical Industry Press: Beijing, China, 2006.
25. Liu, T.Q.; Cao, Z.X.; Wang, H.; Wu, G.L.; Jin, J.J.; Cao, W.Q. A new 2.4 GPa extra-high strength steel with good ductility and high toughness designed by synergistic strengthening of nano-particles and high-density dislocations. *Scr. Mater.* **2020**, *178*, 285–289. [CrossRef]
26. Gladman, T. Precipitation hardening in metals. *Mater. Sci. Technol.* **1999**, *15*, 30–36. [CrossRef]
27. Yoozbashi, M.N.; Yazdani, S.; Wang, T.S. Design of a new nanostructured, high-Si bainitic steel with lower cost production. *Mater. Des.* **2011**, *32*, 3248–3253. [CrossRef]
28. Shirasawa, H.; Tanaka, Y.; Korida, K. Effect of Continuous Annealing Thermal Pattern on Strength and Ductility of Cold Rolled Dual Phase Steel. *Tetsu Hagané* **1988**, *74*, 326–333. [CrossRef]
29. Lesch, C.; Álvarez, P.; Bleck, W.; Sevillano, J.G. Rapid Transformation Annealing: A Novel Method for Grain Refinement of Cold-Rolled Low-Carbon Steels. *Metall. Mater. Trans. A* **2007**, *38*, 1882–1890. [CrossRef]
30. De Knijf, D.; Puype, A.; Föjer, C.; Petrov, R. The influence of ultra-fast annealing prior to quenching and partitioning on the microstructure and mechanical properties. *Mater. Sci. Eng. A* **2015**, *627*, 182–190. [CrossRef]
31. Liu, G.; Zhang, S.; Li, J.; Wang, J.; Meng, Q. Fast-heating for intercritical annealing of cold-rolled quenching and partitioning steel. *Mater. Sci. Eng. A* **2016**, *669*, 387–395. [CrossRef]
32. Yonemura, M.; Nishibata, H.; Nishiura, T.; Ooura, N.; Yoshimoto, Y.; Fujiwara, K.; Kawano, K.; Terai, T.; Inubushi, Y.; Inoue, I.; et al. Fine microstructure formation in steel under ultrafast heating. *Sci. Rep.* **2019**, *9*, 11241. [CrossRef]
33. Xu, D.; Li, J.; Meng, Q.; Liu, Y.; Li, P. Effect of heating rate on microstructure and mechanical properties of TRIP-aided multiphase steel. *J. Alloys Compd.* **2014**, *614*, 94–101. [CrossRef]
34. Liu, G.; Li, T.; Yang, Z.; Zhang, C.; Li, J.; Chen, H. On the role of chemical heterogeneity in phase transformations and mechanical behavior of flash annealed quenching & partitioning steels. *Acta Mater.* **2020**, *201*, 266–277.
35. He, Z.; He, Y.; Ling, Y.; Wu, Q.; Gao, Y.; Li, L. Effect of strain rate on deformation behavior of TRIP steels. *J. Mater. Process. Technol.* **2012**, *212*, 2141–2147. [CrossRef]

Article

The Effect of Different Annealing Strategies on the Microstructure Development and Mechanical Response of Austempered Steels

Eliseo Hernandez-Duran [1,2,3,*], Luca Corallo [1], Tanya Ros-Yanez [4], Felipe Castro-Cerda [2,3] and Roumen H. Petrov [1,2]

1. Research Group Materials Science and Technology, Department of Electromechanical, Systems & Metal Engineering, Ghent University, Tech Lane Science Park Campus A 46, 9052 Ghent, Belgium; luca.corallo@ugent.be (L.C.); roumen.petrov@ugent.be (R.H.P.)
2. Department of Materials Science and Engineering, Delft University of Technology, Mekelweg 2, 2628 CD Delft, The Netherlands; felipe.castro@usach.cl
3. Department of Metallurgy, University of Santiago de Chile, Alameda 3363, Estación Central, 9170022 Santiago, Chile
4. CLEVELAND-CLIFFS INC, Research & Innovation Center, 6180 Research Way, Middletown, OH 45005, USA; tanya.ros@clevelandcliffs.com
* Correspondence: eliseo.hernandez@usach.cl or eliseo.hernandezduran@ugent.be

Abstract: This study focuses on the effect of non-conventional annealing strategies on the microstructure and related mechanical properties of austempered steels. Multistep thermo-cycling (TC) and ultrafast heating (UFH) annealing were carried out and compared with the outcome obtained from a conventionally annealed (CA) 0.3C-2Mn-1.5Si steel. After the annealing path, steel samples were fast cooled and isothermally treated at 400 °C employing the same parameters. It was found that TC and UFH strategies produce an equivalent level of microstructural refinement. Nevertheless, the obtained microstructure via TC has not led to an improvement in the mechanical properties in comparison with the CA steel. On the other hand, the steel grade produced via a combination of ultrafast heating annealing and austempering exhibits enhanced ductility without decreasing the strength level with respect to TC and CA, giving the best strength–ductility balance among the studied steels. The outstanding mechanical response exhibited by the UFH steel is related to the formation of heterogeneous distribution of ferrite, bainite and retained austenite in proportions 0.09–0.78–0.14. The microstructural formation after UFH is discussed in terms of chemical heterogeneities in the parent austenite.

Keywords: austempering; ultrafast heating annealing; thermo-cycling annealing

1. Introduction

A method commonly employed to achieve suitable strength–ductility balance in steels is microstructural grain refinement [1,2]. Phase transformation of austenite into micro and nanosized lath shape BCC (ferrite-martensite-bainite) sub-units can be attained via heat treatment, controlling the temperature of phase transformation [3–5]. Refinement of the parent austenite grain (PAG) size has been proved as an effective strategy towards fine-grained steel grades [2,6–8]. Among the different methods for grain refinement, the addition of microalloying elements [9] and complex thermo-mechanical treatments [10] are well-known routes to achieve a fine distribution of PAGs and improved mechanical properties in high strength steels.

On the other hand, recent trends in advanced high strength steel production account for the development of lean alloy steels with outstanding mechanical performance reached via novel and efficient heat treatments [11,12]. In order to create retained austenite containing multiphase microstructures, most of the thermo-treatments for the new generation of steels take advantage of the decomposition of austenite and carbon partitioning from bainite and/or martensite [12]. Design and study of bainitic and martensitic based TRIP steels are mainly focused on the evaluation of results obtained by manipulation of low temperature heat treatment parameters (in the range from 200 to 500 °C), after a conventional annealing step (heating rate from 10 to 30 °C/s and soaking time at annealing temperature >60 s).

Therefore, unconventional annealing routes could be employed to modify the initial parent austenite phase, resulting in further improvement of the mechanical response of low alloy steels subjected to low temperature thermal paths. Results in thermo-cycling annealing [6–8,13] have shown that multiple annealing and cooling steps, conducing to successive martensite–austenite transformations, are an effective route to obtain a homogeneous distribution of fine-grained PAGs, starting the cycling with a coarse martensitic microstructure.

Another promising annealing route towards the new generation of steels is the ultrafast heating (UFH) [8,14–21]. This strategy represents an optimization of the heat treatment process by employing heating rates $\geq$100 °C/s, reducing the annealing time from several minutes to a window of 1 to 10 s. Thanks to the development of longitudinal and transverse flux induction heating technologies, the ultrafast heating of steel strips is feasible at small and large scales [17,22,23]. Pilot-scale installations for ultrafast heating applications are reported elsewhere [22,23]. The enhanced combination of mechanical properties in lean alloyed UFH steels is developed through the formation of fine-grained heterogeneous microstructures [8,15–18,24,25]. The microstructural grain refinement reached in ultrafast heating experiments is related to several factors including (i) preferential nucleation of austenite [26,27] and interaction between ferrite recrystallization and austenite phase transformation [20,28]; (ii) pinning effect by undissolved cementite carbides [21]; (iii) restricted austenitic grain growth by the high heating rate employed [29]. Moreover, current research on this topic has confirmed that solute heterogeneities in austenite, produced due to the lack of time for homogenization during the annealing step, are responsible for the formation of a complex mixture of constituents upon cooling [15–17,24,30].

This study aims to evaluate and clarify the influence of different annealing strategies on the microstructure development and related mechanical properties of austempered bainitic steels. Annealing treatments carried out here were designed to gain insight into the influence of different microstructural characteristics, produced via modification of the initial parent austenite, on the resulting microstructures and mechanical behavior.

2. Materials and Methods

A lean low carbon steel with composition listed in Table 1 is investigated. The as-received material is a 70% cold-rolled, 1.2 mm thickness steel with a microstructure consisting of 29% of ferrite and 69 ($\pm$3)% of pearlite (Figure 1). Equally distributed bands of ferrite and pearlite were found throughout the thickness of the studied material.

Table 1. Chemical composition, wt.%.

C	Mn	Si	P	S	FE
0.28	1.91	1.44	0.009	0.005	Bal.

The as-received material was subjected to three different annealing strategies, namely conventional (CA), thermo-cycling (TC) and ultrafast heating annealing (UFH). Throughout this manuscript, the heat-treated samples will be referred as CA, TC and UFH based on their annealing history.

Cold-rolled samples of dimensions $10 \times 5 \times 1.2$ mm^3 and $90 \times 20 \times 1.2$ mm^3, with the largest axis parallel to RD, were heat-treated in a Bähr 805A/D dilatometer (TA Instruments, New Castle, DE, USA) and in a Gleeble 1500 thermo-mechanical simulator (Dynamic Systems Inc., Poestenkill, NY, USA), respectively.

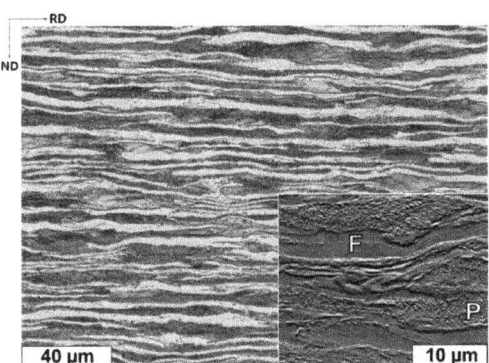

Figure 1. Microstructure of the steel in as-received 70% cold-rolled steel.

The A_{C3} temperature in each annealing treatment was estimated via dilatometric analysis employing the methodology presented in [14]. Samples treated according to the CA treatment were heated at 10 °C/s up to 885 °C, i.e., ≈30 °C above the A_{C3} (≈852 °C), and then soaked for 180 s followed by fast cooling at 160 °C/s. For TC, the first annealing step follows the same parameters as CA, and then three subsequent heating and cooling steps (cycles) were applied. The parameters of each cycle are a constant heating rate of 30 °C/s to 885 °C, soaking time of ≈2 s, and cooling at 160 °C/s to room temperature. The A_{C3} for the last annealing step (step 4) was estimated as 855 °C. The A_{C3} temperature for the samples heated at 500 °C/s in dilatometer was estimated as 892 °C. Nevertheless, this value was obtained in samples heated at 500 °C/s up to the A_{C1} temperature (767 °C), then the heating rate declined to ≈380 °C/s due to the decrease in efficiency of the longitudinal flux induction heating in dilatometer above the curie point and by the formation of paramagnetic austenite. Previous evaluations of the A_{C3} evolution with the heating rate in cold-rolled low alloy steels [31,32] indicated that the A_{C3} temperature shifts slightly when high heating rates are applied. Thomas [31] reported a shift of 1 to 3 °C of the A_{C3} temperature by increasing the heating rate from 100 °C/s to 1000 °C/s in 1020, 1019M and 15B25 cold-rolled steels. Using the Gleeble simulator, UFH samples were heated at 500 °C/s up to 925 °C, approximately 30 °C above the A_{C3} estimated by dilatometric analysis. Then, an isothermal holding step not greater than 0.3 s was employed to avoid chemical homogenization and austenitic grain growth at the annealing temperature. The selected cooling rate, after the annealing step, was 160 °C/s.

Figure 2a,b show the dilatometric curves for CA and TC obtained in samples directly cooled to room temperature and in samples isothermally held at 400 °C. The formation of martensite is clear from the expansion observed below the $M_s^{5\%}$ temperature in the dilatation-change in length v/s temperature curves (Figure 2a,b). In this work, the $M_s^{5\%}$ was defined as the temperature at which a 5% of the total dilatation generated by the martensitic transformation was measured by applying the lever rule method. Since it was not possible to reach a constant heating rate of 500 °C/s in the dilatometer, the M_S temperature for the sample peak annealed at 500 °C/s to 925 °C (UFH) was estimated by means of numerical differentiation of the cooling curves recorded in samples heat-treated using the Gleeble simulator. Figure 2c presents the change in the slope of the cooling curve at low temperature due to the exothermic characteristics of the austenite to martensite transformation. The insert in Figure 2c displays the derivate of the cooling curve.

M_S temperatures of 346 (±5) °C, 322 (±4) °C and 346 (±8) °C were estimated for CA, TC and UFH, respectively. The values presented in parenthesis correspond to the standard deviation of at least 2 measurements. These experimental results are in good agreement with the calculated M_S of 335 °C [33]:

$$M_s(°C) = 692 - 502C^{0.5} - 37Mn - 14Si \text{ (wt.\%)} \qquad (1)$$

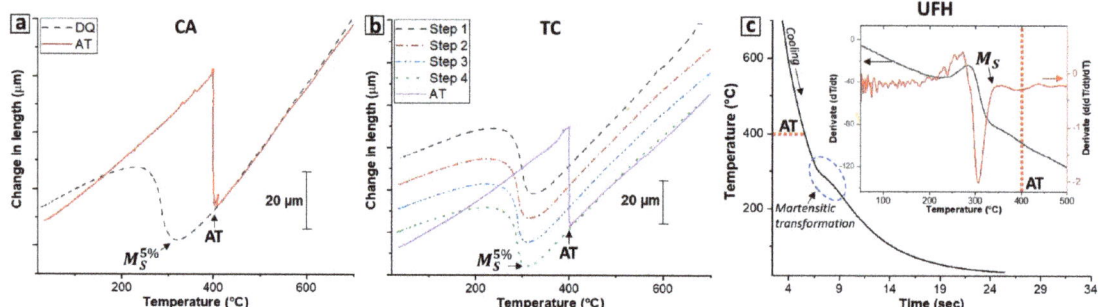

Figure 2. Dilatometric curves obtained during the cooling step, i.e., after the annealing treatments for samples (**a**) CA and (**b**) TC. (**c**) Temperature profile obtained during cooling for the ultrafast heated sample; the insert in (**c**) shows the estimation of the M_S temperature via differentiation of the recorded cooling curve.

On the other hand, the dilatation measured during the isothermal step at 400 °C is to a large extent generated by the transformation of austenite to bainite (Figure 2a,b). Based on the dilatometric results, a set of samples were subjected fast cooling and isothermal at 400 °C for 600 s to induce the stabilization of austenite via carbon redistribution during bainite formation [34]. The austempering process (AT) was performed at 400 °C to avoid the formation of martensite upon cooling. In this way, the analyses of competitive reactions typically observed in Q&P steels [35] such as carbon partitioning from martensite to austenite and/or the tempering of martensite during the isothermal step are excluded in this work.

Figure 3a,b display the temperature record of samples heat-treated in the Gleeble thermomechanical simulator. The temperature was controlled using a K-type thermocouple spot welded to the geometrical center of each sample. Additionally, extra thermocouples were welded at different locations of the sample for measuring possible thermal gradients close to the control thermocouple. Depending on the experimental setup, a small homogeneously treated zone can be obtained in samples heated by Joule effect (electric resistance heating) in the Gleeble simulator. Then, to determine the size of this zone, Vickers hardness measurements were made along the RD direction, on the ND plane. A homogeneous zone of at least 12 mm was determined by employing this method. As schematically presented in Figure 3c, samples used for microstructural and mechanical characterization were extracted from the homogeneously treated zone, which is enclosed by dashed lines. Figure 3c also presents the sample geometry used for tensile testing. Note that the gauge length and shoulders of the dog bone sample fall within the homogeneously treated zone.

The microstructures were characterized by means of light optical microscopy (LOM), scanning electron microscopy in secondary electron mode (SE) and electron backscattered diffraction (EBSD). Samples for microstructural characterization were extracted from the region next to the reduced section of the tensile samples. Metallographic examinations were performed on the RD-ND plane (see Figure 3d). Samples were prepared by grinding and polishing to 0.04 µm colloidal silica suspension (OP-U). LOM micrographs, SE images and EBSD scans were acquired at 280 µm from the sample surface. Image analyses via LOM and SE mode were carried out in samples pre-etched with Nital 2% (2 vol.% HNO_3 in ethanol). A scanning electron microscope FEI Quanta 450 FEG-SEM (ThermoFisher, Hillsboro, OR, USA) was used for microstructural characterization. SE images were acquired employing a working distance of 10 mm and an acceleration voltage of 15 kV. EBSD patterns were acquired using pixels with a hexagonal grid, step size of 120 nm, acceleration voltage of 20 kV, working distance of 14 mm and sample pretilt of 70°. EBSD data acquisition and detector control were operated with EDAX-TSL OIM Data Collection v7.3 software (EDAX AMETEK BV, Tilburg, The Netherlands). The acquired data were post-processed using TSL

OIM Analysis v7 software. The minimum grain size was defined as 5 pixels per grain and grain misorientation angle of 5°.

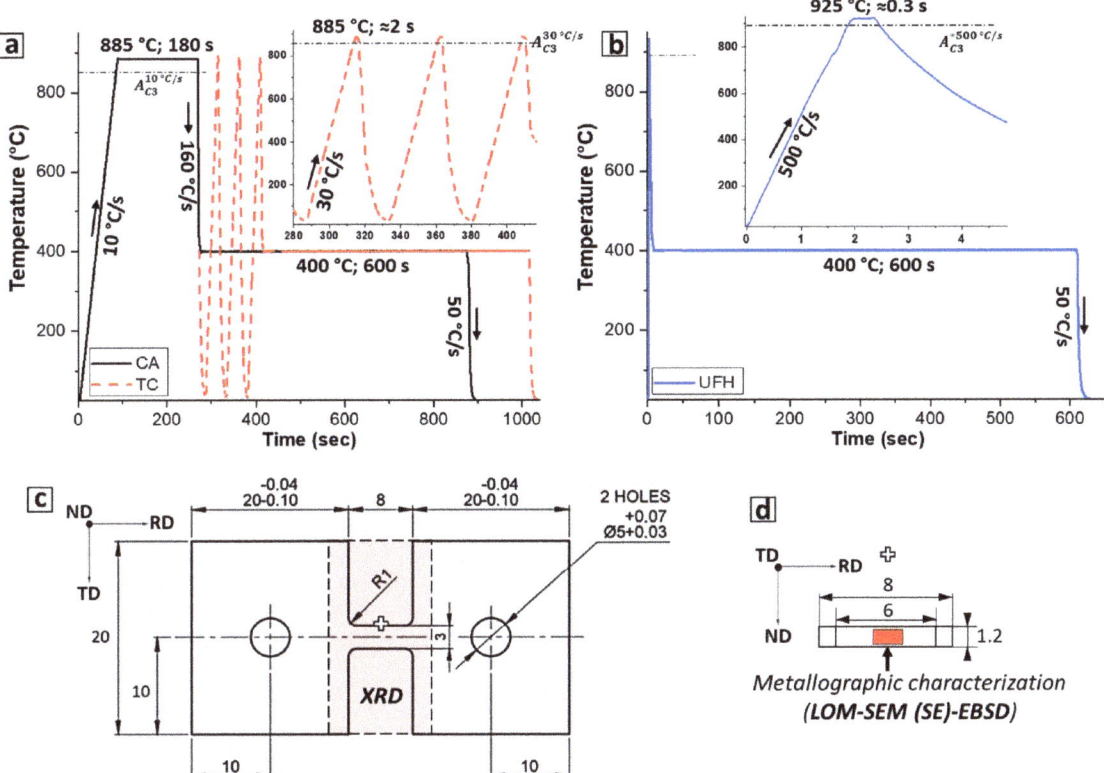

Figure 3. Record of the thermal treatments carried out in the Gleeble simulator: (**a**) CA and TC samples; (**b**) UFH sample. The inserts in (**a**) and (**b**) show the actual temperature recording for the TC and UFH samples, before the austempering step. Horizontal dotted lines denote the A_{C3} temperatures estimated by dilatometry. (**c**) Dog bone geometry and (**d**) area used microstructural characterizations (dimensions in mm). The area enclosed by dashed lines indicates the homogenously heat-treated zone obtained in samples treated using the Gleeble simulator. The cross next to the central region of the gauge length indicates the area used for metallographic characterization.

Taking advantage of the orientation relationship between bainite and parent austenite [36], parent austenite grains (PAGs) were reconstructed from the measured EBSD data using the computer code developed by Gomes et al. [37]. PAG definition was based on 7 square pixels per grain domain and misorientation of 15°.

Quantification of the amount retained austenite (RA) and the carbon content of austenite were estimated by means of X-ray diffraction (XRD) measurements in a Siemens Kristalloflex D5000 diffractometer (Mo-kα source, operation parameters: 40 kV and 40 mA) (Brücker Belgium SA/NV, Kontich, Belgium). Samples cut from the homogenously treated zone were prepared on the RD-TD plane, which is the plane normal to the ND direction (see Figure 3c). A surface layer of ≈300 μm was removed by grinding, followed by repeated polishing and etching steps. XRD patterns were acquired in the 2θ range from 25° to 45° using a step size of 0.03°, dwell time of 20 s and sample holder rotation of 15 rpm. The volume fraction of austenite was determined by the direct comparison method [37] using the integrated area of the $(200)^{BCC}$, $(211)^{BCC}$, $(220)^{FCC}$ and $(311)^{FCC}$

peaks. The retained austenite carbon content was calculated based on the relationship proposed by Roberts [38]:

$$a_\gamma = 3.548 + 0.044 C_\gamma \qquad (2)$$

where a_γ is the lattice parameter (in Å) and C_γ is the austenite carbon content (in wt.%).

Tensile tests were performed in an Instron 5000 device (Instron, Boechout, Belgium) imposing a strain rate of $0.001\ \text{s}^{-1}$. Subsize tensile samples of geometry presented in Figure 3c were strained at constant strain rate up to fracture. Two samples were tested for each austempered condition. The strain evolution during testing was locally measured by 2D-digital image correlation. Image analysis and data evaluation were processed with the Match ID software (Version 2018, MatchID, Ghent, Belgium). An initial gauge length of 6 mm was digitally defined for the strain calculations. Reported yield strength values were based on the 0.2% engineering strain offset. Absorbed energy during uniaxial tensile deformation was calculated as the integrated area under the engineering stress–strain curves. Strain hardening rate was determined as the first derivative of the true stress with respect to the true strain evolution up to necking. Before differentiation of the true stress v/s true strain values, the acquired data points were smoothed using the Locally Weighted Scatterplot Smoothing method (LOWESS).

3. Results

3.1. Microstructure

To evaluate whether ferrite was formed upon cooling, after the annealing steps, an initial microstructural characterization was performed by means of SEM analysis on directly quenched samples (Figure 4). The microstructure of direct quench samples consists predominantly of a lath martensitic (M) matrix, and allotriomorphic ferritic (F) grains are also distinguished (dark gray grains in Figure 4). Ferritic grains of about ≈ 1 µm size are observed at parent austenite grain boundaries in CA (Figure 4a) and TC (Figure 4b,e). Widmanstätten ferrite plates (F_W) [38] were also detected in CA (Figure 4d). A ferrite fraction lower than 1% was obtained after fast cooling for CA, while 2.5 (±0.5)% of ferrite was quantified for TC. The UFH sample mainly consists of martensite and 8.5 (±0.4)% of ferrite with an average grain size of 1.2 (±0.5) µm (Figure 4c). Regions with undissolved spheroidized and lamellar cementite particles (θ) are presented in Figure 4f.

Microstructures produced via a combination of the different annealing strategies and austempering at 400 °C are shown in Figure 5. Inverse pole figures (IPF) for the reconstructed PAGs are presented in Figure 5a–c. The middle row of images (Figure 5d–f) shows combined EBSD image quality and phase maps, where retained austenite grains of film (γ_F) and blocky-type (γ_B) morphologies, are highlighted in green. Bainite and ferrite appear light red. Dark-red to black constituents observed by the image quality (IQ)-phase maps presumably correspond to martensite (M), produced by austenite transformation during the final cooling step [39], after the isothermal holding at 400 °C. The lattice distortion and high dislocation density in martensite decrease the diffraction pattern quality, resulting in a lower and darker IQ scale value than the obtained for the bainitic matrix [39,40]. Based on EBSD-IQ quantification, the amount of martensite was not greater than 1% for all austempered samples. Grain boundaries of misorientation angle between 5–15° and 15–65° are indicated by white and black lines, respectively. TC and UFH annealing led to finer bainitic blocks than those obtained under conventional annealing (see the bainitic block length distribution in Figure 6a). The third row of figures (Figure 5g–i) displays the secondary electron micrographs of the austempered steel grades. A set of parallel bainitic blocks and films of retained austenite are observed in the CA sample (Figure 5g). As presented in the EBSD maps, finer microstructures resulted for the steel samples processed via TC and UFH (Figure 5h,i). Islands with a less etched appearance correspond to partially austenitic–martensitic constituents (M/γ) [39]. Those constituents are clearly distinguished in the EBSD IQ-Phase maps, where M is surrounded by retained austenite grains (Figure 5d–f). The formation of M upon the final cooling step arises due to the heterogeneous distribution of carbon in the residual austenite after bainite transformation during austempering [41–43]. In Figure 5i, undissolved carbides (θ) are also distinguished.

The amount of retained austenite in austempered samples was quantified via XRD as 14%, 15.3%, and 13.8% for the samples CA, TC and UFH, respectively (Table 2). A slightly lower fraction of RA was quantified via EBSD; this is related to non-indexed RA grains with a size smaller than the step size employed for the EBSD data acquisition [18,44].

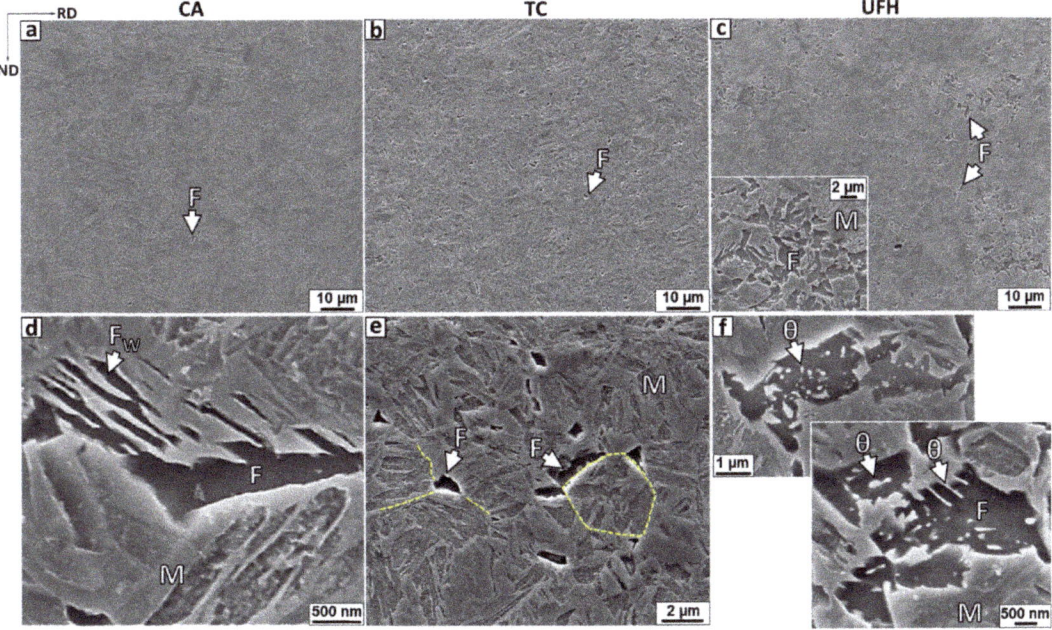

Figure 4. Microstructures of samples directly cooled to room temperature after the annealing step. (**a**–**d**) CA, (**b**–**e**) TC and (**c**–**f**) UFH. In (**e**), prior austenite grain boundaries are highlighted by dashed lines. M: Martensite; F: Ferrite; F_W: Widmanstätten ferrite; θ: Undissolved cementite particles.

The grain size distributions for the austempered steels are presented in Figure 6. A marginal difference and equivalent grain distributions were found for samples treated via TC and UFH, while the conventional annealed steel shows larger bainitic blocks and PAGs. Average bainitic block lengths of 5.3 µm, 3.5 µm and 3.4 µm were obtained for samples CA, TC and UFH, respectively (Figure 6a). PAG reconstructions also revealed that thermo-cycling and ultrafast heating annealing led to grain refinement of the parent austenite and similar grain distributions were obtained after these unconventional types of annealing strategies. Additionally, the narrow distribution of PAGs obtained after TC and UFH is an indication of a more homogeneous distribution of grains (Figure 6b). The average reconstructed PAG sizes for samples CA, TC and UFH are 8.6 µm, 5.7 µm and 5.5 µm, respectively.

Both grain major axis and grain aspect ratio (minimum grain length/maximum grain length) distributions in RA are not greatly influenced by the prior annealing treatment. An average RA grain major axis between 1.5 µm and 1.8 µm was produced after the combination of the different annealing strategies and subsequent austempering (Figure 6c), with the largest distribution of grains for CA. Measured RA grain aspect ratio values display normal distributions with maximum and average close to 0.4, which is related to a rather elongated RA grain shape (Figure 6d).

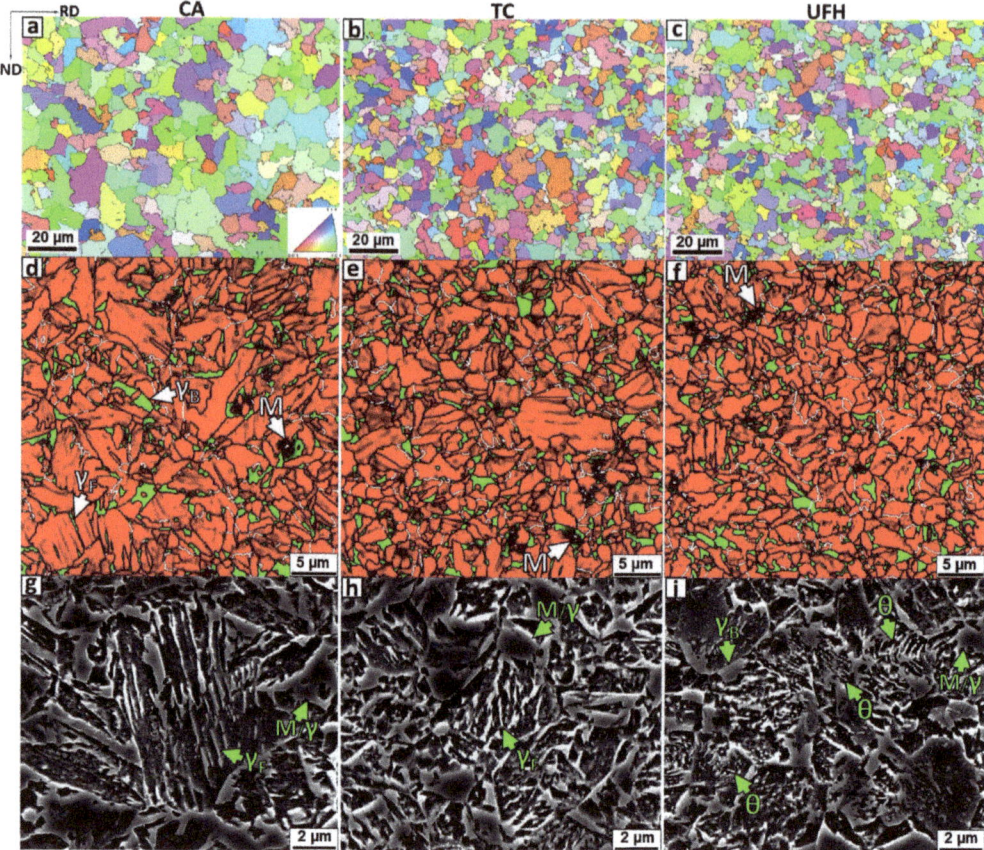

Figure 5. (**a,d,g**) CA, (**b,e,h**) TC and (**c,f,i**) UFH. Microstructures obtained after isothermal holding at 400 °C for 600 s: The first row of images presents the inverse pole figure of the reconstructed PAGs. EBDS IQ-Phase maps and secondary electron images are shown in the second and third row of images, respectively. Retained austenite grains (FCC) appear highlighted in green in the combined IQ-Phase maps, while bainite, ferrite and martensite appear red. White and black lines delineate boundaries of misorientation angle between 5–15° and 15–63°, respectively. M: Martensite; γ: Retained austenite of film (γ_F) and blocky-type (γ_B) morphologies. M/γ: martensite-retained austenite constituent; θ: undissolved cementite particles. Note that the magnification increases from the first to the third row of images.

Table 2. Microconstituents quantification (standard deviation).

Sample	Bainite,% *	Ferrite (SEM),%	Martensite (EBSD), %	RA (EBSD), % (0.2)	RA (XRD), % (0.5)	RA Carbon Content (XRD), wt.%
CA	85.0 (0.5)	<1	<1	11.9	14.0	1.36 (0.02)
TC	82.2 (0.7)	2.5 (0.5)	<1	12.5	15.3	1.33 (0.02)
UFH	77.7 (0.7)	8.5 (0.4)	<1	12.3	13.8	1.40 (0.01)

* Note: Bainite = 100-Ferrite$^{(SEM)}$-Martensite$^{(EBSD)}$-RA$^{(XRD)}$.

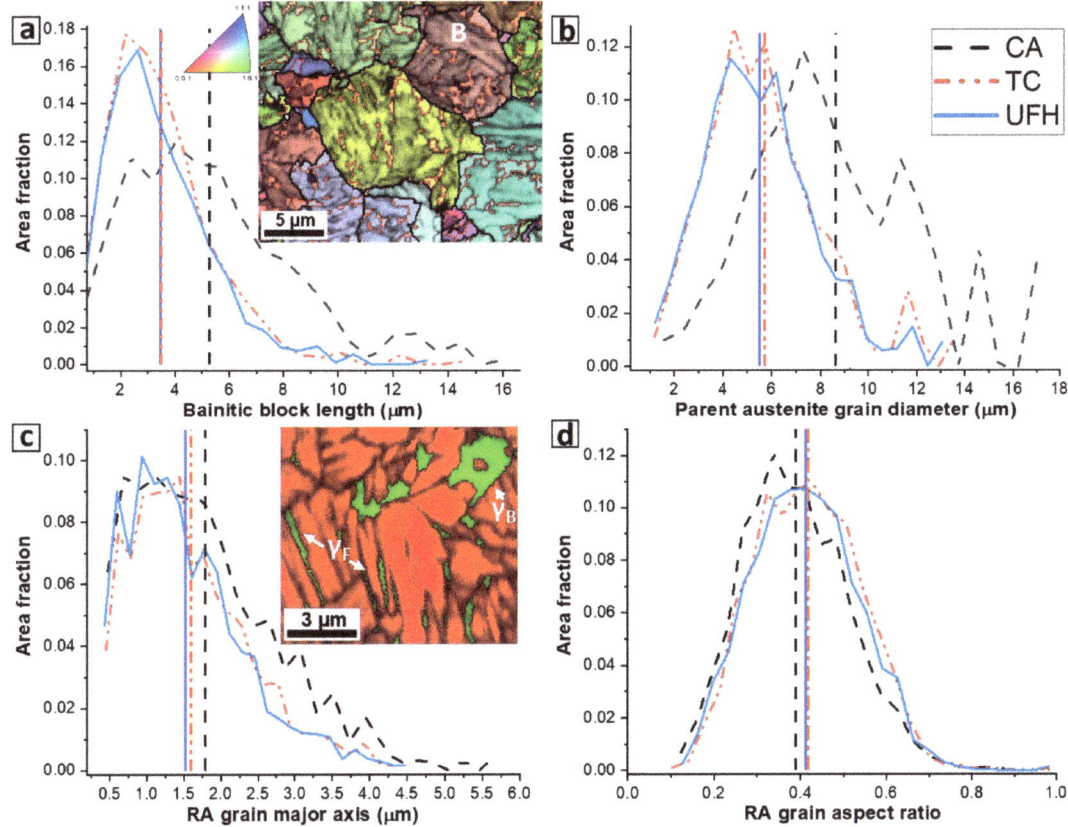

Figure 6. Grain size distribution: (**a**) bainitic block length. (**b**) reconstructed parent austenite grain size diameter. (**c**) retained austenite grain diameter. (**d**) retained austenite grain aspect ratio. Vertical lines denote the average value for each distribution. In (**a**), an IPF of reconstructed PAGs is presented together with the respective IQ map of the "child" bainitic blocks (B) formed after austempering for the CA sample. Retained austenite grains are enclosed by red boundaries. Film (γ_F) and blocky-like (γ_B) RA grains are highlighted in (**c**).

3.2. Textures

Texture analysis was carried out to elucidate the influence of the different annealing strategies on the crystallographic orientation of the transformation products in the studied steels. Figure 7 presents the orientation distribution functions (ODF) for the as-received cold-rolled material and austempered steels. BCC texture is presented at $\varphi_2 = 45°$ section of the Euler space and the main texture components of rolled BCC-Iron (Figure 7a) are presented for comparison. The as-received material (CR) shows a strong ND-RD texture which is typical for cold-rolled ferritic steels [36] and the rotated cube component {001}<110> (Figure 7b). After conventional annealing (Figure 7c), the RD texture fiber disappeared and high intensity is observed along with the ND fiber, with local maxima of orientations concentrated close to the {554}<225> and {111}<112> texture components. Thermo-cycling led to a maximum intensity of 1.64 multiples of random distribution (mrd) (Figure 7d). High intensity is observed close to ND ({554}<225>; {111}<112>) and RD ({112}<110>; {113}<110>) texture components. {001}<110> components are also distinguished after thermo-cycling. Texture obtained after UFH resembles the cold-rolled texture with a strong RD-ND type of texture (Figure 7e). The convex curvature of the ND fiber in the CR sample is maintained in the ultrafast heated bainitic steel. At the same time, the intensity for RD-ND fibers and {001}<110> texture components is lower than that observed in CR.

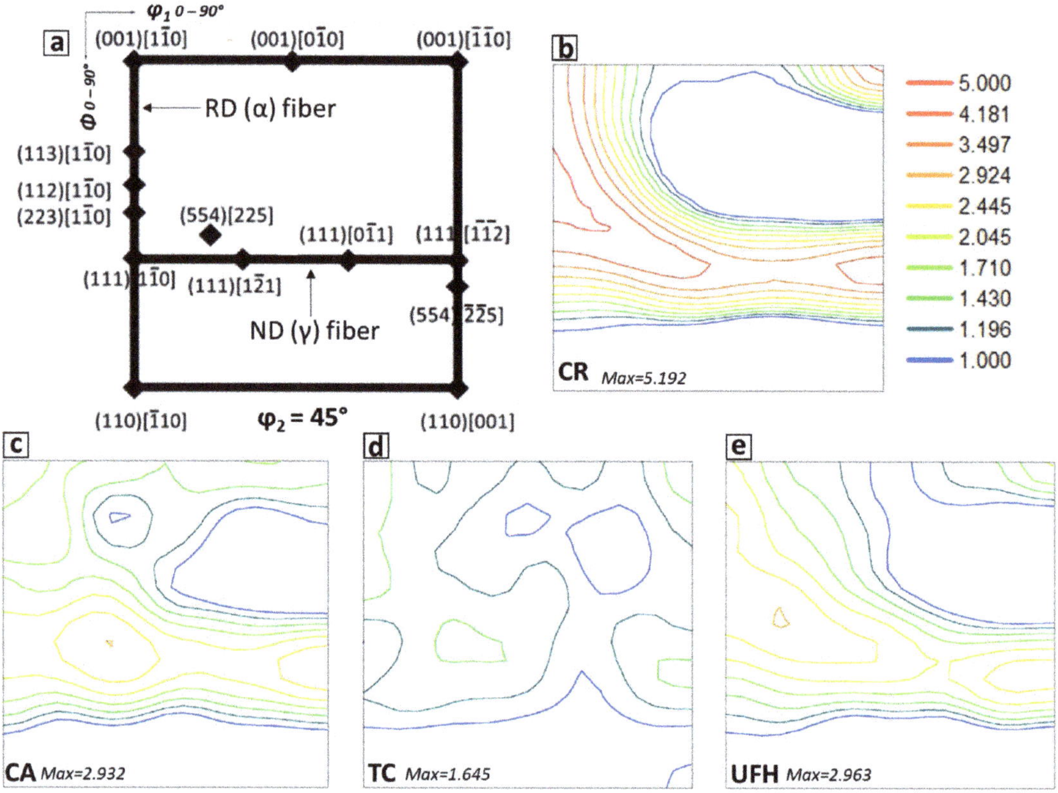

Figure 7. ODF at $\varphi_2 = 45°$ section of the Euler space. (**a**) main texture components of rolled BCC crystals. (**b**) 70% cold-rolled material. (**c**) CA. (**d**) TC. (**e**) UFH. (Scanned area 22,500 µm^2).

3.3. Mechanical Properties

Tensile engineering stress–strain and strain hardening rate v/s true strain curves are shown in Figure 8a,b, respectively. Mechanical property values are summarized in Table 3. Bainitic steels produced in this study display continuous yielding and comparable values of ultimate tensile strength (σ_{UTS}). Yield strength (σ_{ys}) values of 895 MPa, 869 MPa and 862 MPa were measured for samples CA, TC and UFH, respectively. The obtained σ_{UTS} ranged from 1130 to 1135 MPa. Tensile testing, specifically the uniform (ε_u) and total (ε_{total}) elongation values, revealed a considerable difference in ductility for UFH with respect to samples CA and TC. In CA and TC steels, ε_u and ε_{total} are similar with values close to ≈0.15 and 0.25, respectively. The sample UFH shows an ε_u and a ε_{total} of 0.24 and 0.35, respectively. The reported difference in the elongation values represents an increment of 60% in ε_u and 40% in ε_{total} for the sample UFH with respect to CA and TC. Absorbed energy values of 251 MJ/m^3, 268 MJ/m^3 and 375 MJ/m^3 were determined for CA, TC and UFH, respectively.

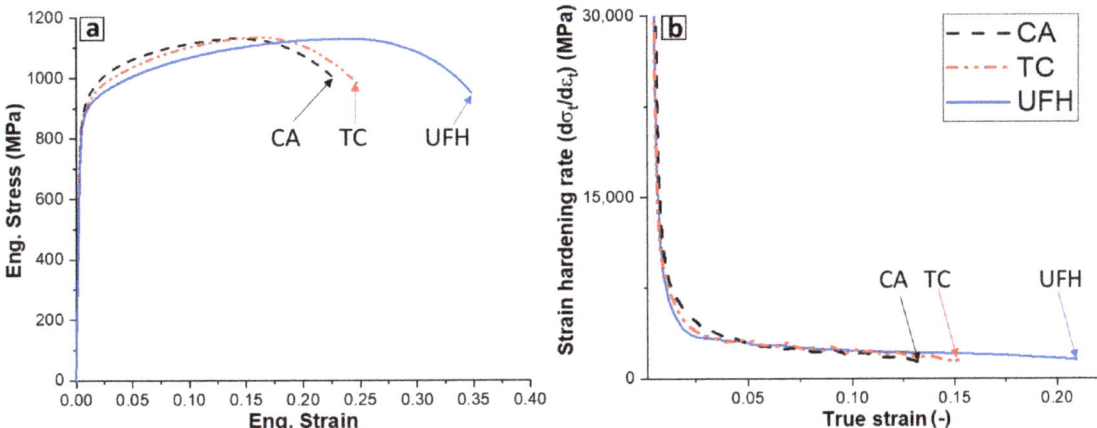

Figure 8. (a) Engineering stress-strain curves and (b) strain hardening rate v/s true strain curves.

Table 3. Mechanical properties (standard deviation).

Sample	σ_{ys}, MPa	σ_{UTS}, MPa	σ_{UTS}/σ_{ys}	ε_u	ε_{total}	Absorbed Energy, MJ/m^3
CA	895 (8)	1131 (2)	1.26 (0.01)	0.14 (0.001)	0.24 (0.005)	251 (6)
TC	869 (8)	1135 (7)	1.31 (0.003)	0.16 (0.003)	0.25 (0.006)	268 (4)
UFH	862 (13)	1130 (6)	1.31 (0.003)	0.24 (0.01)	0.35 (0.004)	375 (1)

The strain hardening behavior of the studied steels is presented in Figure 8b. Below a true strain of 0.05, sample CA displays the highest strain hardening rate and UFH the lowest one. In the true strain range from 0.05 to 0.11, all bainitic steels present a gradual decrease of the strain hardening rate, which is extended up to ≈0.21 for the UFH sample.

4. Discussion

Thermo-cycling and ultrafast heating produced microstructures finer than conventional annealing (Figures 5 and 6). The fine-sized PAGs and product microconstituents obtained via thermo-cycling are the results of multiple reverse transformations martensite-austenite in each cycle. Consecutive nucleation of austenite at prior parent austenite and martensitic grain boundaries [6,8], like blocks and packets, together with an increase in heating rate (10 °C/s to 30 °C/s) and 2 s of holding time, resulted in a measurable refining the grain size in the studied steel. At the same time, multiple reverse transformations randomized texture fibers of the initial material, and localized texture components of low intensity were developed (Figure 7d). Multiple variant selection, related to the transformation from austenite to martensite/bainite [36] (24 variants of the K-S orientation relationship), resulted in the low intensity (multiples of random distribution) observed for the heat-treated samples in comparison with the as-received material. According to this, multiple and subsequent steps of transformation martensite → austenite → martensite are responsible for the weaker texture observed for TC. On the other hand, the conventional annealed sample displays a texture with higher intensity on the {554}<225> and {111}<112> components, being similar to the crystallographic texture observed in recrystallized ferrite [45,46]. The low heating rate employed during conventional annealing (i.e., 10 °C/s) leads to the recrystallization of ferrite during heating, conducing to the transformation of grains with orientations that compose the RD fiber texture (such as {112}<110>) to grains with orientation close to {111}<112> and {554}<225> [45], which are the orientations observed after conventional annealing followed by austempering

(Figure 7c). Also, {111}<112> and {001}<011> orientations components can be obtained as results of transformation from parent austenite grains of brass orientation [47]. Large PAGs and bainitic blocks for CA are the result of the slow heating rate and soaking for 180 s at the annealing temperature, where the selected annealing parameters produce both, isochronal and isothermal austenitic grain growth.

The ODF of the UFH sample (Figure 7e) shows that the general characteristics of the BCC texture are almost the same compared with the as-received material (Figure 7b). This phenomenon can be explained in terms of the texture memory effect hypothesis [46] and similar results have been previously reported for different ultrafast heated steel grades, including Q&P steels [18–20,44]. In this work, evidence of austenite formation and its interaction with non-recrystallized ferrite ($F_{(Non-RX)}$) during the heating process is presented. Figure 9 shows selected $F_{(Non-RX)}$ grains in an intercritical annealed sample heated at 500 °C/s to 800 °C and quenched with no soaking time. A high density of dislocations is indirectly observed in the 1st neighbor kernel average misorientation maps exhibited in Figure 9a,b, proving that ferrite is in a non-recrystallized state during austenite formation. The ODF map presented in Figure 9c supports this observation.

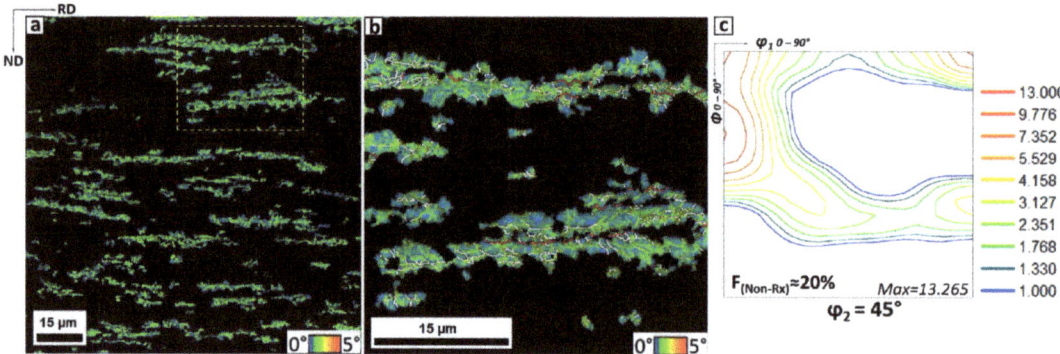

Figure 9. Non-recrystallized ferrite obtained in a sample heated at 500 °C/s to 800 °C followed by direct quenching: (**a**) 1st neighbour kernel average misorientation map. (**b**) Enlarged IQ-KAM map of the area enclosed by the dashed square in (**a**). White and red lines define boundaries with misorientation angles between 5–15° and 15–63°, respectively. (**c**) Orientation distribution function (scanned area 22,500 µm^2).

The ODF of the $F_{(Non-RX)}$ grains shows a convex curvature of the ND fiber and high intensity in the {113}<110> and {112}<110> rolling texture components. Texture characteristics of $F_{(Non-RX)}$ are restored after the transformation of $F_{(Non-RX)}$ → austenite → bainite giving rise to the crystallographic orientation observed for the UFH bainitic steel (Figure 7e), even after heating the sample above the A_{C3} temperature. Those non-recrystallized ferritic regions provide a high density of nucleation sites for austenite [48]. Additionally, the high heating rate and undissolved carbides can effectively suppress the austenitic grain growth upon heating [18,20], resulting in the fine-grained bainitic structure produced after ultrafast heating annealing and austempering.

It is important to note that the four-step thermo-cycling applied in this study, which includes heating up to a temperature range above A_{C3} followed by fast cooling, gives an equivalent grain refinement effect to the obtained through the UFH route, as presented in Figure 6b.

The microstructure produced after a predefined thermal treatment depends on the chemical and morphological characteristics of the parent austenite and subsequent thermal pathways. In this study, a fast-cooling rate of 160 °C/s was employed after the initial annealing step. This approach makes it possible to elucidate the characteristics of the parent austenite based on the microstructure obtained after cooling. Figure 10 shows the microstructure of the as-received ferritic–pearlitic steel (Figure 10a) together with samples

CA (Figure 10b) and UFH (Figure 10c) directly cooled to room temperature (the microstructure of TC is presented in Figure 4b). Clear differences are observed between CA and UFH. In the UFH steel, a banded microstructure that resembles the ferritic–pearlitic bands of the as-received cold-rolled material was obtained. The insert in Figure 10c shows that the darker areas in the optical micrograph are mainly composed of fine-grained ferrite, as it was presented previously in Figure 4c. Contrarily, even distribution of microconstituents was found in CA and TC. The influence of the prior annealing strategies on the produced microstructures is exemplified using schematic continuous cooling transformation diagrams (presented in Figure 10d,e).

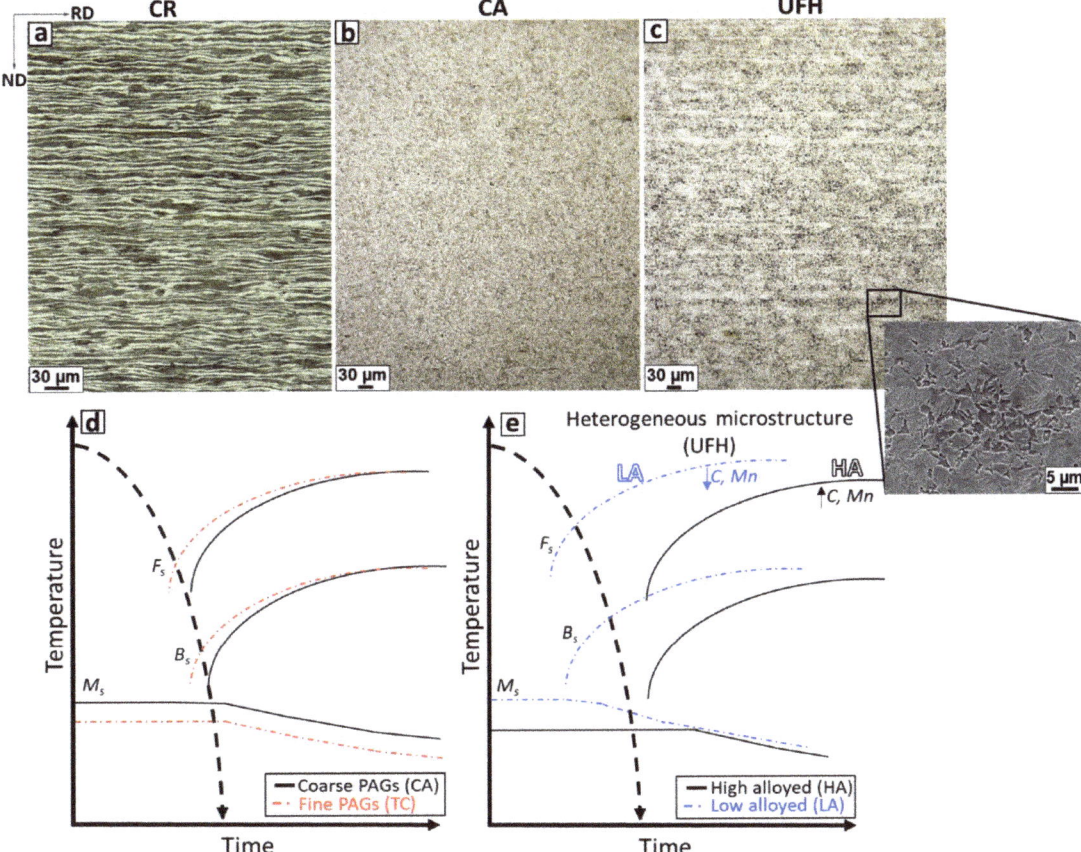

Figure 10. Optical micrographs for: (**a**) as-received material, (**b**) CA and (**c**) UFH samples. As a reference for the reader, the insert in (**c**) shows an enlarged micrograph of the heterogeneous microstructure obtained within the martensitic bands for the UFH sample. (**d**) Schematic CCT diagrams presenting the influence of the PAG size on phase transformations for CA and TC samples. (**e**) Chemical gradients in parent austenite produced via ultrafast heating annealing led to different kinetics of phase transformation (local CCT diagrams) for the UFH sample. The banded microstructure produced after UFH is inherited from the initial cold-rolled material, where LA (ferritic) and HA (pearlitic) are low alloyed and high alloyed regions, respectively.

Conventional annealing produces a homogeneous parent austenite phase of a rather large grain size if it is compared to the grain size distributions of TC and UFH. After cooling, a ferrite fraction lower than 1% was obtained for CA, with ferritic grains nucleated at prior austenite grain boundaries. The decrease in the PAGs size by thermo-cycling annealing led to a higher amount of effective nucleation points [7], resulting in an increased number of ferritic grains obtained after cooling. The decrease in the M_S temperature is also related to the smaller PAGs produced after TC, and this phenomenon has been reported and discussed elsewhere [2,13]. On the other hand, the banded microstructure obtained after ultrafast heating and cooling is linked to the chemical heterogeneity of the parent austenite. The homogenization of manganese and carbon might be constrained during the ultrafast heating annealing [16,30,49]. This is a reason to obtain compositional gradients in the parent austenite, with regions of high and low solute concentration at prior pearlitic (high alloyed region, HA) and deformed ferritic bands (low alloyed region, LA), respectively [50,51]. During the initial stages of nucleation, austenite forms preferentially at prior pearlitic regions [26]. Additionally, austenite can also nucleate at ferrite–ferrite boundaries. Nevertheless, the growth of those nuclei will be controlled through carbon diffusion from the carbon-rich areas [48,52]. Another transformation mechanism that could operate upon fast heating is the massive growth of austenite from proeutectoid ferrite at the last stage of austenite formation [26,27]. As the homogenization of carbon and the alloying elements is likely restricted by the high heating rate and short soaking time employed (<0.3 s), low and high solute regions in parent austenite will transform following different kinetics of phase transformations, as presented in Figure 10e. The results suggest that LA regions decompose to a mixture of ferrite and possible bainite, while prior HA regions are transformed mainly to martensite due to the inhomogeneous distribution of alloying elements in austenite. These results concur with those reported for lean alloy steels subjected to ultrafast heating and fast cooling [15,49,53–55].

According to the mechanical properties, the decrease in the grain size attained via thermo-cycling treatment resulted in equivalent σ_{UTS} and elongation values to the obtained in CA. These observations are in line with previous results in the influence of PAG size and the related mechanical performance of martensitic steels by Hanamura et al. [2]. The results suggest that the decrease in the average bainitic block length from 5.3 μm (CA) to 3.5 μm (TC) and 3.4 μm (UFH) does not play a major role on the overall mechanical behavior of the studied steels. Instead, the combination of ultrafast heating and austempering produced higher uniform and total elongation, resulting in an enhanced strength-ductility balance and superior capacity of energy absorption during tensile testing.

The distribution of microconstituents and corresponding mechanical properties obtained after austempering are summarized in Figure 11. The results indicate that the σ_{UTS} values are insensitive to the processing history and microstructure. This observation agrees with the findings reported by Kumar et al. [56], where a saturation of the strength level was obtained in dual-phase steels with bainite or martensite content higher than 60%.

The resulting mechanical properties obtained via the combination of UFH and austempering agree well with previous findings reported for UFH-Q&P steels [18,25], for which a promising compromise between total elongation and high strength level was found through the formation of ferrite-containing multiphase microstructures. Those results [18,25] suggested that the presence of ferrite does not affect the strength level but effectively contributes towards improving the tensile strain capacity of UFH steels.

In multiphase steels, the fraction, strength, distribution and size of each microstructural constituent define the mechanical behavior [57,58]. The strain hardening is also influenced by stress partitioning and strain accommodation between phases during deformation [59,60]. Additionally, the mechanical stability of retained austenite and its interaction with the surrounding microconstituents play a fundamental role on the strain hardening rate of TRIP-aided steels [59,61,62].

Noticeable differences in the strain hardening rates of the studied steels are observed at the initial stages of deformation, before reaching a strain value of 0.05 (Figure 8b). In an attempt to elucidate the potential effect of the microstructure on the mechanical behavior, the measured strain hardening rates were analyzed by using the modified Crussard-Jaoul analysis [63,64]. Figure 12 shows representative plots of $\ln(d\sigma_t/d\varepsilon_t)$ vs. $\ln(\sigma_t)$ for the studied steels. Three different stages (s_{I-III}) of strain hardening are observed; true stress (ε_t) and true strain (σ_t) values at the transition of each stage are indicated in parenthesis.

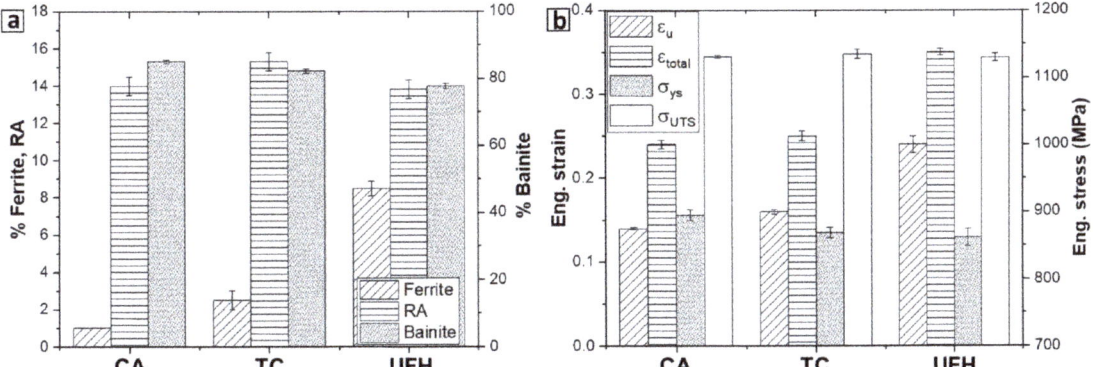

Figure 11. (a) Distribution of microconstituents in austempered samples (the amount of martensite is lower than 1% for all samples). RA: retained austenite. (b) Summary of the mechanical properties measured via uniaxial tensile testing. ε_U: uniform elongation; ε_{total}: total elongation; σ_{ys}: 0.2% yield strength; σ_{UTS}: ultimate tensile strength.

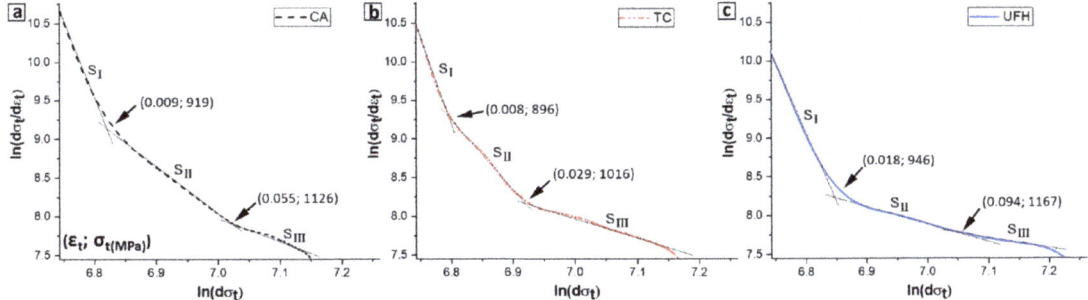

Figure 12. Strain hardening behavior plotted according to the modified C-J analysis in samples (a) CA, (b) TC and (c) UFH. Values indicated in parenthesis at the inflection points of the strain hardening curves correspond to the true strain and true stress, respectively (true strain; true stress).

During stage 1 (s_I), initial yielding and dislocation accumulation in bainitic regions lead to high strain hardening rates for samples CA and TC. At this stage, the accumulation of mobile dislocations at regions near to retained austenite grains takes place [65]. It is expected that the retained austenite grains that compose M/γ islands are among the first to transform due to the constraining effect on strain distribution and locally higher stress levels that arise in the regions surrounding the initial martensitic zones [66]. Retained austenite grains of low mechanical stability might also transform to martensite in this stage. Stage 1 is prolonged to higher levels of stress and strain in the UFH sample due to the homogeneous deformation of soft ferritic grains, which resulted in the lowest strain hardening rate observed at the early stages of deformation [64,67]. In stage 2 (s_{II}), retained austenite grains continuously transform to martensite due to the accumulation of strain. This transformation attenuates the strain hardening rate decreasing by inhibiting the dislocation glide process in regions where newly martensite grains were formed [65,68]. The higher strain hardening rate at the initial stages of deformation for CA and TC samples might be also influenced by a fast rate of austenite transformation upon straining. The C-J plots indicate that most of the austenite transformation proceeds quickly at low strain levels in samples CA and TC (before reaching the stress level corresponding to stage 3). This observation agrees with the results for the kinetics of austenite transformation upon straining in low alloy steels [69–71]. Instead, variations of strain hardening in the UFH sample suggest that austenite transformation is prolonged to higher levels of strain and proceeds at a slower rate than in CA and TC. This is an indication for retained austenite of higher mechanical stability, resulting in improved ductility and energy absorption capacity [70,71]. Liu et al. [16] pointed out that the chemical heterogeneities in retained austenite, generated during ultrafast heating experiments, may play a role on the mechanical behavior of multiphase ultrafast heated steels. The higher carbon and manganese concentration in austenite formed at prior pearlitic colonies may account to improve the mechanical stability of the RA grains [61,62], enhancing the ductility of the UFH steel.

During stage 3 (s_{III}), the deformation of bainite and ferrite continues. Retained austenite grains of higher mechanical stability also transform during this stage. The newly formed martensite islands act like hard particles, producing the redistribution of plastic deformation towards bainitic and ferritic constituents [68].

In addition to this analysis, it should be mentioned that the formation of the heterogeneously banded microstructure produced via UFH might lead to strain/stress gradients between ferrite, bainite and retained austenite (which transforms to martensite upon straining), producing a synergic effect that conduced to the enhancement of ductility without decreasing the strength level, as reported in Refs. [72–74]. Ryu et al. demonstrated that the strain partitioning between microconstituents in low alloy steels drastically influences the stability of retained austenite [75], and this factor could be related to the higher mechanical stability indirectly evaluated for the RA grains in the UFH sample. According to the discussed results and reported mechanical properties for multiphase UFH-Q&P steels [18,25], ferrite grains could effectively contribute to the ductility by decreasing the strain localization, improving the retained austenite stability.

However, the exact quantitative analyses of the influence of the spatial distribution of microconstituents and their contribution to the mechanical behavior, coupled with the evaluation of kinetics and hardening related to the austenite → martensite transformation upon straining remain open for further investigation.

5. Conclusions

In this study, the influence of thermo-cycling and ultrafast heating annealing strategies on microstructure-mechanical properties of austempered steels were evaluated and compared with a steel grade produced via a conventional annealing route. The following conclusions are addressed from the obtained results:

1. Four-step thermo-cycling and ultrafast heating above the A_{C3} led to finer microstructures than conventional annealing. Retained austenite grain distributions were not greatly influenced by the prior annealing treatment.
2. The microstructural refinement attained via thermo-cycling does not show a significant influence on the mechanical response for the studied steels.
3. Ultrafast heating above the A_{C3} followed by fast cooling to room temperature retains a banded microstructure mainly composed of martensite and ferrite. The banded characteristics of the heat-treated material are similar to the observed in the as-received ferritic–pearlitic steel. Microstructural analysis suggested that the banded microstructure developed after heat treatment arises from local chemical heterogeneities in the parent austenite due to insufficient time for diffusion of alloying elements during the UFH process.
4. In contrast to the conventional annealed sample, the grain refined-heterogeneous microstructure produced via a combination of UFH and austempering led to an increase of 40% in total elongation and 50% in energy absorbed measured under uniaxial strain to fracture.
5. The enhancement of ductility of the UFH-bainitic steel is reached without sacrificing the strength level. The obtained results suggest that the formation of heterogeneous

microstructures via ultrafast heating annealing has a greater influence on the mechanical response than the attained level of grain refinement.

Author Contributions: Conceptualization, E.H.-D., F.C.-C., T.R.-Y. and R.H.P.; methodology, E.H.-D. and L.C.; software, E.H.-D. and L.C.; investigation, E.H.-D.; resources, R.H.P.; writing—original draft preparation, E.H.-D.; writing—review and editing, L.C., F.C.-C., T.R.-Y. and R.H.P.; supervision, F.C.-C. and R.H.P.; funding acquisition, E.H.-D., F.C.-C. and R.H.P. All authors have read and agreed to the published version of the manuscript.

Funding: This research was funded by National Agency for Research and Development (ANID-Chile)/Doctorado Nacional/2017-21171319 (E.H.D.) and Project No. 11170104 (F.C.C.).

Institutional Review Board Statement: Not applicable.

Informed Consent Statement: Not applicable.

Data Availability Statement: Not applicable.

Acknowledgments: L. Corallo acknowledges the support of Research Foundation Flanders-FWO, application number 1SC5619N.

Conflicts of Interest: The authors declare no conflict of interest.

References

1. Morrison, W. Effect of grain size on the stress-strain relationship in low-carbon steel. *Trans. ASM* **1966**, *59*, 824–846.
2. Hanamura, T.; Torizuka, S.; Tamura, S.; Enokida, S.; Takech, H. *Effect of Austenite Grain Size on the Mechanical Properties in Air-Cooled 0.1c-5Mn Martensitic Steel*; Trans Tech Publications Ltd.: Stafa-Zurich, Switzerland, 2014; Volume 783–786, pp. 1027–1032. [CrossRef]
3. Tomita, Y.; Okabayashi, K. Heat treatment for improvement in lower temperature mechanical properties of 0.40 pct C-Cr-Mo ultrahigh strength steel. *Metall. Trans. A* **1983**, *14*, 2387–2393. [CrossRef]
4. Garcia-Mateo, C.; Caballero, F.G.; Sourmail, T.; Kuntz, M.; Cornide, J.; Smanio, V.; Elvira, R. Tensile behaviour of a nanocrystalline bainitic steel containing 3wt% silicon. *Mater. Sci. Eng. A* **2012**, *549*, 185–192. [CrossRef]
5. Yokota, T.; Mateo, C.G.; Bhadeshia, H.K.D.H. Formation of nanostmctured steels by phase transformation. *Scr. Mater.* **2004**, *51*, 767–770. [CrossRef]
6. Furuhara, T.; Kikumoto, K.; Saito, H.; Sekine, T.; Ogawa, T.; Morito, S.; Maki, T. Phase transformation from fine-grained austenite. *ISIJ Int.* **2008**, *48*, 1038–1045. [CrossRef]
7. Grange, R.A. Strengthening steel by austenite grain refinemet. *Trans. ASM* **1966**, *59*, 26–48.
8. Grange, R.A. The rapid heat treatment of steel. *Metall. Trans.* **1971**, *2*, 65–78. [CrossRef]
9. Baker, T.N. Microalloyed steels Microalloyed steels. *Ironmak. Steelmak.* **2016**, *43*, 264–307. [CrossRef]
10. Buchmayr, B. Thermomechanical Treatment of Steels—A Real Disruptive Technology Since Decades. *Steel Res. Int.* **2017**, *87*, 1–14. [CrossRef]
11. Zhao, J.; Jiang, Z. Thermomechanical processing of advanced high strength steels. *Prog. Mater. Sci.* **2018**, *94*, 174–242. [CrossRef]
12. Dai, Z.; Chen, H.; Ding, R.; Lu, Q.; Zhang, C.; Yang, Z.; Van Der Zwaag, S. Fundamentals and application of solid-state phase transformations for advanced high strength steels containing metastable retained austenite. *Mater. Sci. Eng. R* **2021**, *143*, 100590. [CrossRef]
13. Celada-Casero, C.; Sietsma, J.; Santofimia, M.J. The role of the austenite grain size in the martensitic transformation in low carbon steels. *Mater. Des.* **2019**, *167*, 107625. [CrossRef]
14. Matlock, D.K.; Kang, S.; De Moor, E.; Speer, J.G. Applications of rapid thermal processing to advanced high strength sheet steel developments. *Mater. Charact.* **2020**, *166*, 110397. [CrossRef]
15. Banis, A.; Bouzouni, M.; Gavalas, E.; Papaefthymiou, S. The formation of a mixed martensitic/bainitic microstructure and the retainment of austenite in a medium-carbon steel during ultra-fast heating. *Mater. Today Commun.* **2021**, *26*, 101994. [CrossRef]
16. Liu, G.; Li, T.; Yang, Z.; Zhang, C.; Li, J.; Chen, H. On the role of chemical heterogeneity in phase transformations and mechanical behavior of flash annealed quenching & partitioning steels. *Acta Mater.* **2020**, *201*, 266–277. [CrossRef]
17. Cola, G.M. Replacing hot stamped, boron, and DP1000 with "room temperature formable" Flash®® Bainite 1500 advanced high strength steel. In Proceedings of the 28th Heat Treating Society Conference, (HEAT Treating 2015), Detroit, MI, USA, 20–22 October 2015; pp. 21–28.
18. De Knijf, D.; Puype, A.; Föjer, C.; Petrov, R. The influence of ultra-fast annealing prior to quenching and partitioning on the microstructure and mechanical properties. *Mater. Sci. Eng. A* **2015**, *627*, 182–190. [CrossRef]
19. Da Costa-Reis, A.; Bracke, L.; Petrov, R.; Kaluba, W.J.; Kestens, L. Grain Refinement and Texture Change in Interstitial Free Steels after Severe Rolling and Ultra-short Annealing. *ISIJ Int.* **2003**, *43*, 1260–1267. [CrossRef]

20. Petrov, R.H.; Sidor, J.; Kestens, L.A.I. Texture formation in high strength low alloy steel reheated with ultrafast heating rates. *Mater. Sci. Forum* **2012**, *702–703*, 798–801. [CrossRef]
21. Petrov, R.; Kestens, L.; Kaluba, W.; Houbaert, Y. Recrystallization and austenite formation in a cold rolled TRIP steel during ultra fast heating. *Steel Grips* **2003**, *289–294*, 221–225.
22. Hudd, R.C.; Lyons, L.K.; De Paepe, A.; Stolz, C.; Collins, J. *The Ultra Rapid Heat Treatment of Low Carbon Strip*; European Commission Project Contract No 7210-MB/818/203/819; Office for Official Publications of the European Communities: Luxembourg, 1998.
23. Griffay, G.; Anderhuber, M.; Klinkenberg, P.; Tusset, V. *New Continuous Annealing Technology with High-Speed Induction Heating Followed by Ultra-Fast Cooling*; European Commission Project Contract No 7210-PR/026; Office for Official Publications of the European Communities: Luxembourg, 2002.
24. Hernandez-Duran, E.I.; Corallo, L.; Ros-Yanez, T.; Castro-Cerda, F.M.; Petrov, R.H. Influence of Mo-Nb-Ti additions and peak annealing temperature on the microstructure and mechanical properties of low alloy steels after ultrafast heating process. *Mater. Sci. Eng. A* **2021**, *808*, 140928. [CrossRef]
25. Dai, J.; Meng, Q.; Zheng, H. An innovative pathway to produce high-performance quenching and partitioning steel through ultra-fast full austenitization annealing. *Mater. Today Commun.* **2020**, *25*, 101272. [CrossRef]
26. Speich, G.; Szirmae, A. Formation of Austenite from Ferrite and Ferrite-Carbide Aggregates. *Trans. Metall. Soc. AIME* **1969**, *245*, 1063–1074.
27. Castro Cerda, F.M.; Sabirov, I.; Goulas, C.; Sietsma, J.; Monsalve, A.; Petrov, R.H. Austenite formation in 0.2% C and 0.45% C steels under conventional and ultrafast heating. *Mater. Des.* **2017**, *116*, 448–460. [CrossRef]
28. Thomas, L.S.; Matlock, D.K. Formation of Banded Microstructures with Rapid Intercritical Annealing of Cold-Rolled Sheet Steel. *Metall. Mater. Trans. A Phys. Metall. Mater. Sci.* **2018**, *49*, 4456–4473. [CrossRef]
29. Mishra, S.; DebRoy, T. Non-isothermal grain growth in metals and alloys. *Mater. Sci. Technol.* **2006**, *22*, 253–278. [CrossRef]
30. Liu, G.; Dai, Z.; Yang, Z.; Zhang, C.; Li, J.; Chen, H. Kinetic transitions and Mn partitioning during austenite growth from a mixture of partitioned cementite and ferrite: Role of heating rate. *J. Mater. Sci. Technol.* **2020**, *49*, 70–80. [CrossRef]
31. Thomas, L. Effect of heating rate on intercritical annealing of low-carbon cold-rolled steel. Ph.D. Thesis, Colorado School of Mines, Golden, CO, USA, 2015.
32. Azizi-Alizamini, H.; Militzer, M.; Poole, W.J. Austenite formation in plain low-carbon steels. *Metall. Mater. Trans. A Phys. Metall. Mater. Sci.* **2011**, *42*, 1544–1557. [CrossRef]
33. Kaar, S.; Steineder, K.; Schneider, R.; Krizan, D.; Sommitsch, C. New Ms-formula for exact microstructural prediction of modern 3rd generation AHSS chemistries. *Scr. Mater.* **2021**, *200*, 113923. [CrossRef]
34. Matas, S.; Hehemann, R.F. The Structure of Bainite in Hypoeutectoid Steels. *Trans. Metall. Soc. AIME* **1961**, *221*, 179–185.
35. Pierce, D.T.; Coughlin, D.R.; Williamson, D.L.; Clarke, K.D.; Clarke, A.J.; Speer, J.G. Characterization of transition carbides in quench and partitioned steel microstructures by Mossbauer spectroscopy and complementary techniques. *Acta Mater.* **2015**, *90*, 417–430. [CrossRef]
36. Takayama, N.; Miyamoto, G.; Furuhara, T. Effects of transformation temperature on variant pairing of bainitic ferrite in low carbon steel. *Acta Mater.* **2012**, *60*, 2387–2396. [CrossRef]
37. Gomes de Araujo, E.; Pirgazi, H.; Sanjari, M.; Mohammadi, M.; Kestens, L.A.I. Automated reconstruction of parent austenite phase based on the optimum orientation relationship. *J. Appl. Crystallogr.* **2021**, *54*, 569–579. [CrossRef]
38. Aaronson, H.I. The proeutectoid ferrite and the proeutectoid cementite reactions. In Proceedings of the the Decomposition of Austenite by Diffusional Processes, Philadelphia, PA, USA, 19 October 1960; pp. 387–546.
39. Navarro-López, A.; Hidalgo, J.; Sietsma, J.; Santofimia, M.J. Characterization of bainitic/martensitic structures formed in isothermal treatments below the Ms temperature. *Mater. Charact.* **2017**, *128*, 248–256. [CrossRef]
40. Hofer, C.; Bliznuk, V.; Verdiere, A.; Petrov, R.; Winkelhofer, F.; Clemens, H.; Primig, S. Correlative microscopy of a carbide-free bainitic steel. *Micron* **2016**, *81*, 1–7. [CrossRef]
41. Scott, C.P.; Drillet, J. A study of the carbon distribution in retained austenite. *Scr. Mater.* **2007**, *56*, 489–492. [CrossRef]
42. Caballero, F.G.; Miller, M.; Garcia-Mateo, C. Slow Bainite: An Opportunity to Determine the Carbon Content of the Bainitic Ferrite during Growth Francisca G. Caballero. *Solid State Phenom.* **2011**, *174*, 111–116. [CrossRef]
43. Clarke, A.J.; Speer, J.G.; Miller, M.K.; Hackenberg, R.E.; Edmonds, D.V.; Matlock, D.K.; Rizzo, F.C.; Clarke, K.D.; De Moor, E. Carbon partitioning to austenite from martensite or bainite during the quench and partition (Q&P) process: A critical assessment. *Acta Mater.* **2008**, *56*, 16–22. [CrossRef]
44. Hernandez-Duran, E.I.; Ros-Yanez, T.; Castro-Cerda, F.M.; Petrov, R.H. The influence of the heating rate on the microstructure and mechanical properties of a peak annealed quenched and partitioned steel. *Mater. Sci. Eng. A* **2020**, *797*, 140061. [CrossRef]
45. Hoelscher, M.; Raabe, D.; Lucke, K. Rolling and recrystallization texture in BCC steels. *Steel Res.* **1991**, *62*, 567–575. [CrossRef]
46. Yoshinaga, N.; Inoue, H.; Kawasaki, K.; Kestens, L.; De Cooman, B.C. Factors affecting texture memory appearing through α → γ → α transformation in IF steels. *Mater. Trans.* **2007**, *48*, 2036–2042. [CrossRef]
47. Butrón-Guillén, M.P.; Jonas, J.J.; Ray, R.K. Effect of austenite pancaking on texture formation in a plain carbon and A Nb microalloyed steel. *Acta Metall. Mater.* **1994**, *42*, 3615–3627. [CrossRef]
48. Zheng, C.; Raabe, D. Interaction between recrystallization and phase transformation during intercritical annealing in a cold-rolled dual-phase steel: A cellular automaton model. *Acta Mater.* **2013**, *61*, 5504–5517. [CrossRef]

49. Albutt, K.; Garber, S. Effect of heating rate on the elevation of the critical temperatures of low-carbon mild steel. *J. Iron Steel Inst.* **1966**, *204*, 1217–1222.
50. Verhoeven, J.D. Review of microsegregation induced banding phenomena in steels. *J. Mater. Eng. Perform.* **2000**, *9*, 286–296. [CrossRef]
51. Grange, R.A. Effect of microstructural banding in steel. *Metall. Trans.* **1971**, *2*, 417–426. [CrossRef]
52. Savran, V.I.; Leeuwen, Y.; Hanlon, D.N.; Kwakernaak, C.; Sloof, W.G.; Sietsma, J. Microstructural features of austenite formation in C35 and C45 alloys. *Metall. Mater. Trans. A Phys. Metall. Mater. Sci.* **2007**, *38*, 946–955. [CrossRef]
53. Lolla, T.; Cola, G.; Narayanan, B.; Alexandrov, B.; Babu, S.S. Development of rapid heating and cooling (flash processing) process to produce advanced high strength steel microstructures. *Mater. Sci. Technol.* **2011**, *27*, 863–875. [CrossRef]
54. Sackl, S.; Zuber, M.; Clemens, H.; Primig, S. Induction Tempering vs. Conventional Tempering of a Heat-Treatable Steel. *Metall. Mater. Trans. A* **2016**, *47*, 3694–3702. [CrossRef]
55. Pedraza, J.; Landa-Mejia, R.; Garcia-Rincon, O.; Garcia, I. The Effect of Rapid Heating and Fast Cooling on the Transformation Behavior and Mechanical Properties of an Advanced High Strength Steel (AHSS). *Metals* **2019**, *545*, 1–12.
56. Kumar, A.; Singh, S.B.; Ray, K.K. Influence of bainite/martensite-content on the tensile properties of low carbon dual-phase steels. *Mater. Sci. Eng. A* **2008**, *474*, 270–282. [CrossRef]
57. Ismail, K.; Perlade, A.; Jacques, P.J.; Pardoen, T.; Brassart, L. Impact of second phase morphology and orientation on the plastic behavior of dual-phase steels. *Int. J. Plast.* **2019**, *118*, 130–146. [CrossRef]
58. Bhattacharyya, A.; Sakaki, T.; Weng, G.J. The Influence of Martensite Shape, Concentration, and Phase Transformation Strain on the Deformation Behavior of Stable Dual-Phase Steels. *Metall. Trans. A* **1993**, *24*, 301–314. [CrossRef]
59. Ennis, B.L.; Jimenez-melero, E.; Atzema, E.H.; Krugla, M. Metastable austenite driven work-hardening behaviour in a TRIP-assisted dual phase steel. *Int. J. Plast.* **2017**, *88*, 126–139. [CrossRef]
60. Calcagnotto, M.; Ponge, D.; Demir, E.; Raabe, D. Orientation gradients and geometrically necessary dislocations in ultrafine grained dual-phase steels studied by 2D and 3D EBSD. *Mater. Sci. Eng. A* **2010**, *527*, 2738–2746. [CrossRef]
61. Jimenez-Melero, E.; van Dijk, N.H.; Zhao, L.; Sietsma, J.; Wright, J.P.; Van Der Zwaag, S. In situ synchrotron study on the interplay between martensite formation, texture evolution and load partitioning in low-alloyed TRIP steels. *Mater. Sci. Eng. A* **2011**, *528*, 6407–6416. [CrossRef]
62. Ebner, S.; Schnitzer, R.; Maawad, E.; Suppan, C.; Hofer, C. Influence of partitioning parameters on the mechanical stability of austenite in a Q&P steel: A comparative in-situ study. *Materialia* **2021**, *15*, 101033. [CrossRef]
63. Tomita, Y.; Okabayashi, K. Tensile Stress-Strain Analysis of Cold Worked Metals and Steels and Dual-Phase Steels. *Metall. Trans. A* **1985**, *16*, 865–872. [CrossRef]
64. Soliman, M.; Palkowski, H. Strain Hardening Dependence on the Structure in Dual-Phase Steels. *Steel Res. Int.* **2021**, *92*, 1–15. [CrossRef]
65. Alharbi, F.; Gazder, A.A.; Kostryzhev, A.; De Cooman, B.C.; Pereloma, E.V. The effect of processing parameters on the microstructure and mechanical properties of low-Si transformation-induced plasticity steels. *J. Mater. Sci.* **2014**, *49*, 2960–2974. [CrossRef]
66. De Knijf, D.; Petrov, R.; Föjer, C.; Kestens, L.A.I. Effect of fresh martensite on the stability of retained austenite in quenching and partitioning steel. *Mater. Sci. Eng. A* **2014**, *615*, 107–115. [CrossRef]
67. Movahed, P.; Kolahgar, S.; Marashi, S.P.H.; Pouranvari, M.; Parvin, N. The effect of intercritical heat treatment temperature on the tensile properties and work hardening behavior of ferrite—Martensite dual phase steel sheets. *Mater. Sci. Eng. A* **2009**, *518*, 1–6. [CrossRef]
68. Timokhina, I.B.; Hodgson, P.D.; Pereloma, E.V Transmission Electron Microscopy Characterization of the Bake-Hardening Behavior of Transformation-Induced Plasticity and Dual-Phase Steels. *Metall. Mater. Trans. A* **2007**, *38*, 2442–2454. [CrossRef]
69. Polatidis, E.; Haidemenopoulos, G.N.; Krizan, D.; Aravas, N.; Panzner, T.; Papadioti, I.; Casati, N.; Van Petegem, S.; Van Swygenhoven, H. The effect of stress triaxiality on the phase transformation in transformation induced plasticity steels: Experimental investigation and modelling the transformation kinetics. *Mater. Sci. Eng. A* **2021**, *800*, 1–10. [CrossRef]
70. Pereloma, E.; Gazder, A.; Timokhina, I. Retained Austenite: Transformation-Induced Plasticity. *Encycl. Iron. Steel. Their Alloy.* **2016**, 3088–3103. [CrossRef]
71. Sugimoto, K.; Kobayashi, M.; Hashimoto, S. Ductility and Strain-Induced Transformation in a High-Strength Transformation-Induced Plasticity-Aided Dual-Phase Steel. *Metall. Trans. A* **1992**, *23*, 3085–3091. [CrossRef]
72. Lyu, H.; Hamid, M.; Ruimi, A.; Zbib, H.M. Stress/strain gradient plasticity model for size effects in heterogeneous nano-microstructures. *Int. J. Plast.* **2017**, *97*, 46–63. [CrossRef]
73. Hassan, S.F.; Al-Wadei, H. Heterogeneous Microstructure of Low-Carbon Microalloyed Steel and Mechanical Properties. *J. Mater. Eng. Perform.* **2020**, *29*, 7045–7051. [CrossRef]
74. Wu, X.; Zhu, Y. Heterogeneous materials: A new class of materials with unprecedented mechanical properties. *Mater. Res. Lett.* **2017**, *5*, 527–532. [CrossRef]
75. Ryu, J.H.; Kim, D.; Kim, S.; Bhadeshia, H.K.D.H. Strain partitioning and mechanical stability of retained austenite. *Scr. Mater.* **2010**, *63*, 297–299. [CrossRef]

Review

A Review of the Boiling Curve with Reference to Steel Quenching

Manuel de J. Barrena-Rodríguez [1,2], Francisco A. Acosta-González [3,*] and María M. Téllez-Rosas [1]

[1] Facultad de Ciencias Químicas, Universidad Autónoma de Coahuila, Saltillo 25280, Mexico; mbarrena85@gmail.com (M.d.J.B.-R.); mauratellez@uadec.edu.mx (M.M.T.-R.)
[2] Universidad Politécnica de la Región Ribereña, Cd. Miguel Alemán 88300, Mexico
[3] Centro de Investigación y de Estudios Avanzados del I.P.N., Unidad Saltillo, Industria Metalúrgica # 1062, Parque Industrial Saltillo-Ramos Arizpe, Ramos Arizpe 25900, Mexico
* Correspondence: andres.acosta@cinvestav.edu.mx; Tel.: +52-844-438-9600

Abstract: This review presents an analysis and discussion about heat transfer phenomena during quenching solid steel from high temperatures. It is shown a description of the boiling curve and the most used methods to characterize heat transfer when using liquid quenchants. The present work points out and criticizes important aspects that are frequently poorly attended in the technical literature about determination and use of the boiling curve and/or the respective heat transfer coefficient for modeling solid phase transformations in metals. Points to review include: effect of initial workpiece temperature on the boiling curve, fluid velocity specification to correlate with heat flux, and the importance of coupling between heat conduction in the workpiece and convection boiling to determine the wall heat flux. Finally, research opportunities in this field are suggested to improve current knowledge and extend quenching modeling accuracy to complex workpieces.

Keywords: boiling curve; quenching severity; boiling and quenching heat transfer; metal quenching heat flow

1. Introduction

Historically, steel heat treatment has evolved from being an ancestral craft to a sophisticated technology that responds to a higher demand of a wide variety of products with increasingly strict quality standards. This technology relies on developments on several knowledge areas like, materials, such as material science, mechanical metallurgy, heat transfer, computation, instrumentation and control theory, robotics, etc. Nonetheless, the core knowledge is represented by the relationship between processing conditions during heat treatment, and microstructure and properties of the workpiece that are obtained after treatment. Of course, processing conditions always includes a thermal cycle ending with a target cooling rate on the workpiece. This cooling rate is very important to determine the final microstructure, properties and physical integrity of the workpiece.

Mathematical modeling has been used with more frequency in the last three decades to understand and quantitatively predict the previously mentioned relationship between processing-microstructure-properties. Modeling or characterizing experimentally the rate of heat transfer during quenching is the basic step for modeling kinetics of solid-state phase transformation in steels, which in turn is used to predict final microstructure, mechanical properties and workpiece internal stresses and deformations.

The motivation of the present work is to point out and criticize important aspects that are frequently underrepresented in the technical literature about determination and use of the boiling curve and/or the respective heat transfer coefficient for metal quenching modeling. Points to review include: effect of initial workpiece temperature on the boiling curve, modification of temperatures that narrow down boiling regimes, uniformity and magnitude of heat flux in production heat treatment, and the effect of the workpiece size and metal properties on the boiling curve. This article summarizes an overview of basic steel quenching concepts, followed by the presentation and analysis of the established theory of heat transfer phenomena during quenching using a vaporizable liquid.

Two classical handbooks on heat treatment and quenching [1,2] were identified as the point to start for this work. In addition, the document reflects opinions emerged from more than 20 years of professional experience (FAAG), in production line, laboratory and computer modeling, on heat transfer in continuous casting of steel. Recent advances on steel quenching have been also reviewed in the context of the present analysis and discussion.

2. Steel Heat Treating Concepts

Heat treatment of alloys or metals is a term to include any thermal cycle from heating a workpiece to a solubilization temperature in the solid state, soaking the workpiece at this temperature to achieve the target solubilization level, and then cooling it down fast enough to develop the metal microstructure that grants the aimed mechanical properties and physical integrity of the workpiece. Solubilization temperature is a function of the chemical composition of the treated alloy; for example, for steels, it is generally from 800 to 950 °C, depending on the steel grade. This temperature is above the eutectoid temperature at which the ferrite phase (Fe_a, Body-Centered Cubic structure) and cementite (Fe_3C) transform into austenite phase (Fe_g, Face-Centered Cubic structure). Because of its crystal structure, austenite is able to dissolve the carbon released from the decomposition of the cementite. Other examples of heat treatable alloys include specific grades of aluminum, titanium and copper alloys. Solubilization temperatures may be estimated from a binary phase diagram. However, since alloys contain several elements, sometimes pseudo-binary phase diagrams or, if available, ternary phase diagrams are preferred.

The cooling rate of the workpiece plays a major role to determine the microstructure and mechanical properties of the alloys. For example, austenite is generally quenched to obtain martensite phase ($Fe_{a'}$, Body-Centered Tetragonal structure). The cooling rate should be high enough to "freeze" the dissolved carbon avoiding its precipitation after cooling. Indeed, martensite is a carbon-supersaturated solid solution, therefore is a non-equilibrium phase which of course does not appear in any equilibrium diagram. Martensite formation is graphically represented by a Temperature-Time-Transformation (TTT) diagram or by a Continuous Cooling Transformation (CCT) diagram. The former represents isothermal transformations, i.e., the workpiece is cooled down at "infinite velocity" from the solubilization temperature to the transformation temperature. Then, it is hold at this temperature until the solid phase transformation ends. On the other hand, CCT diagrams represent what actually happens during cooling at a constant rate. The alloy structure transforms at inconstant temperature. Both diagrams, TTT and CCT, represent time-temperature maps to locate every phase that would form. For example, when steel cools down at a low rate, it will form the equilibrium phases, ferrite and cementite. However, if the cooling rate is higher, bainite may form. This is a mixture of nonlamellar aggregate of carbides and plate-shaped ferrite. If the cooling rate is even higher, the austenite will start to transform to martensite at temperature M_s. As the sample cools down, an increasingly percentage of austenite transforms to martensite until reaching the temperature, M_f, at which the transformation is completed. The values of hardness and tensile strength for this phase are well above than those obtained for bainite or ferrite phases. *Hardenability* refers to the steel's capability to form martensite. The higher the hardenability of a steel, the easier it would be to form martensite since the required minimum cooling rate would be less demanding.

This review is focused on heat transfer phenomena during solid metal quenching, when using liquids as quenchant media. Water, polymer aqueous solutions, and mineral oils are generally used for quenching metals. The choice for a quenching medium for a specific application is based on the required cooling rate, according to the CCT diagram, to form the target phases, but avoiding quality issues like piecework fracture, unacceptable distortion or dimensional variation which would result from thermal and phase transformation stresses. To achieve this, the latent heat and temperature of vaporization, surface tension, and viscosity are the main liquid properties to take into consideration when choosing a quenching medium. Finally, the fluid flow velocity passing over the workpiece determines the rate of heat transfer and therefore has a major impact on the cooling rate.

3. Boiling and Quenching Heat Transfer

This document is focused on quenching workpieces which initial temperature is well above the boiling temperature of the quenching liquid, for example 800 °C. The *boiling curve* is a convenient graphic representation of the heat flux removed from the workpiece surface, or wall surface, as a function of its temperature, T_w. The heat flux is defined as the heat flow removed per unit surface area of the workpiece, q (W/m^2), and therefore it is a local quantity, i.e., it may change along the surface of the workpiece. Differently from gas quenching, liquid quenching includes boiling phenomenon which leads to a parabolic type-function dependence of q with T_w. Such behavior is not obtained from gas quenching, since for this case heat flow is maximum at the beginning of cooling and then decreases exponentially with time. Boiling curves are obtained under natural convection flow (*pool boiling*) or forced convection flow. Both types of flow are applicable to quenching operations. Furthermore, boiling curves can be determined under steady or transient temperature conditions. In the former method, a specimen is heated by a controlled heat source while simultaneously a quenchant flow removes heat from the specimen to reach an equilibrium temperature. This case is mainly found when quenching using sprays, see for example references [3–6]. Abbasi et al. [3] used an electric resistance to supply heat to a metallic sample that simultaneously received a water spray to reach an equilibrium temperature. The authors studied the effect of spray pressure on the removed heat transfer and found that it increases the rate of heat transfer. They attributed this result to an increase in the droplet velocity with spray pressure. Araki et al. [4] also used another arrangement of electrical resistances to supply heat to a thin metal disk which received a controlled spray of water droplets. The disk was heated up to steady temperatures of 240 to 860 °C and received a rate of droplets of 0.5, 1, 2 and 3 droplets per second at velocities between 1.5 to 4 m/s. Bernardin and Mudawar [5] studied film boiling heat transfer of an isolated droplet stream. They also used electric resistances to supply heat to a polished nickel plate to reach equilibrium temperatures up to 400 °C. The authors presented empirical correlations of the heat flux removed by these continuous stream of monodispersed water droplets. All these studies used a low mass flux of droplets impacting the surface (<1 kg/m^{2}s), and also the droplets velocities were below 5 m/s. More recently, Hernández-Bocanegra et al. [6] developed a new steady state system which is able to study actual production air-mist and spray jets. It uses electromagnetic heating to supply heat to a platinum sample 8 mm diameter by 2.5 mm thickness. Using a properly designed coil, a 5 kW high frequency generator was able to maintain the sample at equilibrium temperatures ranging from 200 to 1200 °C while simultaneously receiving a high mass flux spray. Droplets impacted the platinum disk at velocities between 10 and 30 m/s, which represents high pressure spray quenching conditions. On the other hand, transient temperature experiments consist in preheating an instrumented specimen in a muffle and then transporting it to a quenching rig to generate the corresponding cooling curve until ambient temperature. Reference [7] details a concept on thermal equilibrium establishment which is useful to determine the required soaking (heating) time for the workpiece to reach the target austenitization temperature, and references [8,9] describe some examples of experimental cooling curves. Kobasko [7] proposed a universal correlation to calculate the heating (soaking) time of any steel part. This time is directly proportional to Kondratjev form factor, K, is inversely proportional to thermal diffusivity of a material and Kondratjev number, K_n, and depends on the accuracy of the temperature measurement. Reference [1] (pp. 69–128) describes the technique to measure cooling curves for steel quenching characterization. Several standard probes, instrumented with thermocouples, have been developed to obtain such curves. Some of them are the SAE 5145 steel Grossmann probe, French probe, Beck hollow spherical copper probe, spherical silver probe, JIS silver probe, and Liscic-NANMAC steel probe. This method has been also applied to study secondary cooling of continuous casting strands. Stewart et al. [8] implemented a test apparatus using a stainless-steel specimen cooled from 1200 °C to ambient temperature. Li et al. [9] also used an instrumented stainless-steel plate to obtain cooling curves from several initial temperatures, 400, 550, 700, 800, 900 and 1000 °C. They found that the respective boiling curves depend on the initial temperature. This document focuses on the boiling curve obtained under transient temperature conditions.

3.1. Boiling Curve Determination

Boiling curves are obtained from *thermal analysis*. This is an experimental technique which uses embedded sub-superficial thermocouples within the workpiece, as it is shown in Figure 1. A typical distance from wall surface to thermocouple tip is 1 to 3 mm. The temperature-time data recorded during quenching are used to solve the so-called *Inverse Heat Conduction Problem* (IHCP) [10]. This method solves iteratively the heat conduction differential equation to find the proper boundary condition consisting in the heat flux evolution function $q(t)$ that minimizes the sum of absolute differences between the measured, T_m, and computed, T_c, temperatures, $\sum_{i=1}^{N} |T_m - T_c|_i$, where N is the number of data points collected during the quenching test. The computed temperatures correspond to the thermocouple tip position at the recorded times during cooling. The final solution of the heat conduction equation also provides the temperature evolution at any location within the solid, in particular it provides the temperature at the wall surface, T_w. Therefore, the solution of the IHCP leads to the q vs. T_w plot. Sometimes, the wall temperature is replaced by the *wall superheat* defined by the difference $T_w - T_{sat}$, where T_{sat} is the *saturation temperature* of the liquid, i.e., its boiling temperature. The boiling curve would be the same, except for the shift of the temperature scale. Figure 1 shows a graphic summary of the boiling curve determination from thermal analysis using the solution of the IHCP.

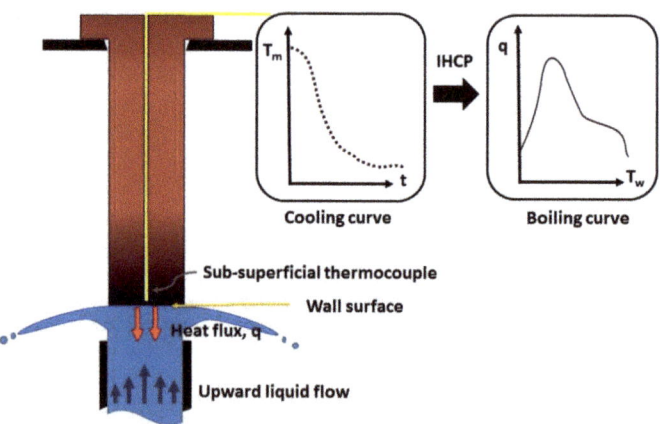

Figure 1. Graphic summary of the boiling curve determination from a cooling curve and using the solution of the IHCP. The workpiece is an instrumented steel cylinder for the end-quenched Jominy test.

3.2. Effect of Initial Temperature of the Solid Workpiece

Figure 2a shows typical boiling curves when quenching metallic workpieces from several initial temperatures, $T_1, T_2, \ldots,$ and T_6. As it was mentioned previously, reference [9] presents an example using several initial temperatures between 400 to 1000 °C and quenching stainless-steel plates using a water spray. At any initial temperature, the heat flux is low but increases as the surface temperature decreases until reaching a maximum value, so called *critical heat flux* (CHF). Thereafter, heat flux keeps decreasing during cooling.

At this point, it is convenient to say that Babu et al. [11] proposed a normalized boiling curve to represent all curves for different soaking temperatures into a single curve. In spite of its usefulness, this method has not gained further visibility in the research community. Figure 2b shows schematically the boiling phenomena occurring on the surface of a workpiece during its quenching by immersion. They are described as follows: At higher initial temperatures, a continuous vapor film forms on the solid surface, which acts as a thermal resistance for heat flow. The presence of this vapor blanket explains the quasi-plateaus observed in the curves corresponding to initial temperatures, $T_w \geq T_3$. Heat flux is kept at a relatively low value; however, when the temperature drops, the vapor film collapses under the liquid pressure, and heat flux increases as a result of a direct contact between the solid surface and the liquid. However, for the curves with $T_w < T_3$ the plateau does not appear, suggesting the absence of an initial stable vapor film. *Leidenfrost* temperature, T_L, is defined as the minimum wall temperature where the *stable vapor film regime* prevails. Below this temperature, *nucleate boiling regime* appears, in which swarms of bubbles continuously form and detach from the wall surface. This process is maintained until wall temperature is too low to promote boiling. This is the *onset of nucleate boiling* temperature, T_{ONB}, and is where *single phase convection regime* appears. The above explanation does not clarify why heat flux is very low at the starting high temperature. This initial low heat flux has mainly attributed to the measurement delay from the thermocouple (*time constant*). This is a characteristic delay that depends on the wire diameter, the thinner the wire the shorter the delay. Rabin and Rittel [12] showed that the time response of a solid-embedded thermocouple is far from being similar to the corresponding response of a fluid-immersed thermocouple. The authors found that the solid-embedded thermocouple has a significantly faster time response at the initiation of the process, but it requires a much longer time to reach a steady state temperature. They also claim that the thermal diffusivity of the thermocouple should be at least one order of magnitude higher than that of the measured domain in order to obtain meaningful results in transient measurements. This is difficult to obtain for quenching metals, since both, thermocouple and workpiece have mutually similar thermal diffusivity values.

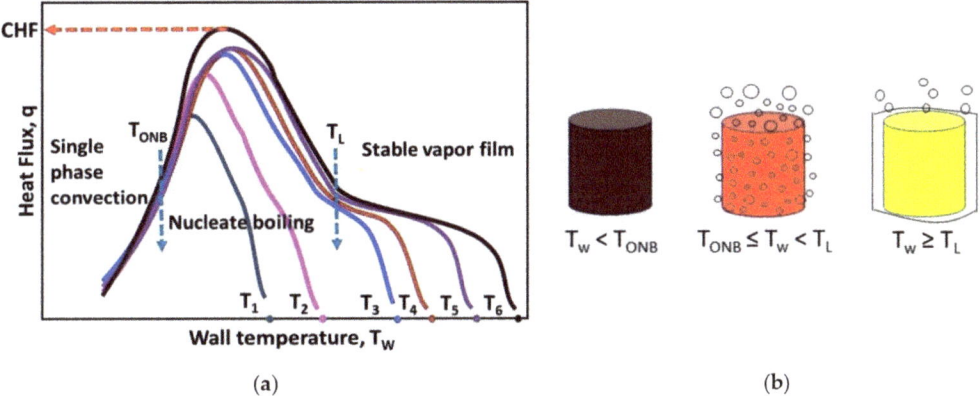

Figure 2. Schematic representation of (**a**) boiling curves corresponding to the continuous cooling of probes from several initial temperatures, $T_1, T_2, \ldots, T_6$ and (**b**) heat transfer regimes during quenching by immersion.

3.3. Effect of Quenchant Liquid

The quenchant liquid properties that have a major impact on the quenching process are: the boiling latent heat and temperature, viscosity, and surface tension. Boiling temperature, also called saturation temperature, T_{sat}, has an effect on the transition temperature from the nucleate boiling regime to the single-phase convection regime, i.e., over the onset of nucleate boiling temperature, T_{ONB}. This temperature is always larger than the saturation temperature, especially when quenching under forced convection. In this case, the flowing liquid spends a very short time in contact with the wall surface, preventing its boiling, even if the wall temperature is many degrees above T_{sat}. In contrast, for natural convection, T_{ONB} exceeds only few degrees the value of T_{sat}. The wall superheating to reach the nucleate boiling regime, $\Delta T_w = T_{ONB} - T_{sat}$, is therefore dependent on the fluid dynamics of the quenchant. This means that under constant fluid flow conditions, a liquid which has a larger value for T_{sat} will have a larger T_{ONB} than the corresponding value for a liquid having a smaller T_{sat}. This is the case for water and oil, two commonly used quenchants. Water saturation temperature at 1 atm is 100 °C, while the boiling temperature of a typical mineral oil is 280 °C. Therefore, T_{ONB} for oil is larger than the corresponding value for water. T_{ONB} is particularly important because at wall temperatures above it, vapor bubbles form on the wall surface, creating an obstacle for a direct liquid-solid contact. Vapor thermal conductivity is one order of magnitude smaller than that of liquid. Therefore, the rate of heat conduction through the vapor bubbles is considerably smaller than heat conduction through the liquid. This is represented schematically in Figure 3a,b, where the vertical arrows represent the heat flux from the wall. As a result, heat flux from a wall is more uniform when single phase convection is present, that is, when $T_w < T_{ONB}$. Heat flux uniformity is especially important to minimize thermal and phase transformation stresses. When martensite phase forms during quenching, its higher specific volume with respect to austenite phase generates stresses in the workpiece. Quenching is a controlled cooling process to form martensite as uniform as possible, avoiding that the corresponding stresses may lead to unacceptable workpiece distortion or even crack formation. Martensite forms at a temperature, M_s, that depends on the steel grade. Therefore, it is advisable that $T_{ONB} > M_s$. This is the main reason oil is used as a quenchant rather than water for treating steel grades that are more susceptible to crack formation and distortion during quenching.

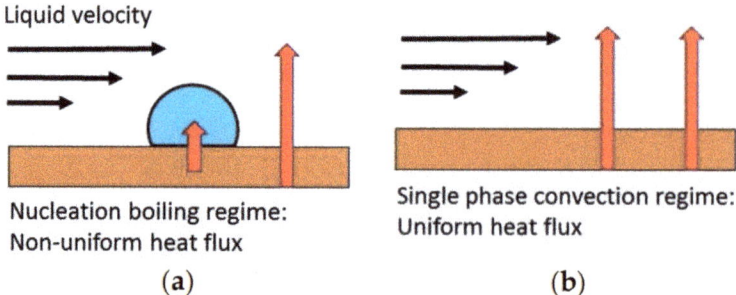

Figure 3. Schematic comparison of heat transfer distribution from a workpiece when quenching under: (**a**) nucleation boiling regime, in which heat flux through vapor bubbles is smaller than heat flux to the liquid and (**b**) single phase convection regime, where heat flux in uniform through the wall surface.

The latent heat for vaporization also plays an important role to remove heat during quenching. The higher the latent heat is, the slower the rate of bubble formation becomes. This, in turn, influences the temperature range for each boiling regime to occur.

Liquid viscosity, as defined by Newton's viscosity law, is a physical property that measures the fluid resistance to flow under a given shear stress. This property determines the velocity profile in the fluid within the boundary layer. This layer is a thin region located just over the wall and it may be formed by liquid, a mixture liquid-vapor, or vapor phase, depending on the present boiling regime. Liquid viscosity has an effect in the single-phase convection and nucleate boiling regimes. A low viscosity value promotes a higher fluid velocity and therefore an improved rate of heat transfer in the single-phase regime. In addition, a low viscosity value favors a faster vapor bubble growth and detachment from the wall; this improves heat flux in the nucleate boiling regime. Generalization of the previous ideas should be taken with caution since liquid viscosity is a strong function of temperature. Viscosity changes during quenching should be considered carefully to properly model the heat transfer process.

Surface tension, γ_{LG}, is a property of an interface liquid-gas that measures the energy per unit surface area. The higher the surface tension, the larger is the energy associated with this free surface and the more difficult is to increase it. Surface tension plays a role in the wetting of the wall surface and in the rate of bubble formation and growth. This is discussed below in Section 3.5.

3.4. Effect of Liquid Velocity and Wall Temperature

Before presenting the effect of liquid velocity and wall temperature it is useful to recall the connection between the heat transfer coefficient, h, and the boiling curve, which is given by the Newton convection equation expressed as,

$$q = h\left(T_f - T_w\right) \quad (1)$$

where q is the heat flux at the wall, as defined previously and T_f is the bulk liquid temperature. Therefore, h can be obtained from Equation (1) and using the boiling curve data (q vs. T_w). Readily, we can plot the corresponding h vs. T_w curve. This curve is also a parabolic-type function having a maximum value at a temperature which is generally close to that for the CHF.

Liquid velocity plays an important role in heat transfer from a wall surface to the fluid. The laminar boundary layer theory shows that the local heat flux from a flat wall, at a distance x from the leading edge of the plate, is given by the following equation [13],

$$q = \left\{0.332\, k\, Pr^{1/3}\left[\frac{V_\infty}{\nu x}\right]^{1/2}\right\}\left(T_f - T_w\right) \quad (2)$$

where k and ν are the thermal conductivity and the kinematic viscosity of the fluid, respectively; Pr is the Prandtl number ($=\nu/\alpha$, kinematic viscosity divided by the thermal diffusivity of fluid), and V_∞ is the fluid velocity relative to the wall, and T_w is the wall temperature. The term that appears inside the curly brackets represents the local heat transfer coefficient for a single-phase fluid flowing parallel to the wall. Equation (2) shows that heat flux increases with the square root of the fluid velocity. During quenching, the previous equation is only valid under the single-phase convection regime. Nucleate boiling regime introduces bubble formation, growth and detachment from the wall surface and Equation (2) is no longer valid. However, heat flux in this regime also increases with fluid velocity. Finally, the stable vapor blanket regime leads also to a single phase flowing over the wall surface. However, the combined heat flux is the sum of two contributions: a convection contribution given by Equation (2) and a radiation term given by the following equation.

$$q_r = \sigma F \varepsilon\left(T_f^4 - T_w^4\right) = \left\{\sigma F \varepsilon\left(T_f^2 + T_w^2\right)\left(T_f + T_w\right)\right\}\left(T_f - T_w\right) \quad (3)$$

where $s = 5.669 \times 10^{-8}$ W/m^{2}K^4 is the Stefan–Boltzmann constant, F is the view factor which is commonly equal to one for quenching workpieces, and e is the emissivity of the wall surface. Analogous to Equation (2), the term inside the curly brackets represents the *radiative heat transfer coefficient*. It is seen that this coefficient depends on the wall absolute temperature rised to the third power. Table 1 presents some examples of combined heat transfer coefficients as a function of the fluid velocity [14–16]. They were determined from thermal analysis and the solution of the IHCP. Data from refs. [14,16] were obtained using the ivf SmartQuench® system which follows specifications of ISO 9950. It includes an Inconel 600 instrumented probe that is heated up in a resistance furnace to the test initial temperature. Then, the probe is immersed into a preheated oil bath and the cooling curve is registered. Reference [15] reports the use of steel cylindrical probes, (Ø28 mm × 56 mm height and Ø44 mm × 88 mm) to study the effect of both, fluid velocity and probe diameter on the heat flux. The cylinders were thermally isolated in the basal and top surfaces, to force heat to flow in the radial direction. Heat transfer coefficient values were estimated from data of references [14,15] using Equation (1). In all cases, the heat transfer coefficient increases with fluid velocity. The table shows only the peak values for the heat transfer coefficient and misses the detailed variation of h with wall temperature, T_w, and also with the initial temperature, T_o.

Table 1. Experimentally determined heat transfer coefficients as a function of fluid velocity.

Reference Comments	Fluid Velocity [1] Heat Transfer Coefficient (W/m²K)			
Ref. [14] [2] Oil Q8 Bellini FNT® at 66 °C Probe initial temperatures: 750, 800, 850 and 900 °C	0 rpm 4246	1000 rpm 4459	2000 rpm 4671	3000 rpm 4883
Ref. [15] [2,3] Mineral oil at 30 °C Probe initial temperature: 870 °C	0 m/s, 28 mm 2697	0 m/s, 44 mm 3848	0.33 m/s, 28 mm 4348	0.33 m/s, 44 mm 5635
Ref. [16] Oil Quenchway 125B at 130 °C Probe initial temperature 850 °C	0 m/s 3000	0.2 m/s 3100	0.35 m/s 3500	0.5 m/s 3900

[1] Units may be rpm or m/s as it is indicated. [2] Computed values from CHF, T_{max} and T_f data. [3] 28 and 44 mm are the diameters of cylindrical probes, steel grade AISI 1040.

3.5. Effect of Wall Surface Roughness and Wettability

Essentially, wall surface roughness is represented by the average length of mili or micro scale bumps and batches forming a net of gaps over the wall surface. When a workpiece is submerged into a liquid, air remains trapped in these gaps. During the nucleate boiling regime, vapor bubbles will grow from these "air seeds". In contrast, a smooth surface is basically a gap-free surface which requires a higher superheating, as compared to a rough surface, to promote bubble nucleation and grow. Surface roughness can be controlled by machining specific patterns on the wall surface, or it can be a consequence of solid deposits formed on the surface, for example salt deposition when using hard water for quenching, or when quenching a workpiece covered with an oxide layer, which formed during its previous heating in the furnace. The wall surface roughness has an effect on single phase convection since improves the specific surface area, increasing the heat flux; although in most heat-treating cases, this effect is marginal. In contrast, surface roughness plays a major role during nucleate boiling regime and its effect on the boiling curve is explained including the surface wettability as follows.

Wettability is a property that measures the ability to extend the liquid-solid interface surface. A higher wettability means a larger liquid-solid interface area which leads to a higher heat flux, when remaining all other variables as constants. Wettability can be measured using the contact angle, θ, which is determined from the Young equation and is represented by the following expression.

$$\gamma_{LS} + \gamma_{LG}cos\theta - \gamma_{GS} = 0, \tag{4}$$

where γ_{LS}, γ_{LG}, γ_{GS} are the surface tensions at the liquid-solid, liquid-gas and gas-solid interfaces, respectively. In the range $0 \leq \theta \leq 180°$, $cos\theta$ increases when θ decreases. Therefore, a decrease in surface tension leads to a decrease in the contact angle, improving wall wettability. This means that decreasing surface tension, for example using tensoactives, improves wettability and therefore increases the heat flux during quenching in the single-phase convection regime. This is not the case at high wall temperature, above T_L. Under stable vapor blanket regime, liquid properties are no longer heat flux controlling but the vapor properties are. Therefore, under this regime, wettability has no effect. In the intermediate regime, nucleation boiling, wettability has a specific effect on the rate of bubble nucleation and growth. Vapor bubbles are formed in the gaps that are already occupied by air. When wettability increases, liquid penetrates deeper into every gap leaving a smaller "air seed" for vapor bubble nucleation. This would make more difficult for a bubble embryo to grow. The wettability effect is to increase heat flux as a result of improving direct contact between liquid and the solid wall.

Wettability has been also modified by applying and electric voltage [17]. The authors report laboratory results showing that *interfacial electrowetting* (EW) fields can disrupt the stable vapor film. They attributed this phenomenon to electrostatic attraction of the liquid molecules to the wall surface. As a result of improving direct contact between liquid-wall, heat flux is increased. Experiments include the use of stagnant water bath to quench metallic spheres from 400 °C.

3.6. Effect of Solid Properties and Workpiece Size

The previous discussion has been based on the considerable number of studies on the effect of liquid and wall surface characteristics on the boiling curve. However, the role of solid properties and workpiece size on the heat flux curve have not received so much attention. This may be justified since convection boiling has developed itself from steady heat transfer applications, as for example studies on heat exchangers. In these cases, there is a constant temperature profile in the solid, which reduces to a uniform temperature when the *characteristic length* of the solid is small enough. This length, L, can be estimated when the *Biot number*, Bi, is less than 0.1, according to the following equation:

$$Bi = \frac{hL}{k} < 0.1 \qquad (5)$$

where h is the heat transfer coefficient on the wall surface and k is the thermal conductivity of the solid workpiece. Heat treatment of metallic workpieces is a transient process which is frequently applied to workpieces that have a characteristic length such that Biot number is larger than 0.1. Under this scenario, the role of solid properties and workpiece dimensions on the boiling curve should not be underestimated.

Thermo-physical properties include density (ρ), thermal conductivity (k) and specific heat (C_p) which are generally temperature dependent. Moreover, treated alloys show solid phase transformations during quenching that release the corresponding latent heat. This energy has an effect on the boiling curve and has to be considered carefully when using thermal analysis and the solution of the IHCP to determine the boiling curve. Neglecting this contribution to the heat flux may lead to a wrong interpretation of quenching experimental data. The latent heat for transformation from solid "a" to solid "b", ΔH_{a-b}, can be considered in a heat transfer analysis by adding it to the specific heat within the transformation temperature range, ΔT_{ab}, to define the *apparent specific heat* according to the following equation.

$$C_{p,\,app} = C_{p,a} + \frac{\Delta H_{a-b}}{\Delta T_{ab}} \qquad (6)$$

where $C_{p,a}$ represents the specific heat of solid phase "a". The previous equation has been successfully used in metals solidification heat transfer studies [18,19], and it has been implemented also to analyze cooling curves from steel quenching experiments [20,21].

The boiling curve is the result of coupling convective boiling phenomena with the heat conduction process in the solid. Heat has to be transported from the solid bulk to the interface where the fluid will remove it. Obviously, larger workpieces spend more time to cool down than smaller ones, so it is easy to understand that temperature vs. time or heat flux vs. time curves will change. However, the effect of workpiece size on the boiling curve, heat flux vs. wall temperature, is not that obvious. For example, Sanchez-Sarmiento et al. [22] reported a model for residual stresses in spring steel quenching. The authors determined by thermal analysis, and using the solution of the IHCP, the heat transfer coefficient as a function of wall temperature for AISI 5160H steel cylinders of

13.45 and 20.56 mm diameter quenched in an aqueous solution and in an oil. Under the same liquid and wall temperatures, they reported significant differences in the maxima heat transfer coefficients for these probes. The authors did not offer a fundamental explanation of these findings but it is clear that the cylinder diameter played a role. This result shows the importance to couple heat transfer between liquid and solid to properly represent heat flux during quenching.

3.7. Link between Heat Transfer Coefficient, H, and Quenching Severity, H

Quenching severity is a quantity that has been widely used in heat treatment engineering to design quenching systems, and it represents the ability of a quenching medium to extract heat from a hot workpiece. It is expressed as the *Grossman H-value*, which commonly varies for a steel workpiece between 7.9 (oil, no agitation) to 196 m^{-1} (brine, strong agitation) [23]. This quantity is related to the heat transfer coefficient, h, by the following expression.

$$H = \frac{h}{k} \tag{7}$$

where k is the thermal conductivity of the solid metal. It should be mentioned that since both, h and k are temperature dependent, then so H is. Moreover, since h depends on the type of quenchant liquid and fluid velocity, H is also dependent on these variables. In heat treatment practice, H is considered a constant average value for each specific quenching condition.

The ultimate quenching goal of practical importance is measured by the relative depth of martensite phase that was formed during cooling. This relative depth is defined as the ratio of the actual depth of martensite layer divided by the total characteristic length of the workpiece, for example the radius for a cylindrical workpiece. Steel hardenability is closely related to this relative depth, which depends on both, H and the steel grade. In order to characterize alloy hardenability in terms only of steel grade, a widely known laboratory method known as the end-quench Jominy test [24] was proposed by Walter E. Jominy and A.L. Boegehold in 1937 to measure steel hardenability, and it is summarized in the following section.

3.8. Hardenability Determination, the End-Quench Jominy Test

The end-quench Jominy test uses a standard cylindrical probe 100 mm long × 25.4 mm diameter which is heated up to an austenization temperature and then is set over an open vertical tube from where water flows upward at a controlled constant rate. The liquid impacts the basal face of the cylinder, as it is shown in Figure 1, and heat is removed from the probe in the downward direction by convection boiling and radiation. The lateral surface of the cylinder is not isolated therefore heat flows by natural convection and radiation. However, it can be shown, that the lateral heat flux is much smaller than the basal heat flux, and therefore heat flows essentially in one dimension (1D). The resulting steel structure changes from the basal surface all along the cylinder. Martensite forms at the basal surface and its local proportion decreases further away from this point. Hardness depends on the obtained phases; therefore, hardenability is determined from a hardness profile measured along the cylindrical probe. Since the probe and test setup dimensions and the water flowrate are standard, the hardness curve depends only on the steel grade.

This laboratory test has been the focus of numerous works aiming to understand the relationship between heat transfer, kinetics of phase transformation and mechanical properties after quenching. A particularly useful analysis has been presented by Smoljan et al. [25], who developed a model to simulate the hardness of quenched and tempered steel workpieces. In these studies, heat transfer is characterized using at least one thermocouple, as it is shown in Figure 1. The cooling curve data are analyzed using the solution of the IHCP to determine the heat flux at the wall and, more important, the temperature field evolution in the workpiece. This computed temperature is used to calculate the cooling rate map within the solid probe and then the kinetics of phase transformation and the corresponding mechanical properties. Thermal analysis and mathematical modeling have proved to be useful methods to improve our understanding on the relationship, heat transfer, microstructure evolution and properties. However, when dealing with design, optimization and quality issues during production heat treatment, our knowledge of such a relationship may not allow us to predict quenching phenomena with enough accuracy, unless a detailed data base were available. Section 4.3 presents a summary of these data.

4. Discussion of Poorly Attended Aspects of the Boiling Curve

This section presents fundamental aspects of the boiling curve that are important for heat treatment of steel and alloys, but they are no analyzed neither discussed in the open literature. A reason for this lack of attention may be that heat transfer conditions during metal quenching differ from those for most studied cases in convective boiling, where the wall temperature is maintained constant, and/or the wall superheat is below 400 °C. Therefore, the number of scientific papers focused on studying heat transfer during metal quenching is considerably smaller. It should be recalled that heat transfer conditions in heat treatment of alloys involve very high initial temperatures, therefore all regimes may be simultaneously present in a single workpiece during quenching. Furthermore, the transient nature of quenching establishes a coupled heat transfer between the conduction in the metal workpiece and the convective boiling flow. Finally, liquid velocity over the wall has been taken to extreme values, for example using high pressure sprays, to achieve intensive cooling during quenching.

4.1. Initial Wall Heat Flux

The effect of initial temperature of the solid wall on the boiling curve was described in Section 3.2. It was pointed out that, independently from the initial temperature, heat flux starts from a low value and then increases when T_w decreases. This leads to a family of q vs. T_w curves rather than to a single curve, as is shown schematically in Figure 2a. If the measured low initial heat flux is attributed to the thermocouple time constant, we must ask how to make sure that the initial heat flux is actually low. An answer can be inferred from studying quenching during continuous casting of steel. Motomochi-Espinoza and Acosta-González [26] reported a heat transfer analysis to design a laboratory rig representing heat flow during secondary cooling in continuous casting of thin slab. They pointed out that the steel entering the secondary cooling system has a high temperature gradient across its solidified shell (for example 12 mm thickness), which is associated with an estimated heat flux ~0.9 MW/m^2. This heat flux is considerably larger than the initial values reported for laboratory or plant quenching tests. The difference between a solidified shell and a workpiece is the initial temperature gradient. Treated workpieces are soaked in furnaces to reach a uniform temperature. Further, once the workpiece is removed from the furnace, essentially no thermal gradient develops during its transportation period from the furnace to the quenching tank. Therefore, the workpiece starts cooling from a homogeneous temperature that needs time to develop and promote heat conduction. This is a plausible factor influencing the observed low initial heat flux values in boiling curves. In the next section, we will include a discussion of the fluid velocity as another a factor that becomes important to increase the initial heat flux by suppressing the stable vapor blanket. The advantage of this factor is that it is suitable for control in a quenching production line.

The main remark of the previous analysis is that initial solid temperature plays an important role to determine the whole q vs. T_w curve and therefore the proper boiling curve should be considered in any mathematical model of a phase transformation in solid state metal. For example, some studies of heat treatment of alloys [27–30] present models to predict the kinetics of phase transformation during quenching assuming a boiling curve, as a boundary condition to solve the heat conduction equation. This curve was commonly reported in an independent work and which workpiece was heated at a specific initial temperature. Unless the initial temperature in the actual study were the same as in the heat conduction study where the boiling curve was taken from, the results of modeling phase transformation would be meaningless.

4.2. Definition of Fluid Velocity and Its Effect on the Boiling Curve

Liquid velocity refers to the relative velocity between the workpiece and the liquid. Since it is known that fluid velocity changes from point to point, then which velocity value is commonly considered to report the effect of liquid velocity on the boiling curve? The answer is not unique, but there are several conventional values. For example, for pool boiling quenching, a zero liquid velocity is reported in spite of the natural flow promoted by the buoyant force and the rising bubbles around the workpiece. Some authors report the liquid velocity during forced convection in agitated tanks by propeller driven flow, as the shaft rotation speed, in revolutions per minute (RPM). In other cases, forced convection is also obtained from a circulating flow through a quenching tank. A liquid flow rate enters the tank through an inlet tube and leaves the tank out through an outlet located somewhere far away. In this case, fluid velocity is taken as the average liquid velocity in a given cross section area, for example, at the inlet tube. Recently [31,32], it has been proposed to use the isothermal shear stress on the wall to correlate with boiling heat flux, inspired by the Reynolds–Colburn analogy, rather than using the liquid velocity itself. In spite there is no universal agreement on which liquid velocity to report, it is well known that increasing liquid velocity improves heat transfer from the workpiece as was shown by Totten and Lally from observations in a physical model of a quenching tank [33]. This fact has led to an important technological development called intensive quenching [34] where high pressure sprays are used to promote a very high cooling rate, leading to maximum surface compressive stresses, and avoiding workpiece surface cracking. Figure 4 shows schematically the effect of the liquid velocity on a boiling curve. Notice that fluid velocity has an effect on the CHF and on the transition temperatures, T_L and T_{ONB}. This is associated to a higher fluid velocity in the boundary layer. At a high velocity, the contact time between liquid and solid is very short and therefore liquid cannot reach its boiling temperature. Then, single phase convection takes place improving the magnitude of heat flux. This regime can also improve uniformity on the heat flux as explained in Figure 3a,b. At a higher wall temperature, liquid flowing at velocity v_2 is able to remove the vapor blanket as compared to liquid flowing slower. Therefore, Leidenfrost temperature increases with liquid velocity.

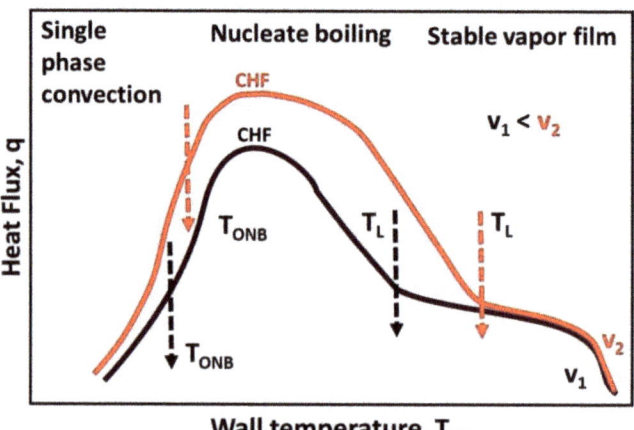

Figure 4. Schematic representation on the effect of liquid velocity on the boiling curve. Notice that Leidenfrost (T_L) and onset of nucleate boiling (T_{ONB}) temperatures change with fluid velocity, where velocity v_1 is smaller than velocity v_2.

4.3. The Need for a Comprehensive Data Base for Heat Treatment Analysis

Ideally, thermophysical properties of alloys and quenchant liquids, mechanical properties of phases and microconstituents that are present in solid alloys, and a full data set of empirical parameters for phase transformation kinetic models would be enough to feed computerized models of heat transfer phenomena, phase transformation kinetics and mechanical behavior to predict the resulting microstructure and properties distribution in any workpiece subjected to a specific heat treatment. However, our current knowledge on convection boiling is not complete enough to predict, from first principles, the thermal evolution of actual workpieces during immersion quenching. An accurate prediction of nucleation, growth and detachment of multiple vapor bubbles from the wall surface requires to solve the multiphase fluid flow equations for turbulent flow conditions. Turbulent flow itself is a developing area that is currently best modeled by *Direct Numerical Simulation* (DNS). This method needs very fine grids and extremely small time steps to obtain accurate numerical solutions, demanding prohibitive computer effort for quenching simulation. Therefore, analysis of heat transfer during quenching is more conveniently carried out using alternative simplifying assumptions. These assumptions may be: (1) Isothermal fluid flow, (2) conjugate heat transfer, (3) interpenetrated liquid-vapor phases, or (4) using experimental thermal analysis and the solution of the IHCP.

The approach by isothermal fluid flow includes only the solution of the continuity and momentum equation in the liquid under steady and turbulent conditions [31,35]. The temperature evolution of the solid workpiece is not considered in this analysis. The fluid flow field is computed in the whole quenching tank that contains one or several workpieces under treatment. Velocity distribution is used to infer, qualitatively, the cooling rate of the workpiece [31] or use it to estimate a heat transfer coefficient from separate heat transfer calculations [35]. The higher the liquid velocity passing on the workpiece is, the higher the rate of heat removal would be. This approach is useful to study the effect that changes on the flow direction and magnitude have on the uniformity of heat removal from the workpiece.

The method based on conjugate heat transfer includes the calculation of both, fluid-dynamics of the quenchant and temperature evolution in the workpiece. No heat transfer coefficient is required in this approach since the heat flux at the wall surface is computed by specifying that both, heat flux and temperature are continuous at the wall surface, and thereby are computed. In this method, no boiling is considered since the liquid flows isothermally. This approach demands more computer effort than the purely isothermal flow model, however, it allows to compute cooling rates in the solid. These cooling rates are the basis to predict microstructure evolution and properties.

A more elaborate approach is the multiphase flow method represented by the interpenetrated liquid-vapor phases. The solid temperature can be computed or assumed equal to a fixed value, and the fluid flow calculations include boiling of liquid. In this approach, the rate of vaporization is computed but the interface of individual bubbles is not represented explicitly. Rather, volume fractions of vapor and liquid are determined for every control volume, that is for every position in the fluid. This is why the method takes the name interpenetrated phases. Bubble size should be specified but only to compute the rate of mass, momentum and energy interchange with the liquid, which is important to determine buoyant and drag forces and vaporization-condensation rates. In this regard, empirical coefficients for the rates of vaporization and condensation should be known. This approach requires more computer effort than the two previous methods but a greater detail in the quenching rate distribution over the wall surface can be obtained.

Finally, the most used method—both simple and trustworthy—is thermal analysis using the solution of the IHCP. As was mentioned above, a metallic probe is instrumented with one or several sub-superficial thermocouples. The instrumented probe is heated up to the homogenization temperature and then removed from the furnace and quenched. Each cooling curve is used in the solution of the Inverse Heat Conduction Problem to determine the heat flux at the wall as a function of time. The heat transfer coefficient can be computed from this heat flux using Equation (2). In essence, this procedure measures the temperature evolution at one, or more points, within the workpiece and then uses the solution of the IHCP to compute the temperature evolution in the rest of the workpiece. The main drawback of this approach is the need to generate thermal analysis for every needed quenching condition. For example, when using different quenchants, different alloys, different workpiece sizes, quenching under different liquid velocities, etc. Extrapolation of the obtained results for the studied cases is not recommended for other new cases.

There are standard instrumented probes [7] that are employed to characterize the quenching power of liquids. These probes are made of Inconel, stainless-steel or silver, which do not suffer a phase transformation in the solid-state during quenching. This is convenient to analyze the respective cooling curve data since no latent heat of phase transformation should be considered. An extension of this idea is to use instrumented probes made of the same steel grade under heat treatment [36]. These probes are processed in the quenching tank as if they were the quenching workpieces, emulating the actual production conditions.

4.4. Research Opportunities

The previous overview allows to identify the following research opportunities:

Mathematical modeling of heat transfer. Heat transfer coefficient is the key variable to simulate the thermal field evolution in a steel workpiece. However, overwhelmingly of technical papers considers that this coefficient is only a function of wall temperature, fluid velocity and its properties, and at best surface roughness. All these variables correspond to the region from the wall surface to the liquid. We do not agree with this idea. The resulting heat transfer coefficient also depends on the steel thermophysical properties and on the thermal gradients in the workpiece. This is because the net heat flux at the wall surface is the result of coupling heat conduction and convection boiling. There are few authors who report, implicitly, such an idea. For example, in ref. [2], heat transfer coefficient is also a function of position on the workpiece surface and time. The authors do not explain additional factors, other than fluid flow, affecting this dependence. We believe that such a dependence with position and/or time is not a proper choice because it hides the real physics behind it. For this reason, there is opportunity to develop mathematical models considering coupling between heat conduction in the steel and convection boiling in the fluid. The main physical ingredients controlling this coupling, remain to be investigated.

Experimental fluid flow and heat transfer. Laboratory measurements of heat flux are generally concentrated on the solid cooling curves. However, simultaneous fluid flow may be also characterized. There is a challenge of observing and recording the fluid flow field in very short periods, during cooling of the probe. An alternative strategy is to compute the actual convective boiling flow field rather than measure it. Therefore, analysis would include measured cooling curves in the solid and computed fluid flow field.

5. Conclusions

The analysis and criticism of the technical literature on heat transfer phenomena during solid steel quenching has led to the following conclusions.

1. The low initial heat flux value in the boiling curve has been historically associated to the time constant of the thermocouple. Furthermore, at high wall temperatures, a stable vapor blanket may form acting as a thermal resistance to limit the heat flux. However, the measured heat flux in presence of this vapor blanket is larger than the initial heat flux. Therefore, the proposed argument based on the initial uniform temperature in the solid is a plausible explanation for such a low initial heat flux. Heat flux is promoted in the solid only after there is a thermal gradient in this material.
2. It is known that heat flux removed from a quenching workpiece increases with fluid velocity. However, in spite of its importance, velocity has been reported as an average value and using different definitions. For example, rotation speed of a propel (RPM), average liquid velocity through the cross-section area of the inlet tube, or a local

value which is proportional to the isothermal shear stress and pressure on the wall surface. This makes a comparison between results from different researchers difficult. In addition, a generalization of these results to integrate a data base becomes complex.
3. Ideally, heat flux during quenching should be high and uniform above the workpiece surface. This would promote a uniform phase transformation throughout the whole workpiece minimizing thermal and phase formation stresses. According to the literature analysis carried out by the present authors, the best heat transfer regime to achieve such a uniformity is the single-phase convection, not the nucleation boiling regime, as it may be though because of the CHF. During single phase convection, liquid is always in contact with the wall surface minimizing heat flux fluctuations. Furthermore, the heat flux can be increased by augmenting the fluid velocity. For example, using a high-pressure spray. This increment in the liquid velocity would also increase the value of T_{ONB}, avoiding bubble formation on the wall.
4. Mathematical modeling of the quenching process has been based on the solution of the conservation of mass, momentum and energy differential equations. Quenching modeling has been useful to understand the involved phenomena. However, there is no unique formulation to represent this complex process and a number of alternatives have been developed. From the literature analysis, we found models that can be classified into these categories: (1) Isothermal fluid flow, (2) conjugate heat transfer, (3) interpenetrated liquid-vapor phases, or (4) using experimental thermal analysis and the solution of the IHCP. The ultimate users should select the approach method, according to their particular objectives and their technical capabilities.

Author Contributions: Writing-original draft preparation, M.d.J.B.-R.; supervision, writing-review and editing, F.A.A.-G.; writing-review and editing, M.M.T.-R. All authors have read and agreed to the published version of the manuscript.

Funding: This research received no external funding.

Institutional Review Board Statement: Not applicable.

Informed Consent Statement: Not applicable.

Data Availability Statement: The used data can be found in the respective references.

Acknowledgments: The authors express gratitude to their respective Institutions for the support received during this pandemic, Covid-19.

Conflicts of Interest: The authors declare no conflict of interest.

References

1. Totten, G.E.; Bates, C.E.; Clinton, N.A. *Handbook of Quenchants and Quenching Technology*; ASM International: Materials Park, OH, USA, 1993.
2. Simsir, C. Modeling and Simulation of Steel Heat Treatment-Prediction of Microstructure, Distortion, Residual Stresses, and Cracking. In *ASM Handbook Steel Heat Treating Technologies*; Dosset, J., Totten, G.E., Eds.; ASM International: Materials Park, OH, USA, 2014; Volume 4B.
3. Abbasi, B.; Kim, J.; Marshal, A. Dynamic pressure based prediction of spray cooling heat transfer coefficients. *Int. J. Multiph. Flow* **2010**, *36*, 491–502. [CrossRef]
4. Araki, K.; Yoshinobu, S.; Nakatani, Y.; Moriyama, A. Stationary measurement for heat transfer coefficient in droplet-cooling of hot metal. *Trans. ISIJ* **1982**, *22*, 952–958. [CrossRef]
5. Bernardin, J.D.; Mudawar, I. Film boiling heat transfer of droplet steams and sprays. *Int. J. Heat Mass Transf.* **1997**, *40*, 2579–2593. [CrossRef]
6. Hernández-Bocanegra, C.A.; Castillejos-Escobar, A.H.; Acosta-González, F.A.; Zhou, X.; Thomas, B.G. Measurement of heat flux in dense air-mist cooling: Part I—A novel steady-state technique. *Exp. Therm. Fluid Sci.* **2013**, *44*, 147–160. [CrossRef]
7. Kobasko, N.I. Thermal Equilibrium and Universal Correlation for Its Heating-Cooling Time Evaluation. *Theor. Phys. Lett.* **2021**, *29*, 205–219. [CrossRef]
8. Stewart, I.; Massingham, J.D.; Hagers, J.J. Heat transfer coefficient effects on spray cooling. *Iron Steel Eng.* **1996**, *73*, 17–23.
9. Li, D.; Wells, M.A.; Cockcroft, S.L.; Caron, E. Effect of Sample Start Temperature during Transient Boiling Water Heat Transfer. *Metall. Mater. Trans. B* **2007**, *38*, 901–910. [CrossRef]

10. Beck, J.V. Nonlinear estimation applied to the nonlinear inverse heat conduction problem. *Int. J. Heat Mass Transf.* **1970**, *13*, 703–716. [CrossRef]
11. Babu, K.; Kumar, T.S.P. Mathematical Modeling of Surface Heat Flux During Quenching. *Metall. Mater. Trans. B* **2010**, *41*, 214–224. [CrossRef]
12. Rabin, Y.; Rittel, D. A Model for the Time Response of Solid-embedded Thermocouples. *Exp. Mech.* **1999**, *39*, 132–136. [CrossRef]
13. Incropera, P.I.; Dewitt, D.P. *Fundamentals of Heat and Mass Transfer*, 3rd ed.; John Wiley and Sons: New York, NY, USA, 1990; p. 393, ISBN 0-471-61246-4.
14. Passarella, D.N.; Aparicio, A.; Varas, F.; Ortega, E.B. Heat Transfer Coefficient Determination of Quenching Process. In *Mecánica Computacional, Proceedings of the Asociación Argentina de Mecánica Computacional, San Carlos de Bariloche, Argentina, 23–26 September 2014*; Bertolino, G., Cantero, M., Storti, M., Teruel, Y.F., Eds.; Asociación Argentina de Mecánica Computacional: Santa Fe, Argentina, 2014; Volume XXXVIII, pp. 2009–2021.
15. Fernandes, P.; Prabhu, K.N. Effect of section size and agitation on heat transfer during quenching of AISI 1040 steel. *J. Mater. Process Technol.* **2007**, *183*, 1–5. [CrossRef]
16. Aronsson Rinby, C.; Sahlin, A. Compilation and Validation of Heat Transfer Coefficients of Quenching Oils. Master's Thesis, Chalmers University of Technology, Gothenburg, Sweden, 2012.
17. Shahriari, A.; Hermes, M.; Bahadur, V. Electrical control and enhancement of boiling heat transfer during quenching. *Appl. Phys. Lett.* **2018**, *108*, 091607. [CrossRef]
18. Muhieddine, M.; Canot, E.; March, R. Various Approaches for Solving Problems in Heat Conduction with Phase Change. *Int. J. Finite Vol.* **2009**, *6*, 1–19.
19. Camporredondo, S.J.E.; Castillejos, E.A.H.; Acosta, G.F.A.; Gutiérrez, M.E.P.; Herrera, G.M.A. Analysis of Thin-Slab Casting by the Compact-Strip Process: Part I. Heat Extraction and Solidification. *Metall. Mater. Trans. B* **2004**, *35*, 541–560. [CrossRef]
20. Durand, D.; Durban, C.; Girot, F. Coupled Phenomena and Modeling of Material Properties in Quench Hardening Following Inductive Heating of the Surface. In Proceedings of the 17th ASM Heat Treating Society Conference including the 1st International Induction Heat Treating Symposium, Indianapolis, IN, USA, 15–18 September 1997; Milam, D., Poteet, D.A., Pfaffman, G.D., Rudnev, V., Muehlbauer, A., Albert, W.B., Eds.; ASM International: Materials Park, OH, USA, 1997; pp. 855–863. [CrossRef]
21. Lopez-Garcia, R.D.; Garcia-Pastor, F.A.; Castro-Roman, M.J.; Alfaro-Lopez, E.; Acosta-Gonzalez, F.A. Effect of Immersion Routes on the Quenching Distortion of a Long Steel Component Using a Finite Element Model. *Trans. Indian Inst. Met.* **2016**, *69*, 1645–1656. [CrossRef]
22. Sánchez-Sarmiento, G.; Castro, M.; Totten, G.E.; Harvis, L.; Webster, G.; Cabré, M.F. Modeling Residual Stresses in Spring Steel Quenching. In Proceedings of the 21st ASM Heat Treating Society Conference, Indianapolis, IN, USA, 5–8 November 2001; Liščić, B., Tensi, H.M., Shrivastava, S., Specht, F., Eds.; ASM International: Materials Park, OH, USA, 2001; pp. 1–10.
23. Poirier, D.R.; Geiger, G.H. Correlations and Data for Heat Transfer Coefficients. In *Transport Phenomena in Materials Processing*, 2nd ed.; The Minerals Metals & Materials Society: Warrendale, PA, USA; Springer International Publishing: Cham, Switzerland, 2016. [CrossRef]
24. *Standard Test Methods for Determining Hardenability of Steel*; ASTM International: West Conshohocken, PA, USA, 2007; pp. 18–26.
25. Smoljan, B.; Iljkić, D.; Totten, G.E. Mathematical Modeling and Simulation of Hardness of Quenched and Tempered Steel. *Metall. Mater. Trans. B* **2015**, *46*, 2666–2673. [CrossRef]
26. Motomochi-Espinoza, P.A.; Acosta-González, F.A. A Method to Design Laboratory Tests Aimed to Determine the Heat Flux during Steel Processing (in Spanish). In Proceedings of the Memorias del 37° Congreso Internacional de Metalurgia y Materiales, Saltillo, Coahuila, Mexico, 11–13 November 2015; Tecnológico Nacional de México: Mexico City, Mexico, 2015; pp. 223–9540.
27. Arimoto, K.; Li, G.; Arvind, A.; Wu, W.T. The Modeling of Heat Treating Process. In *Proceedings of the 18th Conference of Heat Treating, 12–15 October 1998*; Wallis, R.A., Walton, H.W., Eds.; ASM International: Materials Park, OH, USA, 1998; pp. 23–30. [CrossRef]
28. Rohde, J.; Thuvander, A.; Melander, A. Using thermodynamic information in numerical simulation of distortion due to heat treatment. In Proceedings of the 5th ASM Heat Treatment and Surface Engineering Conference in Europe Incorporating the 3rd International Conference on Heat Treatment with Atmospheres, Gothenburg, Sweden, 7–9 June 2000; Mittemeijer, E.J., Grosch, J., Eds.; ASM International: Materials Park, OH, USA, 2000; pp. 21–29. [CrossRef]
29. Esfahani, A.K.; Babaei, M.; Sarrami-Foroushani, S. A numerical model coupling phase transformation to predict microstructure evolution and residual stress during quenching of 1045 steel. *Math. Comput. Simul.* **2021**, *179*, 1–22. [CrossRef]
30. Agarwal, P.K.; Brimacombe, J.K. Mathematical Model of Heat Flow and Austenite-Pearlite Transformation in Eutectoid Carbon Steel Rods for Wire. *Metall. Mater. Trans. B* **1981**, *12*, 121–133. [CrossRef]
31. Barrena-Rodríguez, M.J.; González-Melo, M.A.; Acosta-González, F.A.; Alfaro-López, E.; García-Pastor, F.A. An Efficient Fluid-Dynamic Analysis to Improve Industrial Quenching Systems. *Metals* **2017**, *7*, 190. [CrossRef]
32. González-Melo, M.A.; Acosta-González, F.A. Efficient Prediction of Heat Transfer during Quenching based on a Modified Reynolds-Colburn Analogy. In Proceedings of the Thermal Processing in Motion including the 4th International Conference on Heat Treatment and Surface Engineering in Automotive Applications, Spartanburg, SC, USA, 5–7 June 2018; Ferguson, L., Frame, L., Goldstein, R., Guisbert, D., Scott MacKenzie, D., Eds.; ASM International: Materials Park, OH, USA, 2018; pp. 91–97.
33. Totten, G.E.; Lally, K.S. Proper Agitation Dictates Quench Success Part 1. *Heat Treat.* **1992**, *9*, 12–17.

34. Aronov, M.A.; Kobasko, N.I.; Powell, J.A.; Young, J. Practical application of intensive quenching technology for steel parts and real time quench tank mapping. In Proceedings of the 19th ASM Heat Treating Society Conference, Cincinnati, OH, USA, 1–4 November 1999; Krauss, G., Midea, S.J., Pfaffmann, G.D., Eds.; ASM International: Materials Park, OH, USA, 1999; pp. 1–9.
35. MacKenzie, S.; Li, Z.; Ferguson, B.L. Effect of Quenchant Flow on the Distortion of Carburized Automotive Pinion Gears. In Proceedings of the 5th International Conference on Quenching and Control Distortion including European Conference on Heat Treatment 2007, Berlin, Germany, 25–27 April 2007; International Federation for Heat Treatment and Surface Engineering: Winterthur, Switzerland, 2007; pp. 1–11.
36. Kumar, T.S.P. Influence of Steel Grade on Surface Cooling Rates and Heat Flux during Quenching. *J. Mater. Eng. Perform.* **2013**, *22*, 1848–1854. [CrossRef]

Article

Precipitation Criterion for Inhibiting Austenite Grain Coarsening during Carburization of Al-Containing 20Cr Gear Steels

Huasong Liu [1], Yannan Dong [1], Hongguang Zheng [2], Xiangchun Liu [2], Peng Lan [1], Haiyan Tang [1] and Jiaquan Zhang [1,*]

[1] School of Metallurgical and Ecological Engineering, University of Science and Technology Beijing, Beijing 100083, China; liuhuasong_ustb@163.com (H.L.); ustb_dyn0801@163.com (Y.D.); lanpeng@ustb.edu.cn (P.L.); tanghaiyan@metall.ustb.edu.cn (H.T.)
[2] Central Research Institute of Baoshan Iron and Steel Co., Ltd., Shanghai 201900, China; zhenghongguang@baosteel.com (H.Z.); liuxiangchun@baosteel.com (X.L.)
* Correspondence: jqzhang@metall.ustb.edu.cn

Citation: Liu, H.; Dong, Y.; Zheng, H.; Liu, X.; Lan, P.; Tang, H.; Zhang, J. Precipitation Criterion for Inhibiting Austenite Grain Coarsening during Carburization of Al-Containing 20Cr Gear Steels. *Metals* 2021, *11*, 504. https://doi.org/10.3390/met11030504

Academic Editor: Andrea Di Schino

Received: 10 February 2021
Accepted: 15 March 2021
Published: 18 March 2021

Publisher's Note: MDPI stays neutral with regard to jurisdictional claims in published maps and institutional affiliations.

Copyright: © 2021 by the authors. Licensee MDPI, Basel, Switzerland. This article is an open access article distributed under the terms and conditions of the Creative Commons Attribution (CC BY) license (https://creativecommons.org/licenses/by/4.0/).

Abstract: AlN precipitates are frequently adopted to pin the austenite grain boundaries for the high-temperature carburization of special gear steels. For these steels, the grain coarsening criterion in the carburizing process is required when encountering the composition optimization for the crack-sensitive steels. In this work, the quantitative influence of the Al and N content on the grain size after carburization is studied through pseudocarburizing experiments based on 20Cr steel. According to the grain structure feature and the kinetic theory, the abnormal grain growth is demonstrated as the mode of austenite grain coarsening in carburization. The AlN precipitate, which provides the dominant pinning force, is ripened in this process and the particle size can be estimated by the Lifshitz–Slyosov–Wagner theory. Both the mass fraction and the pinning strength of AlN precipitate show significant influence on the grain growth behavior with the critical values indicating the grain coarsening. These criteria correspond to the conditions of abnormal grain growth when bearing the Zener pinning, which has been analyzed by the multiple phase-field simulation. Accordingly, the models to predict the austenite grain coarsening in carburization were constructed. The prediction is validated by the additional experiments, resulting in accuracies of 92% and 75% for the two models, respectively. Finally, one of the models is applied to optimize the Al and N contents of commercial steel.

Keywords: gear steel; AlN precipitate; carburization; austenite grain size; Zener pinning; precipitation criterion

1. Introduction

Gear steels include carbon structural steel (such as Q235), low-alloy high-strength structural steel (such as Q345), high-quality carbon structural steel (such as 45 steel), alloy structural steel (such as 40MnB and 42CrMo), and structural steel with guaranteed hardenability (such as 20CrMnTi and 20CrMo), etc. For the heavy-duty gear steels including 20CrMnTi, Cr-Mo and Cr-Ni-Mo series, carburizing and quenching treatments are required before usage. In the carburizing process, the austenite grains are prone to grow up, which affects the tensile strength, elongation, impact toughness [1–3], heat treatment distortion [4], and fatigue crack resistance [5,6] of the gears. To inhibit austenite grain coarsening, the presence of certain nanoscale particles is needed to achieve the pinning on the austenite grain boundary (GB) [7]. Common gear steels use AlN as the pinning particles, of which the carburizing temperature is in the range of 930–980 °C. A higher carburizing temperature requires the addition of Nb or Nb–Ti microalloying elements [8,9].

Although high-temperature carburizing has the advantages of increasing the carburized layer thickness and shortening the carburizing time, the commercial gear steels mostly use AlN to pin the austenite GB at present. The adequate acid-soluble Al and N in steel can guarantee the control of austenite grain size during carburizing. However, for the gear steel grades containing peritectic or hypo-peritectic composition, the excessive addition of Al or N may cause the surface crack problem in continuous casting production. For instance, in several Chinese plants, the production of 20Cr, 20CrMo, and 20CrMnTi has suffered frequent surface or corner cracks for a long time. Reducing the content of Al or N is conducive to improving the third brittle zone of the slab surface [10]. Hence the composition design has to balance the requirements of carburization and the suppression of continuous casting cracks. For this purpose, the quantitative relationship between the Al and N content and the austenite grain size after carburization is required. Work has been carried out to investigate this relationship. Militzer et al. studied the austenite grain growth kinetics in Al-killed plain carbon steels and found that the grain growth depended strongly on the degree of AlN precipitation [11]. Pous-Romero et al. studied the austenite grain growth in a nuclear pressure vessel steel of which the pinning particle was determined as AlN [12]. Two regimes of grain growth were reported to be produced due to the existence or dissolution of precipitates. Other experimental works also manifested that high contents of Al and N were beneficial for obtaining the fine austenite grain structure and avoiding abnormal grain coarsening [13]. However, the precise criterion of composition or AlN precipitation has not been reported to date, resulting in uncertainty in the reduction of Al or N content for crack-sensitive steels.

The present work aims to reveal the criterion of austenite grain size control in the carburization of Al-bearing gear steels. Six compositions with various Al and N contents were designed based on the commercial 20Cr steel, and the pseudocarburizing experiments were carried out in the temperature range of 977–1019 °C. Combining with the AlN precipitation behavior, the dependence of grain growth behavior on the precipitate condition was revealed. After determining the grain coarsening mode in carburization, the critical condition for abnormal grain growth was obtained by the multiple phase-field (MPH) simulation. Finally, the models for predicting the grain coarsening were proposed and validated by additional experiments.

2. Materials and Methods
2.1. Material

A typical gear steel grade 20Cr was investigated in this work, which suffers great crack sensitivity in the continuous casting procedure. To investigate the effect on the grain growth behavior of Al and N contents, six steel compositions were designed, based on the commercial 20Cr steel. The raw materials were melted in a 180 Kg vacuum induction furnace and cast into an ingot with a section of 170×170 mm^2. Then the cooled casts were hot-forged to the bars with a section of 34×34 mm^2, which started at 1200 °C and air-cooled to room temperature after the deformation. Finally, the pseudocarburizing test specimens were machined from the bars with a size of $12 \times 12 \times 15$ mm^3 of which the long axis was parallel to the deformation direction. The precise composition of each bar was measured at a quarter of the thickness, and the results are listed in Table 1. It was found that the contents of main solutes except Al and N were rather stable, and the contents of Ti and Nb were all trace except for the S3. The metallographic structure of S1 in the cross-section is shown in Figure 1, which was composed of polygonal ferrite and pearlite at room temperature.

2.2. Experiments

The pseudocarburizing experiments were carried out in a high-temperature resistance furnace. Since the practical carburizing temperature is in the range of 950–980 °C for 20Cr, the experimental carburizing temperatures were set as 977, 998, and 1019 °C, which covered the highest carburizing temperature of the current Al-containing gear steels. The carburizing time was chosen as 4 h. Once finished carburizing, the samples were taken out from the furnace and water-quenched immediately. The furnace temperature was calibrated by the platinum and rhodium 10-platinum thermocouple, and argon gas was continuously blown into the furnace to reduce oxidation during the experiment.

Table 1. Chemical compositions of the experimental steels (wt.%).

Steel	C	Si	Mn	S	Cr	Ni	Ti	Nb	Al	N
S1	0.21	0.32	0.88	0.002	1.23	0.16	<0.005	<0.0005	0.028	0.0183
S2	0.22	0.31	0.88	0.002	1.22	0.16	0.0023	<0.0005	0.019	0.015
S3	0.21	0.3	0.88	0.002	1.22	0.15	0.013	0.0007	0.029	0.0115
S4	0.21	0.3	0.85	0.002	1.2	0.16	<0.0005	<0.0005	0.014	0.015
S5	0.21	0.32	0.87	0.002	1.2	0.17	0.0005	<0.0005	0.01	0.011
S6	0.22	0.31	0.89	0.002	1.21	0.15	0.0012	0.0005	0.018	0.0065

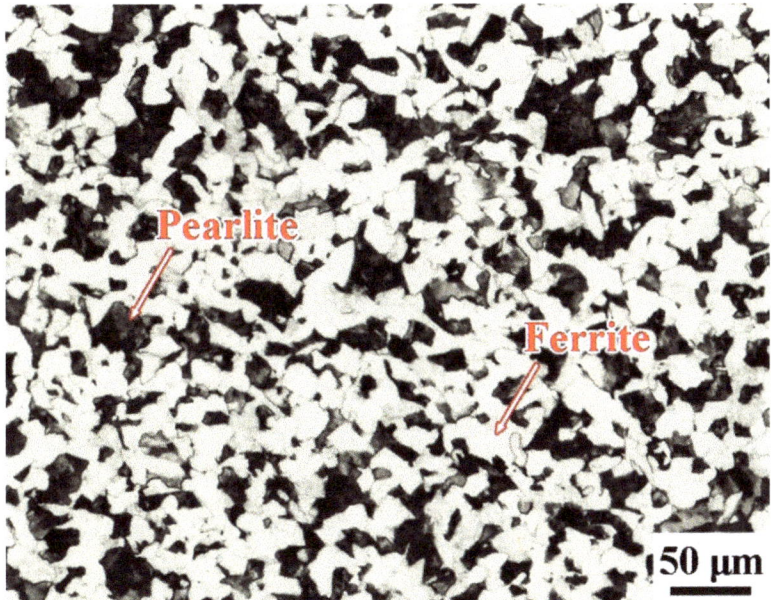

Figure 1. The microstructure of S1 at room temperature.

To further avoid the influence of surface oxidation, the middle section of each sample was used as the surface for microstructure characterization. The section was polished to 0.5 μm and etched by the saturated picric acid aqueous solution to reveal the original austenite GBs. The prior austenite grain structure (PAGS) was photographed and collected by an optical microscope (UOP, Chongqing, China), and the grain size was measured by the area equivalent diameter. Each sample was measured to at least 1500 (when the grains were fine) or 500 (when the grains were coarse) grains. Using the carbon replication method, the morphology and composition of the precipitates in some samples were detected by a transmission electron microscope (TEM, Oxford Instruments, Abingdon, UK) and an attaching energy dispersive spectrometer (EDS, Oxford Instruments, Abingdon, UK).

2.3. Precipitation Model

The mass fraction of precipitation was calculated by the commercial software Thermo-Calc using the TCFE-v8.1 database (2017, Thermo-Calc Software, Stockholm, Sweden), and the mean radius of AlN particles was estimated through the classical Lifshitz–Slyosov–Wagner (LSW) theory [14]

$$r_p = \left(r_0^3 + \frac{8\sigma_I C D V_m^2}{9RTV_B}t\right)^{1/3} \quad (1)$$

where σ_I is the surface energy of the particle–matrix interface, V_m the precipitate molar volume, C the equilibrium mole fraction of solute Al in the matrix, D the bulk diffusion coefficient of solute Al in austenite, V_B the mole volume of solute Al, R the gas constant, T and t the absolute temperature and ripening time, respectively. The initial radius of the AlN particle r_0 was set as 7.5 nm in the calculation, which had very little influence on the result. The mole fraction C was obtained by the solubility product K_γ of AlN in austenite.

2.4. MPH Method

A two-dimensional polycrystalline phase-field model was selected to analyze the occurrence condition of abnormal grain growth. This model uses the order parameter φ_i to describe the probability that a spatial point belongs to the grain orientation i, where $\varphi_i = 1$ (0) means it is located in the grain (outside the grain), and $\varphi_i = 0\sim1$ means it is located at the GB. For a system containing less than Q grains, i.e., $i = 1, 2 \ldots Q$, the order parameter of each point must satisfy $\sum_1^Q \varphi_i = 1$. That is, each spatial point can only locate either inside a grain or at the GB of multiple grains. The order parameter evolution is calculated by [15]

$$\begin{aligned}
\frac{\partial \varphi_i}{\partial t} &= -\frac{2}{n}\sum_{j \neq i}^n L_{ij}\left(\frac{\delta F}{\delta \varphi_i} - \frac{\delta F}{\delta \varphi_j}\right) \\
\frac{\delta F}{\delta \varphi_j} &= \sum_{i \neq j}^n \left(\frac{\varepsilon_{ij}^2}{2}\nabla^2 \varphi_i + \omega_{ij}\varphi_i\right) \\
\omega_{ij} &= \frac{4\sigma_{ij}}{W_{ij}}, \varepsilon_{ij}^2 = \frac{8W_{ij}\sigma_{ij}}{\pi^2}, L_{ij} = \frac{\pi^2 m_{ij}}{8W_{ij}}
\end{aligned} \quad (2)$$

where n is the total number of grain orientations existing at a certain point, σ_{ij}, m_{ij}, and W_{ij} the GB energy, GB mobility, and GB thickness between grains i and j, respectively. In this simulation, the anisotropy of these parameters is all ignored.

The Zener pinning of the precipitate particle is required in the simulation of abnormal grain growth, which is introduced by modifying the GB mobility [16]

$$m_{ij} = \begin{cases} m_{ij}^0 \exp\left(-\frac{0.12 P_z}{|\Delta G_{ij}| - P_z}\right) & \text{if } |\Delta G_{ij}| > P_z \\ 0.01 m_{ij}^0 & \text{else} \end{cases} \quad (3)$$

where m_{ij}^0 is the GB mobility without pinning, P_z the particle pinning force. The driving force of the GB between grains i and j is obtained by the grain diameters d_i and d_j [17]

$$|\Delta G_{ij}| = 2\sigma_{ij}\left|\frac{1}{d_i} - \frac{1}{d_j}\right| \quad (4)$$

The initial grain structure in MPH simulation was generated by randomly placing the circular nuclei with the size distribution obeying the log-normal function

$$f(d) = \frac{1}{ds\sqrt{2\pi}}\exp\left[-\frac{(\ln d - \mu)^2}{2s^2}\right] \quad (5)$$

where d is the grain diameter, μ and s the mean and standard deviation of $\ln d$, respectively. The MPH simulations were carried out on a personal computer with Matlab (2016b, Mathworks, Natick, MA, USA) language, and the number of grid points is 801 × 801. All the parameters used in this simulation and LSW calculation are given in Table 2.

Table 2. Parameters in the multiple phase-field (MPH) simulation and Lifshitz−Slyosov−Wagner (LSW) calculation.

Parameter	Value	Reference
Grid space Δx	$\exp(\mu)/20$	-
Time step Δt	$0.9\Delta x^2/(4L_{ij}\varepsilon_{ij}^2)$	-
W_{ij}	$6\Delta x$	-
σ_{ij}	0.5, J/m^2	-
m_{ij}^0	0.5×10^{-12}, m^4/(J·s)	-
σ_I	0.75, J/m^2	[18]
D	$0.000251\exp(-253{,}400/RT)$, m^2/s	[19]
V_m	1.33×10^{-5}, m^3/mol	[18]
V_B	1.05×10^{-5}, m^3/mol	1
K_γ	$\exp(4.5985-11{,}568/T)$	[19]

[1] The mole volume of solute Al is estimated by the density of Al at 600 °C.

3. Results

3.1. Initial Austenite Grain Structure

The PAGS after holding 300 s at a carburizing temperature was regarded as the initial grain structure of carburization. Taking S2 as an instance, the initial grain structure of 977 °C carburizing is shown in Figure 2a. The austenite grains were uniform and fine before coarsening. Figure 2b gives the frequency distribution of grain size of S1, S2, and S4 when holding at different temperatures for 300 s. Although the carburizing temperatures were different, the initial grain structures were close in these cases. The size distributions could be well fitted by the log-normal distribution function in Equation (5). It should be pointed out that, when the contents of Al and N were very low, the coarse grains were formed even after holding 300 s at high temperatures, such as S6 carburizing at 1019 °C. According to the other cases, the average grain size at the start of each carburization could be determined as about 12 μm.

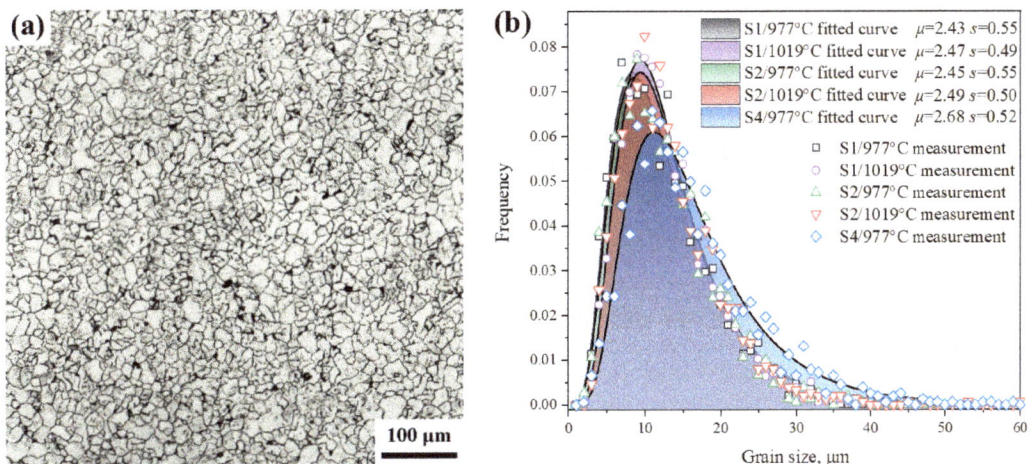

Figure 2. (**a**) Prior austenite grain structure (PAGS) after holding 300 s at 977 °C of S2; (**b**) Grain size-frequency distributions after holding 300 s at various temperatures of S1, S2, and S4.

3.2. PAGS after Pseudocarburization

The PAGSs after 977 °C pseudocarburization are shown in Figure 3. With the same carburizing scheme, quite different PAGSs were developed. Due to the high contents of

Al and N in S1, S2, and S3, the austenite grains remained fine after carburizing, which nearly maintained the initial state shown in Figure 2a. For the other steels, of which the Al or N content was low, very coarse grains were formed. In these cases, as shown in Figure 3b,c, the grain coarsening was quite inhomogeneous. As a result, the coarse- and fine-grain regions were divided apparently, which was an indication of the abnormal grain growth [20]. With the decrease of Al or N content, the coarsening of PAGS became so severe that the fraction of the coarse-grain region increased. A similar phenomenon existed with the increase of carburizing temperature.

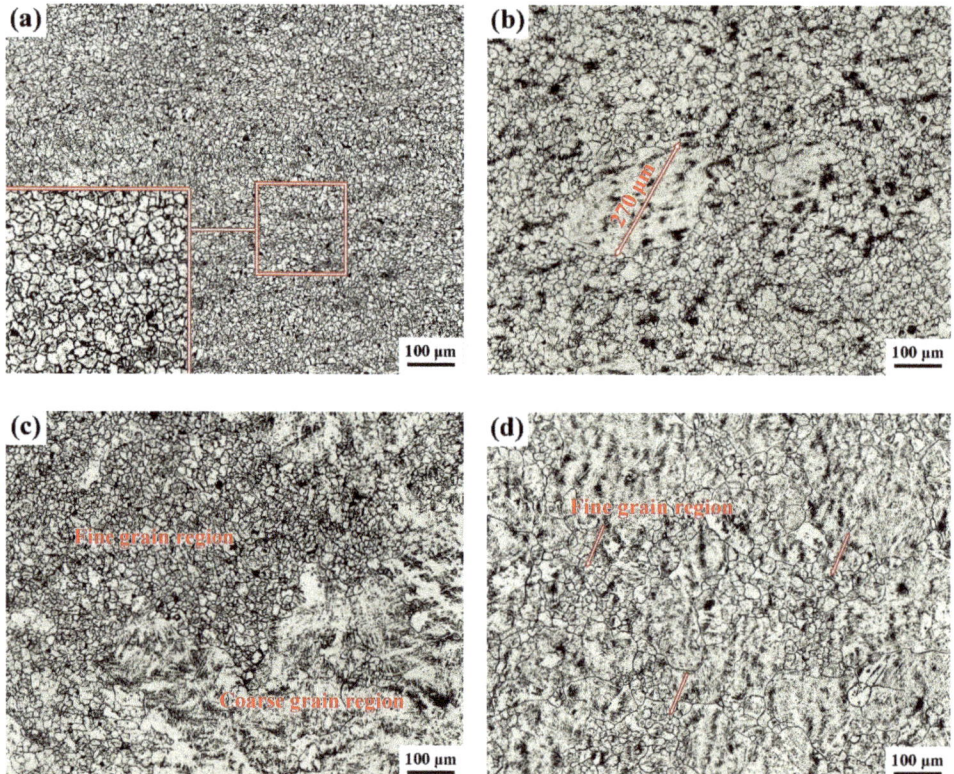

Figure 3. PAGSs after holding for 4 h at 977 °C of (**a**) S2, (**b**) S4, (**c**) S5, and (**d**) S6.

3.3. AlN Precipitation

Figure 4 shows the TEM micrograph and the EDS spectrum of the precipitate particles in S3 and S4 after the pseudocarburization at 977 °C. Due to the presence of Ti, the precipitates in S3 always contained Ti, of which the larger particles should be Al-rich (Ti,Al)(C,N) composite precipitates, while the fine particles were TiN or Ti-rich composite precipitates. However, bearing the trace content of Ti or Nb in other steels, the precipitates in S4 were mainly AlN particles and a few large MnS particles. The detection demonstrated that, during the high-temperature carburizing, the precipitates had undergone significant Ostwald ripening, resulting in the coarse particles more than 100 nm in size and the fine particles with sizes of only tens of nm. In this process, due to the higher N content and a certain amount of Ti in S3, the number of fine particles was higher than that of S4.

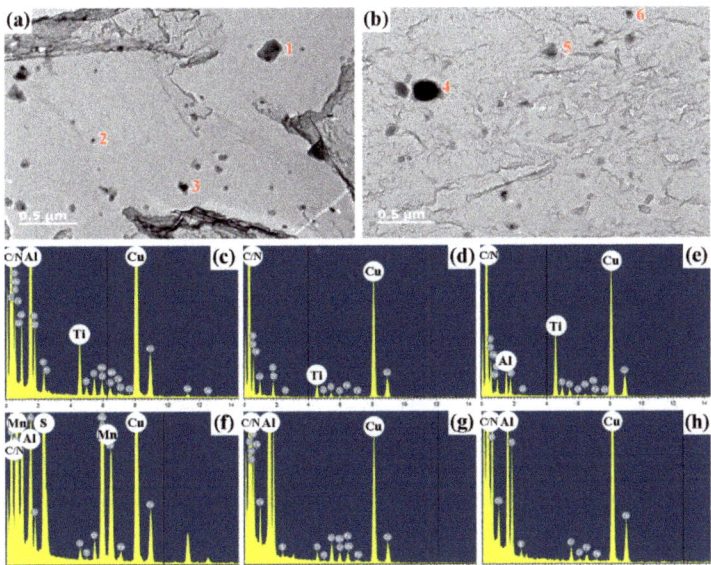

Figure 4. TEM micrograph of the precipitates after holding for 4 h at 977 °C of (**a**) S3 and (**b**) S4 and the corresponding EDS spectra of (**c**–**e**) particles 1–3 and (**f**–**h**) 4–6.

The calculated mass fraction of S1 equilibrium precipitation is shown in Figure 5a. In the range of 950–1050 °C, AlN was the main precipitate and only a small amount of TiN and MnS existed. So, the dominant pinning particle at the pseudocarburizing temperatures should be AlN as expected. Figure 5b plots the equilibrium mass fraction of this precipitate in various steels, and the sum fraction of AlN and TiN was added for S3 owing to the existence of Ti. As the contents of Al and N reduced, the initial precipitation temperature and the precipitation amount of AlN both decreased. The evolution of AlN particle radius during the carburization was then calculated by the LSW theory, and the results of carburizing at 977 and 1019 °C are shown in Figure 6. Interestingly, it was revealed that the influence of Al and N contents on the particle size was complicated and no monotonous relationship could be found. According to the TEM detections, the average AlN particle radius in S4 after the carburization at 977 °C was about 27.5 nm, which was very close to the calculation result of the kinetic model.

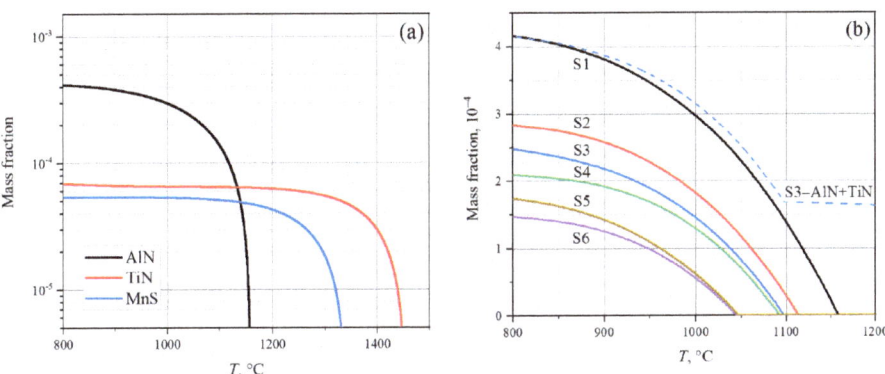

Figure 5. (**a**) Calculated equilibrium precipitation in S1; (**b**) equilibrium mass fractions of AlN in different steels.

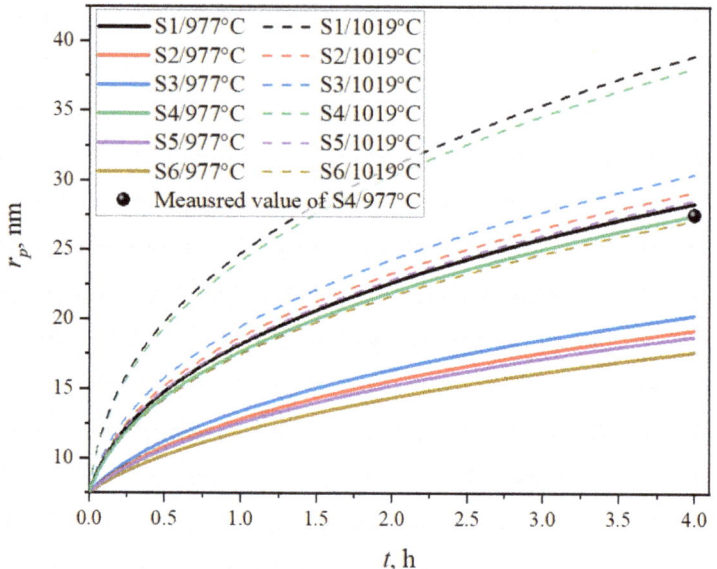

Figure 6. Calculated average radiuses of AlN particles as functions of time during carburizing at 977 and 1019 °C in various steels.

3.4. Relationship between the Precipitation and PAGS

The average grain size after carburization of each case is listed in Table 3. In these cases, the PAGSs of S1, S2, and S3 maintained the fine and uniform structure as shown in Figure 3a, while in those of S4, S5, and S6 appeared the huge grains as Figure 3b–d revealed. It was noted that the average grain size could only partly reflect the coarsening of PAGS. Taking S4 as an instance, although the huge grains were formed after carburizing at 977 and 998 °C, the average grain sizes remained close to the initial one because the number of coarse grains was far less compared to the number of fine grains.

Table 3. The average grain size of each case after pseudocarburization (μm).

Carburizing T (°C)	S1	S2	S3	S4	S5	S6
977	11.72	13	12.46	16	33.91	22.24
998	13.36	12.7	13.16	12.43	42.47	27.36
1019	11.46	13.69	13.23	33.03	57.5	41.78

The influence of precipitate amount on the PAGS after carburization was examined. The corresponding relationship is shown in Figure 7a. Considering the average grain size could not reflect the grain coarsening, the black square and red circular dots marked the fine PAGS and the PAGS containing coarse grains, respectively. A critical mass fraction between these two structures was revealed. Once the mass fraction was below the critical value, the huge grains appeared, and the average grain size increased as the precipitate amount decreased. Given the content of Ti in steels, especially in S3, the influence of TiN precipitation was also considered in Figure 7a through the sum fraction of TiN and AlN. However, when considering the TiN precipitation, the critical mass fraction for grain coarsening could be determined as about 1.56×10^{-4}. If using f_p/r_p (f_p is the volume fraction of precipitate) as the pinning strength of AlN particles, Figure 7b shows the relationship between pinning strength and PAGS feature. A similar result with Figure 7a is obtained, where the critical f_p/r_p is about 14,574 m^{-1}. Below this critical value, there exists an abnormal case with fine PAGS as circled in Figure 7b. For this case, the actual pinning strength should be higher than the present value, because it belongs to S3 which should consider the pinning of TiN. The critical pinning strength corresponds to the carburizing case at 977 °C of S4. The indication can also be found in the associated PAGS in Figure 3b, where the matrix is still fine grains and the coarse grains are few and isolated.

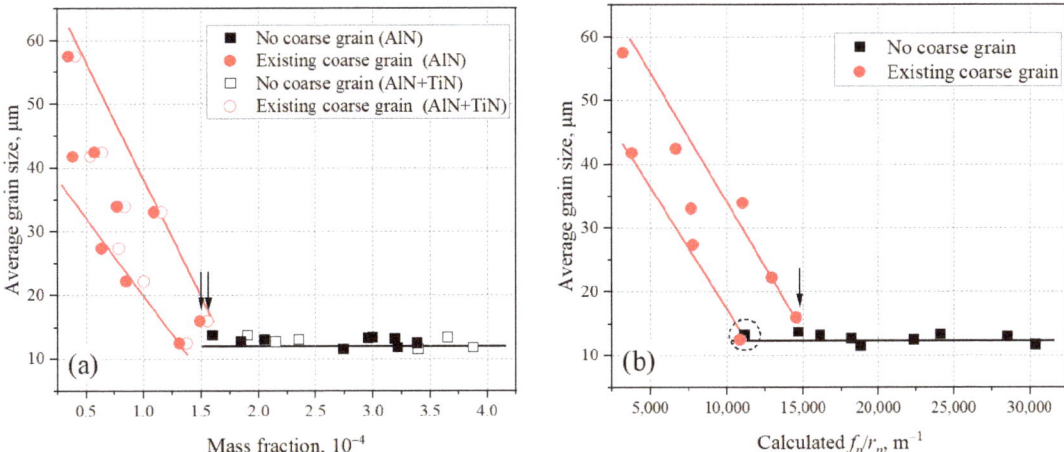

Figure 7. Relationships between the PAGS feature after carburization and the equilibrium mass fraction of precipitation (**a**) and the pinning strength f_p/r_p (**b**). In image (**b**), the circled dot corresponds to the case of S3 carburizing at 1019 °C.

4. Discussion

4.1. Grain Coarsening Mode in Carburization

The grain size distributions of S2 and S4 before and after the carburizing treatments are shown in Figure 8. It is seen that regardless of the occurrence of grain coarsening, the portion of fine grains remained almost unchanged during the carburization. The failure of the grain size control was the appearance of large grains, of which the size could reach dozens of times that of fine grains. This is an indication that the coarsening of PAGS should be carried out in an abnormal grain growth regime in carburization. The moving rate of austenite GB can be described as

$$v = m_0(P_0 - P_z - P_d) \tag{6}$$

where m_0 is the intrinsic GB mobility, P_0 the intrinsic driving force of GB migration. The solute dragging force P_d is related to the GB velocity v as [21]

$$P_d = \frac{avc}{1 + b^2 v^2} \tag{7}$$

where a and b are the constants related to the binding energy and diffusion coefficient of the solute element at the GB, and c the bulk concentration of solute. When v is very low, P_d is almost a linear function of v. Therefore, Equation (6) is rewritten as

$$v = \frac{m_0}{1 + acm_0}(P_0 - P_z) = m(P_0 - P_z) \tag{8}$$

where m is the equivalent GB mobility considering the solute dragging which corresponds to the m_{ij}^0 in Equation (3).

When grain growth occurs, two distinct regimes exist, i.e., normal grain growth and abnormal grain growth. Almost all grains can have obvious GB motion in the former regime, while only part of the grains can grow up in the latter one. Many works have confirmed that abnormal grain growth can be induced by the effect of particle pinning [16,17,20,22]. Besides, under the dragging force shown in Equation (7), the transition from the low-speed motion to the high-speed motion of GB was also expected to lead to abnormal grain growth [23]. However, the required driving force for this transition is extremely large and no experimental report has been found for the austenite grain in steel. So this paper was concerned only about the low-speed region of GB motion to obtain the equivalent mobility as shown in Equation (8). That means, only the particle pinning work is expected to cause abnormal grain growth. The driving force of GB migration is calculated by

$$P_0 = 2\sigma\kappa \qquad (9)$$

where σ is the isotropous expression of σ_{ij}, κ the GB curvature. In a three-dimensional polycrystalline system, κ can be expressed as a function of the adjacent grain sizes $2|1/d_1 - 1/d_2|$ [24]. For a system with an average grain size of d_m, assuming the maximum grain size is $3d_m$, the driving force range of GB migration is about $(0, 8\sigma/3d_m)$ and the average driving force is about $2\sigma/3d_m$ [25]. Considering features of the grain growth regimes and according to Equation (8), when normal growth occurs, the Zener pinning force should be within $(0, 2\sigma/3d_m)$ so that most of the GBs can move. When abnormal growth occurs, the pinning force should be within $(2\sigma/3d_m, 8\sigma/3d_m)$, then only the GBs with a large size difference in neighboring grains can move. Nearly no grain growth occurs when the pinning force exceeds $8\sigma/3d_m$.

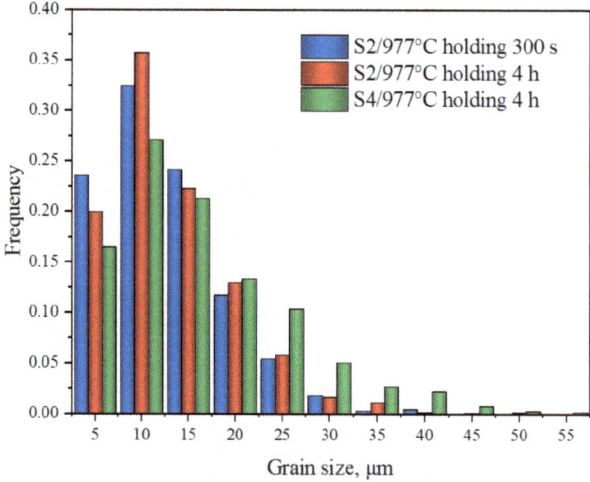

Figure 8. Comparison of austenite grain size-frequency distributions of S2 at the start of carburization and S2 and S4 after carburization at 977 °C.

During carburizing, the pinning work of precipitates has four possible situations: (1) the initial pinning force falls into $(0, 2\sigma/3d_m)$; (2) the initial pinning force falls into $(2\sigma/3d_m, 8\sigma/3d_m)$; (3) the pinning force is greater than $8\sigma/3d_m$ initially but falls into $(2\sigma/3d_m, 8\sigma/3d_m)$ after the dissolution and ripening; (4) the pinning force is always greater than $8\sigma/3d_m$. Among these situations, the austenite grains will grow up normally in case (1) or have an abnormal growth in cases (2) and (3). We should note that the occurrence of grain growth does not mean the failure of grain size control. After the normal grain growth, in case the mean grain size is less than the permissible value, the control of grain size is successful. However, for the gear steel that needs to be carburized, the precipitation in the steel aims to prevent the austenite grain growth which means the excessive precipitates are always designed. Therefore, the failure of grain size control should arise from the insufficient precipitates or the dissolution and Ostwald ripening of precipitates which reduces the pinning force during carburizing. Then the pinning force is less than $8\sigma/3d_m$ and drops to the adjacent region of abnormal grain growth. Once the abnormal growth occurs, the coarsened grains will continue to grow up and develop huge sizes far exceeding the values permissible by the users. Therefore, the austenite grains can have two states in the carburizing process, i.e., no growth and abnormal growth, and the latter is the manifestation of grain coarsening. In our experiments, with the decreasing contents of Al and N, the pinning force would also decrease from the region associated with no grain growth. This is why the grain size variation had a discontinuity in Figure 7. The criterion between the no grain growth and the grain coarsening behaviors corresponds to the critical condition of abnormal grain growth.

4.2. Condition for Abnormal Grain Growth

To achieve the precise condition of abnormal grain growth, the MPF method was employed to simulate the various grain growth behaviors. According to Equation (8), the occurrence of abnormal growth is determined by the competition between P_z and P_d, which is independent of mobility. The particle pinning force can be calculated by

$$P_z = \xi \frac{\sigma \cdot f_p}{r_p} \tag{10}$$

where ξ is a dimensionless constant, determined by the particle shape, the geometric characteristics of the interface between the particle and the matrix, the coherence of the interface, and the connection properties [26]. Equations (9) and (10) show that both P_0 and P_z are linearly related to σ, so the occurrence of abnormal grain growth is also independent of the GB energy. In the MPH simulation, the mobility without pinning work m_{ij}^0 was assumed as 0.5×10^{-12} m^4/(J·s), and the GB energy σ was set as 0.5 J/m^2. When constructing the initial grain structure, the standard deviation was set as 0.33, and the average size was within 5–25 µm. It should be pointed out that, in the present model, if using the measured standard deviation of 0.52, the abnormal growth cannot occur. This may be because, with the wide grain size distribution, the coarsening grains are so many that the growth cannot continue due to the encounters, which indicates that the present particle pinning model of Equations (3) and (4) can be improved in the future.

The generated initial grain structure with an average size d_0 of 10 µm is shown in Figure 9a, which is similar to the actual initial grain structure in Figure 2a. After the simulations for 4 h carburizing, the relationship between the pinning force and the grain growth regime is achieved in Figure 10. It can be seen that under different initial sizes, three regimes of grain growth, i.e., normal growth, abnormal growth, and no growth can occur according to the pinning force. The upper bound condition of the abnormal grain growth was well fitted by the power function shown in Figure 10. Three simulated grain structures are given in Figure 9b,c for instance. The fine and uniform grain structure was obtained in the no grain growth regime, while the mixed one after abnormal grain growth was similar to the PAGS after carburization shown in Figure 3c,d.

4.3. Prediction of Grain Coarsening in Carburization

Figure 7a reveals that under experimental conditions, the precipitate mass fraction could be used as the critical condition of abnormal grain growth. The reason is, when the carburizing temperatures were close, the difference between the average sizes of AlN particles was not too much. Figure 11 shows the relationship between the mass fraction of AlN precipitate and the calculated AlN pinning strength after carburizing for 2 and 4 h. Although the particle sizes were different, the function between mass fraction and pinning strength was close to a linear one. Then the critical condition for abnormal grain growth could be presented by the former. Therefore, with the close carburizing temperature and time, the critical mass fraction of AlN precipitate can be used to avoid the occurrence of abnormal grain growth as

$$f_{p,m} > 1.56 \times 10^{-4} \tag{11}$$

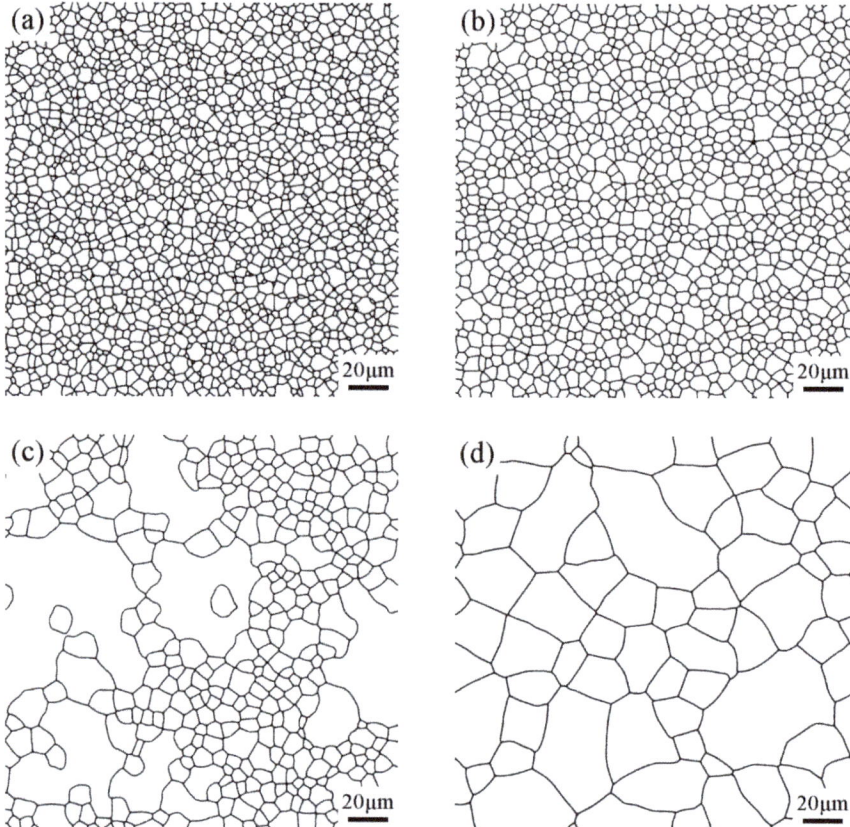

Figure 9. The initial grain structure in MPH simulation (**a**) and the final grain structures after simulated carburization with d_0 of 10 µm. Images (**b**–**d**) correspond to the cases of (**b**–**d**) in Figure 10.

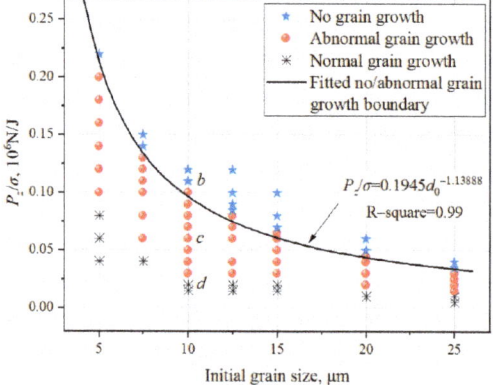

Figure 10. Various grain growth regimes as functions of initial grain size and pinning force.

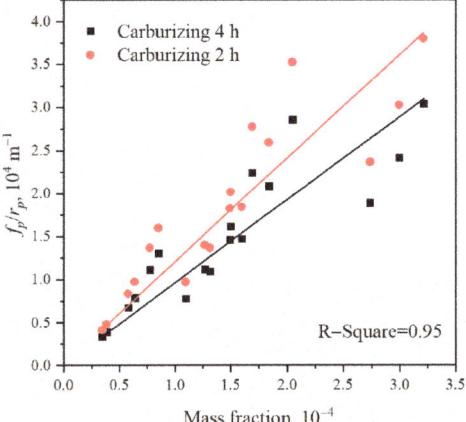

Figure 11. Relationship between the calculated pinning strength and the mass fraction of AlN precipitate.

However, in case the carburizing temperature or time changes much, the variation of AlN particle size needs to be considered. After the MPH simulation results as shown in Figure 10, the boundary of pinning force for abnormal grain growth can be predicted by

$$P_z = 0.1945 \cdot d_0^{-1.13888} \cdot \sigma \tag{12}$$

where d_0 is in the unit of m. Note that, the driving force of GB in a two-dimensional system is half of that in the actual three-dimensional system, as shown in Equations (4) and (9). Hence the critical pinning force in this equation needs to be doubled when applied in a three-dimensional system. With the initial grain size of 12 μm, the calculated critical P_z/σ is 156,377 m^{-1}. Combined with the critical pinning strength shown in Figure 7b, the dimensionless constant ξ can be determined from Equation (10) for the pinning of AlN particles on austenite GB. The obtained value of 10.73 is between 3/2 and 12 as recommended in the literature [27,28]. Then the critical condition for inhibiting the grain coarsening in carburization can be obtained as

$$\frac{f_p}{r_p} > 0.03625 \cdot d_0^{-1.13888} \tag{13}$$

The occurrence of abnormal grain growth depends on the competition between P_z and P_d of grain structure. If ignoring the variation of standard deviation in grain size distribution and assuming the dimensionless constant is independent of the composition, Equation (13) should be applicable for all the steels with AlN precipitate pinning the austenite GB. Unlike Equation (11), this model involves the initial grain size and the prediction of precipitate size. The former is altered by the chemical composition and the microstructure before the reverse transformation, and the latter may have an error in the estimation. So only satisfying Equation (13) may be dangerous for the abnormal grain growth. That means the pinning strength f_p/r_p should be significantly greater than the right-hand value of Equation (13) to ensure the avoidance of abnormal grain growth. After Equation (13), we also noticed that increasing the initial grain size was beneficial for controlling the austenite grain size during carburization. This may be a potential approach to raise the carburizing temperature or reduce the Al and N contents in gear steels, even though the initial grain size after austenitization is difficult to alter at present.

4.4. Model Validation and Application

To verify the accuracy of Equations (11) and (13), a series of additional experiments have been conducted. The carburizing temperature, time, and features of the final PAGS are listed in Table 4. Based on the two models, the predictions were compared with the measurements in Figure 12, and several PAGSs of the verified cases are shown in Figure 13. It was seen that the accuracy rates of Equations (11) and (13) were 92% and 75%, respectively. Although the former was higher, we should note that the failure cases of Equation (13) all appeared around the critical value of f_p/r_p, which may be within the sensitive range as mentioned before. However, Equation (11) failed to predict the grain growth behavior of S1 when carburizing at 1035 °C, of which the AlN mass fraction was much larger than the threshold. In Equation (11), the ability of grain growth inhibition of S1 was the highest, while in Equation (13), it was greatly weakened due to the high carburizing temperatures. The result indicates that, if the carburizing temperature and time are close to the investigated ones in this work, Equation (11) should be preferred, because in this model, the avoidance of the estimation of precipitate ripening and the initial grain size can reduce the potential error of prediction as Figure 12 has shown. Otherwise, the variation of precipitate particle size must be considered because the mean size can be altered significantly. Then, Equation (13) is suggested. Moreover, according to the validations, the sensitive range of pinning strength can be determined to consider the change of critical pinning strength owing to the variation of initial grain size or the calculation of precipitate ripening. The upper limit of this range is suggested to be $0.03625 \cdot d_0^{-1.13888} + 4000$ m^{-1}.

Table 4. The pseudocarburizing conditions, the calculated states of precipitates, and features of the final PAGSs of the verified experiments.

Steel	T (°C)	t (hour)	$f_{p,m}$	r_p (nm)	f_p/r_p (m^{-1})	Existing Coarse Grains
S4	950	4	0.00017	22	20,300	No
S4	950	8	0.00017	28	16,216	No
S4	977	2	0.00015	22	18,239	Yes
S4	998	2	0.00013	26	13,666	Yes
S2	998	6	0.00019	27	18,209	No
S2	1019	6	0.00016	33	12,851	No
S2	1035	4	0.00014	34	10,964	Yes
S2	1050	2	0.00012	31	10,130	Yes
S1	998	8	0.0003	42	19,181	No
S1	1019	8	0.00027	49	14,956	No
S1	1035	4	0.00025	44	15,364	Yes
S1	1050	2	0.00023	39	15,759	No

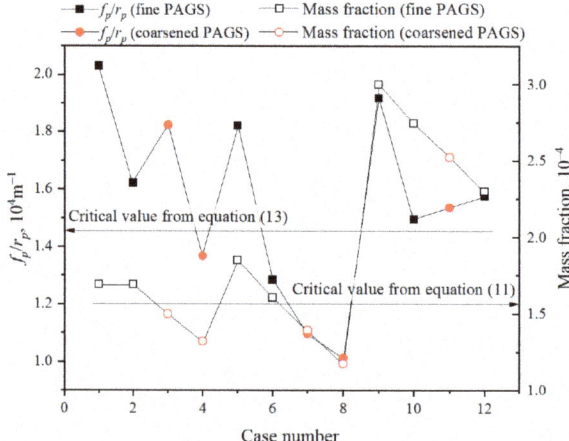

Figure 12. Calculated pinning strengths f_p/r_p (left, solid) and mass fractions (right, hollow) of the verified experiments. The cases are in the same sequence as in Table 4, and the black squares and red circles represent the fine and coarsened PAGSs, respectively.

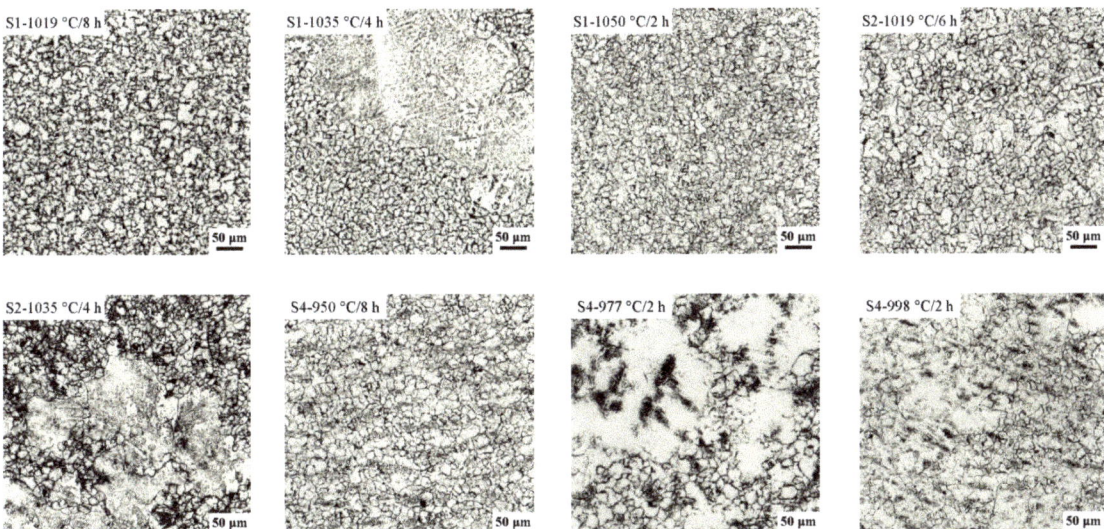

Figure 13. Several PAGSs of the verified experiments.

Using the models, we can propound the suggestion of composition modification for similar steel grades. For instance, a steel plant in China has suffered surface cracking in the continuous casting of 20Cr for a long time. The main composition of this steel is similar to the one in Table 1 and the contents of Al and N are listed in Table 5. The variation ranges of these elements allowed by the user are provided as well. It is noticed that the N content in this steel is rather high. Since the practical carburizing schedule of this steel is holding for 4 h at 960 °C, we choose Equation (11) to optimize the Al and N contents. According to the calculation of Thermo-Calc, the mass fraction of AlN at 960 °C associated with different Al and N contents is shown in Figure 14. As indicated, the AlN mass fraction of the original composition is much higher than the criterion value. We also note, however, that in the production, the steel composition and the process parameters may have slight fluctuations. So the design of Al and N contents should consider the safety distance from the critical AlN mass fraction. Using the safety distance of AlN mass fraction as 5×10^{-5}, the suggested composition is given in Table 5, which is expected to improve the hot ductility of the slab surface in continuous casting.

Table 5. The user-allowed composition range (wt.%), the original and modified composition, and the estimated third brittle zone of the commercial 20Cr steel.

Element	Allowed Range	Original	Modified
Al	0.023–0.042	0.035	0.025
N	0.011–0.021	0.018	0.012
Third brittle zone [1]		709–848 °C	718–812 °C

[1] The third brittle zone was estimated by the model from Schwerdtfeger et al. [10] with 40% as the threshold of reduction of area. Cooling rate and strain rate using 0.5 °C/s and 0.001 s^{-1}, respectively.

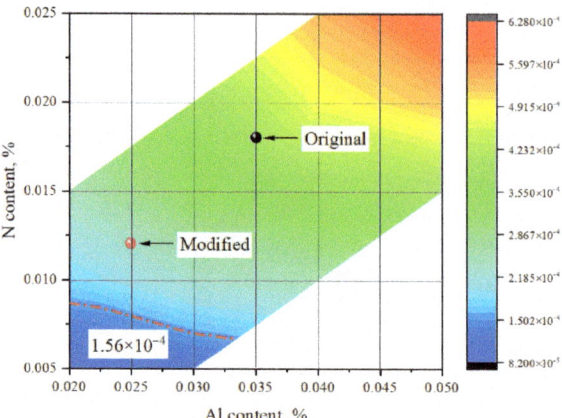

Figure 14. Calculated mass fraction of AlN precipitation at 960 °C with the variations of Al and N contents.

5. Conclusions

Aiming at the criterion for inhibiting grain coarsening in carburization, the influence of Al and N content on the austenite grain growth during carburization was investigated in this work through a series of pseudocarburizing experiments. Several conclusions can be summarized as follows.

1. The initial grain structures of carburization are close in various compositions and carburizing temperatures, of which the average size is about 12 μm. After the carburization, nearly no growth occurs if the austenite grains remain fine and uniform. However, for the cases of existing coarse grains, the fine- and coarse-grain regions are divided clearly and the coarse-grain size can be dozens of times the fine-grain size.
2. AlN precipitates provide the dominant pinning to austenite GB in the experimental steels. During the carburization, the precipitates occur the significant ripening, developing the coarse particles more than 100 nm in size and the fine particles with sizes of only tens of nm. In this process, the average particle size variation can be well simulated by the LSW theory.
3. The mass fraction and the pinning strength (defined as f_p/r_p) can determine the coarsening of austenite grain structure during carburization. In both the relationships, a critical mass fraction or a critical pinning strength was revealed, below which the PAGS will be coarsened.
4. Abnormal grain growth is the failure mode of grain size control in the carburization of gear steels. Based on the MPH simulation, the relationship between Zener pinning force and grain growth regime was constructed, and the threshold of pinning force for the abnormal grain growth was described by a power function of initial grain size. Combined with the experiment result, the dimensionless constant for the pinning of AlN on austenite GB was determined as 10.73.
5. Two models for predicting the austenite grain coarsening in carburization were constructed through the condition of abnormal grain growth. One of them concerns the mass fraction of AlN precipitates, and the other involves the variation of particle size. According to the verified experiments, the accuracies of the two models are 92% and 75%, respectively. The model which indicates the critical AlN mass fraction was used to optimize the composition of a commercial 20Cr steel associated with frequent surface cracks in production. Considering the safety distance beyond the criterion, the suggestion for reducing Al and N contents was propounded.

6. Prospects

When seeking the criterion of austenite grain coarsening, it is obvious that the determination of precipitation status, the calculation of pinning force, and the determination of pinning force conditions for abnormal grain growth act as the critical factors concerning the model accuracy. Hence, we trust that the three directions can facilitate the accuracy of the criterion in future research. First, the precise calculation about the precipitate dissolution and ripening process to reflect the distribution of particles. Second, the modified estimation of pinning force to consider the significant size difference between particles. Third, the advanced approach of introducing the pinning force in MPH simulation to produce abnormal grain growth more realistically than the present way.

Author Contributions: Conceptualization, H.L. and J.Z.; methodology, H.L. and P.L.; software, Y.D.; validation, H.L. and Y.D.; formal analysis, H.Z.; investigation, H.L. and Y.D.; resources, H.Z. and X.L.; writing—original draft preparation, H.L.; writing—review and editing, H.L. and J.Z.; visualization, P.L.; supervision, P.L. and J.Z.; funding acquisition, H.Z., H.T and P.L. All authors have read and agreed to the published version of the manuscript.

Funding: This research was funded by National Natural Science Foundation of China, grant number U1860111 and 51874033; and Fundamental Research Funds for the Central University, grant number FRF-TP-19-017A3.

Institutional Review Board Statement: Not applicable.

Informed Consent Statement: Not applicable.

Data Availability Statement: The main data had been provided in the paper already. Any other raw/processed data required to reproduce the findings of this study are available from the corresponding author upon request.

Conflicts of Interest: The authors declare no conflict of interest.

References

1. Wang, M.Q.; Shi, J.; Hui, W.J.; Dong, H. Microstructure and mechanical properties of V-Nb microalloyed steel for heavy-duty gear. *Trans. Mater. Heat Treat.* **2007**, *28*, 18–21.
2. Białobrzeska, B.; Konat, Ł.; Jasiński, R. The influence of austenite grain size on the mechanical properties of low-alloy steel with boron. *Metals* **2017**, *7*, 26. [CrossRef]
3. Moravec, J.; Novakova, I.; Sobotka, J.; Neumann, H. Determination of grain growth kinetics and assessment of welding effect on properties of S700MC steel in the HAZ of welded joints. *Metals* **2019**, *9*, 707. [CrossRef]
4. An, J.M.; Qin, M.; Ding, Y. Effect of austenite grain size on heat treatment distortion automotive gear steels. *Heat Treat.* **2013**, *28*, 48–51.
5. Ma, L.; Wang, M.Q.; Shi, J.; Hui, W.J.; Dong, H. Rolling contact fatigue of microalloying case carburized gear steels. *Chin. J. Mater. Res.* **2009**, *23*, 251–256.
6. Ma, L.; Wang, M.Q.; Shi, J.; Hui, W.J.; Dong, H. Influence of niobium microalloying on rotating bending fatigue properties of case carburized steels. *Mater. Sci. Eng. A* **2008**, *498*, 258–265. [CrossRef]
7. Yan, B.; Liu, Y.; Wang, Z.; Liu, C.; Si, Y.; Li, H.; Yu, J. The effect of precipitate evolution on austenite grain growth in RAFM steel. *Materials* **2017**, *10*, 1017. [CrossRef] [PubMed]
8. Enloe, C.M.; Findley, K.O.; Speer, J.G. Austenite grain growth and precipitate evolution in a carburizing steel with combined niobium and molybdenum additions. *Metall. Mater. Trans. A* **2015**, *46*, 5308–5328. [CrossRef]
9. Alogab, K.A.; Matlock, D.K.; Speer, J.G.; Kleebe, H.J. The influence of niobium microalloying on austenite grain coarsening behavior of Ti-modified SAE 8620 steel. *ISIJ Int.* **2007**, *47*, 307–316. [CrossRef]
10. Schwerdtfeger, K.; Spitzer, K.H. Application of reduction of area–temperature diagrams to the prediction of surface crack formation in continuous casting of steel. *ISIJ Int.* **2009**, *49*, 512–520. [CrossRef]
11. Militzer, M.; Hawbolt, E.B.; Meadowcroft, T.R.; Giumelli, A. Austenite grain growth kinetics in Al-killed plain carbon steels. *Metall. Mater. Trans. A* **1996**, *27*, 3399–3409. [CrossRef]
12. Pous-Romero, H.; Lonardelli, I.; Cogswell, D.; Bhadeshia, H.K.D.H. Austenite grain growth in a nuclear pressure vessel steel. *Mater. Sci. Eng. A* **2013**, *567*, 72–79. [CrossRef]
13. Li, S.J.; Fan, Y.D.; Huang, S.Y. Effect of Al, Ti, Nb microalloying on mixed grain size of case-hardened steel ZF7. *Special Steel* **2013**, *34*, 52–54.
14. Ardell, A.J. The effect of volume fraction on particle coarsening: Theoretical considerations. *Acta Metall.* **1972**, *20*, 61–71. [CrossRef]

15. Kim, S.G.; Kim, D.I.; Kim, W.T.; Park, Y.B. Computer simulations of two-dimensional and three-dimensional ideal grain growth. *Phys. Rev. E* **2006**, *74*, 061605. [CrossRef] [PubMed]
16. Apel, M.; Böttger, B.; Rudnizki, J.; Schaffnit, P.; Steinbach, I. Grain growth simulations including particle pinning using the multiphase-field concept. *ISIJ Int.* **2009**, *49*, 1024–1029. [CrossRef]
17. Kim, J.M.; Min, G.; Shim, J.H.; Lee, K.J. Effect of time-dependent pinning pressure on abnormal grain growth: Phase field simulation. *Met. Mater. Int.* **2018**, *24*, 549–559. [CrossRef]
18. Kang, Y.; Yu, H.; Fu, J.; Wang, K.; Wang, Z. Morphology and precipitation kinetics of AlN in hot strip of low carbon steel produced by compact strip production. *Mater. Sci. Eng. A* **2003**, *351*, 265–271. [CrossRef]
19. Cheng, L.M.; Hawbolt, E.B.; Meadowcroft, T.R. Modeling of dissolution, growth, and coarsening of aluminum nitride in low-carbon steels. *Metall. Mater. Trans. A* **2000**, *31*, 1907–1916. [CrossRef]
20. Wang, F.; Davis, C.; Strangwood, M. Grain growth behaviour on reheating Al–Nb-containing steel in the homogenised condition. *Mater. Sci. Technol.* **2018**, *34*, 587–595. [CrossRef]
21. Cahn, J.W. The impurity-drag effect in grain boundary motion. *Acta Metall.* **1962**, *10*, 789–798. [CrossRef]
22. Rudnizki, J.; Zeislmair, B.; Prahl, U.; Bleck, W. Prediction of abnormal grain growth during high temperature treatment. *Comput. Mater. Sci.* **2010**, *49*, 209–216. [CrossRef]
23. Kim, S.G.; Park, Y.B. Grain boundary segregation, solute drag and abnormal grain growth. *Acta Mater.* **2008**, *56*, 3739–3753. [CrossRef]
24. Di Nunzio, P.E. A discrete approach to grain growth based on pair interactions. *Acta Mater.* **2001**, *49*, 3635–3643. [CrossRef]
25. Rios, P.R.; Fonseca, G.S. Geometrical models for grain, grain boundary and grain edge average curvature in an Al-1mass%Mn alloy. *Scripta Mater.* **2005**, *52*, 893–897. [CrossRef]
26. Dépinoy, S.; Marini, B.; Toffolon-Masclet, C.; Roch, F.; Gourgues-Lorenzon, A.F. Austenite grain growth in a 2.25Cr-1Mo vanadium-free steel accounting for Zener pinning and solute drag: Experimental study and modeling. *Metall. Mater. Trans. A* **2017**, *48*, 2289–2300. [CrossRef]
27. Nes, E.; Ryum, N.; Hunderi, O. On the Zener drag. *Acta Metall.* **1985**, *33*, 11–22. [CrossRef]
28. Rios, P.R. Overview no. 62: A theory for grain boundary pinning by particles. *Acta Metall.* **1987**, *35*, 2805–2814. [CrossRef]

Review

Heuristic Design of Advanced Martensitic Steels That Are Highly Resistant to Hydrogen Embrittlement by ε-Carbide

Michio Shimotomai

MOTP LLP, Masago 3-17-4-1303, Mihama Ward, Chiba 261-0011, Japan; shimotomai317@gmail.com

Abstract: Many advanced steels are based on tempered martensitic microstructures. Their mechanical strength is characterized by fine sub-grain structures with a high density of free dislocations and metallic carbides and/or nitrides. However, the strength for practical use has been limited mostly to below 1400 MPa, owing to delayed fractures that are caused by hydrogen. A literature survey suggests that ε-carbide in the tempered martensite is effective for strengthening. A preliminary experimental survey of the hydrogen absorption and hydrogen embrittlement of a tempered martensitic steel with ε-carbide precipitates suggested that the proper use of carbides in steels can promote a high resistance to hydrogen embrittlement. Based on the surveys, martensitic steels that are highly resistant to hydrogen embrittlement and that have high strength and toughness are proposed. The heuristic design of the steels includes alloying elements necessary to stabilize the ε-carbide and procedures to introduce inoculants for the controlled nucleation of ε-carbide.

Keywords: steel; martensitic steel; ε-carbide; tempering; hydrogen embrittlement; mechanical strength; inoculant; materials design

Citation: Shimotomai, M. Heuristic Design of Advanced Martensitic Steels That Are Highly Resistant to Hydrogen Embrittlement by ε-Carbide. *Metals* **2021**, *11*, 370. https://doi.org/10.3390/met11020270

Academic Editor: Andrea Di Schino

Received: 14 December 2020
Accepted: 19 February 2021
Published: 23 February 2021

Publisher's Note: MDPI stays neutral with regard to jurisdictional claims in published maps and institutional affiliations.

Copyright: © 2021 by the author. Licensee MDPI, Basel, Switzerland. This article is an open access article distributed under the terms and conditions of the Creative Commons Attribution (CC BY) license (https://creativecommons.org/licenses/by/4.0/).

1. Introduction

High-strength low-carbon martensitic steels yield low-cost environmentally efficient materials. Wider use of these materials could improve energy savings and reduce the carbon footprint of many products. The use of such steels in martensitic condition provides a good weldability and a high strength, which makes them attractive materials for structural applications. However, an increase in steel strength enhances hydrogen embrittlement (HE). For example, the strength of steel bolts for automobiles has been limited to 1400 MPa because of HE. In hydrogen-powered vehicles, steel with a high resistance to HE is required for cost and weight reductions [1]. A promising approach to develop steels with a high strength and a low HE susceptibility may be through the use of ε-carbide precipitates in steels. In this work, a conceptual design of advanced steels with a high resistance to HE is proposed based on a literature survey and on experimental results of the influence of ε-carbide on the HE susceptibility of steel.

The paper is organized as follows. Section 2 presents a critical literature review on hydrogen absorption of ε-carbide and steel precipitation strengthening. Section 3 presents preliminary experimental results on the influence of ε-carbide in martensitic steel on the HE susceptibility. In Section 4, a steel design with a high strength that is compatible with a low HE susceptibility is proposed. The design includes alloying elements that are required to stabilize ε-carbide and inoculants to nucleate ε-carbide precipitates. The final section provides concluding remarks.

2. Literature Survey

2.1. ε-Carbide and Hydrogen Embrittlement

Hughes et al. [2] first related the delayed fracture tests of a low-alloy steel in corrosive environments to ε-carbide precipitates in the steel microstructure. Their conjecture was based on Berg's paper [3], which indicated that ε-carbide would have a high affinity for hydrogen. This report is critically reviewed below.

Berg thought that ε-carbide in steel might hold and bind hydrogen. If so, the ε-carbide could reduce negative effects of hydrogen in steels, such as fish-scaling and embrittlement. At that time, it was assumed that ε-carbide was a hexagonal carbide transient to χ-carbide (Hägg carbide) with a composition of Fe_2C. In addition, the X-ray powder diffraction analysis ability was limited. He aimed to prepare the carbide starting from an iron catalyst following the method by Hofer et al. [4]. A mixture of oxides, Fe_2O_3 (142 gm), CuO (12.5 gm), and KOH (0.38 gm) was heated to 200 °C in a stream of H_2 of commercial purity for 1 week, and then heated to 170 °C in a stream of CO of commercial purity for 2 weeks. The CO gas contained 1% H_2 as an impurity gas. The mixture was enclosed in four porcelain tubes with perforated end disks and placed in a row in a quartz tube. The gases were forced to flow through the tubes. The first product of carburization in CO was supposed to be carbonyl hydride that was then converted into carbide. The contents of the four porcelain tubes were analyzed separately for C and H. The content of Fe was not analyzed. The chemical analysis yielded a close coincidence of the percentage of C and H on the atomic weight basis. This was the basis of Berg's conclusion that the carbide held hydrogen up to the composition of Fe_2HC. The hydrogen atoms incorporated into the carbide were thought to have originated from the impure CO gas.

Berg, however, did not identify the reaction product of X-ray structural analysis, nor was it identified by the temperature dependence of the magnetization. Therefore, it remains doubtful if his carbide was a single-phase Fe_2C. In fact, the iron carbide prepared by Hofer et al. with the same method was a magnetic mixture of hexagonal Fe_2C (60%) and χ-carbide (40%) [4]. Because the Curie temperature of the former is 380 °C and the latter 250 °C, Berg could have discriminated the two phases in his sample. Furthermore, his chemical reaction product could have contained iron carbonyl or metallic iron. More importantly, the hexagonal Fe_2C, which he had intended to prepare, was later confirmed to be monoclinic Fe_5C_2 by a precise X-ray analysis [5]. Today, the ε-carbides precipitated in steels are established as hexagonal $Fe_{2.4}C$ by the precise counting of Fe and C atoms in ε-carbide precipitates using atom-probe tomography [6,7].

In summary, Berg carried out an imperfect synthesis of a carbide that is different from the ε-carbide precipitated in steel. At best, he prepared a carbide, which is nowadays termed monoclinic Fe_5C_2. Accordingly, his conclusion that the carbide holds hydrogen up to Fe_2HC lacks experimental verification. No direct evidence has been presented for Fe_5C_2 precipitation by tempering of martensitic steels [8]. This literature survey indicates that Hughes's discussion [2] on HE of steel that relies on the Berg study has no sound foundations.

Recently, hydrogen trapping in a high-strength steel was reported by Zhu et al. [9]. Hydrogen was charged electrochemically at room temperature into a tempered martensitic steel with ε-carbide precipitates. The electrolyte was an aqueous solution of 0.5 M H_2SO_4, and the current density was 30 mA/cm^2. The influence of hydrogen charging on the tensile elongation was tested. The thermal desorption spectrum of the steel charged for 5 min revealed two peaks: a strong one at ~90 °C and a weak one at ~400 °C. The former was correlated to diffusive hydrogen (0.62 wppm) in the steel lattice and the latter to non-diffusive hydrogen trapped at the ε-carbide (0.31 wppm). It was concluded that the ε-carbide precipitates played a limited role in the alleviation of HE. They also measured concentration profiles of C and H across an ε-carbide precipitate with three-dimensional atom probe tomography. The values were ~11 at % for C and ~0.4 at % for H. This implies that the ε-carbide in their experiment may be formulated as $Fe_{2.4}CH_{0.04}$. It is not clear if this value represents the full hydrogen-trapping capacity of ε-carbide.

2.2. Precipitation Strengthening by ε-Carbide

Leslie was the first to note the possibility of precipitation strengthening of ferrite by ε-carbide, although he termed it as "metastable carbide" [10]. Leslie studied the aging of Fe–0.014% C, Fe–0.45% Mn–0.017% C, and Fe–3.25% Si–0.029% C alloys at 60–200 °C, after quenching from 730 °C. He noted two distinct configurations of the precipitated carbides, long chains of carbide particles on dislocation lines and single particles in the matrix. The main changes in mechanical properties during the quench-aging resulted from the closely spaced carbides that precipitated within the matrix. He suggested Orowan's equation [11] as the underlying mechanism. The matrix nucleation sites for the ε-carbide remained undefined.

A search for the role of alloying elements on the nucleation of ε-carbide in Fe–C alloys has been reported for titanium, vanadium, and chromium [12]. Fine precipitates of TiC of 8~30 nm in size were found to act as preferential nucleation sites. The TiC particles formed during the alloy preparation and remained undissolved during subsequent solution treatment at 720 °C. The roles of V- and Cr-carbides remained unclear because of the limited additions of vanadium and chromium to the Fe–C alloys.

Isolated cases of the role of ε-carbide in engineering steels have been reported. Eglin steel is an ultra-high-strength steel alloy that was developed at Eglin Air Force Base in the early 2000s and was patented in 2009 [13]. The steel has strength levels like AerMet100, AF1410, and HP9–4–30 but is produced at a reduced cost because of the reduction or elimination of expensive alloying elements, such as nickel and cobalt. A comprehensive study followed to correlate the mechanical properties and microstructural evolution in the heat-affected zone of the Eglin steel [14]. ε-carbide particles were found in an auto-tempered lath of martensite in the coarse grains of the heat–affected zone (HAZ) of Eglin steel. The ε-carbide particles improved the HAZ toughness significantly.

Abrahams (author in [14]) invented a "low alloy high performance steel". The patent application was filed in 2016 and patented in the United States in 2019 [15]. The steel composition was 0.24%~0.32% C, 2.00%~3.00% Cr, 0.50%~1.50% Mo, 0.05%~0.35% V, 1.00% Mn or less, ~3.00% Ni or less, and ~1.50% Si or less, with a balance of Fe. The hardened and tempered article had a high impact toughness and other favorable physical properties such as an ultimate tensile strength, yield strength, elongation to failure, and hardness. The steel is believed to have been strengthened by nanoscale ε-carbides within a primarily martensitic matrix. The carbide was 100~150 nm long and ~10 nm wide. The carbide particles have a feathery rod-like shape, which are semi-coherent to the matrix. According to the inventor, without being bound by any theory, ε-carbide is believed to provide strength while preserving the dynamic toughness.

Xia et al. developed a low-carbon high-silicon martensitic steel in which ε-carbide plays a role [16]. The chemical composition is 0.21% C–1.8% Si–1.1% Mn–0.7% Cr–0.14% Ni–0.19% Mo. The steel has a tensile strength of 1548 MPa, an impact toughness of 120 J/cm^2, and a fracture toughness (K_{IC}) of 94.8 MPa m$^{1/2}$ when it is austenitized at 900 °C for 1 h followed by water quenching and then tempered at 320 °C for 1 h. Its microstructure is characterized by lath martensite, retained austenite films, and ε-carbide. The ε-carbides are present as short flakes in the martensite matrix and enhance the steel toughness.

At the end of Section 2.2, a comment is provided on low-temperature-tempered (LTT) martensitic steels, typically 4130 and 4340 steels. A review on this technology has been given by Krauss [17]. The low-temperature tempering of quenched martensite between 150 and 200 °C improves its toughness but maintains its hardness and strength at a high level. This increase in strength is attributed to the effect of increasing the density of fine-transition carbides, believed to be orthorhombic η-carbide, on the strain-hardening behavior of the martensitic microstructures. The chemical composition of the η-carbide was assumed to be between that of Fe_2C and Fe_3C. However, controversy existed between orthorhombic η-carbide and hexagonal ε-carbide. Thompson investigated this question using imaging and electron diffraction techniques with a transmission electron microscope [18] and found that the distinction between orthorhombic and hexagonal carbides appeared unlikely; η-carbide and ε-carbide that were observed by electron diffraction patterns were nearly the same. Therefore, fine transition carbides in LTT martensitic steels are ε-carbides. The HE susceptibility of LTT martensitic steels was outside of the research interest, whereas that of high-temperature-tempered LTT steels (tempered at ~600 °C) has been reported because of their application in oil and gas engineering in environments that contain H_2S [17].

3. Experiments on Hydrogen Absorption

The literature suggests that ε-carbide precipitation is a promising tool to increase steel strength and toughness if the size and dispersion are controlled. However, the influence of ε-carbide on HE has not been studied extensively. If ε-carbide increases the storage capacity of hydrogen of the steel, it will extend the time required to provide sufficient hydrogen atoms for the formation of hairline cracks.

To evaluate the HE susceptibility of steels, an easy and precise testing method has been proposed, namely immersion hydrogen charging with ammonium thiocyanate (NH_4SCN) solution [19]. By adjusting the solution concentration, diffusible hydrogen atoms that intrude on the steel specimen surface are controlled. The advantage of this solution is the reduced dissolution of a specimen, owing to its high pH of 4 to 5. For comparison, aqueous solutions of HCl for hydrogen charging tests are usually pH 1.

In this study, a quench-tempered martensitic steel was tested. Its chemical composition is Fe–0.21% C–0.22% Si–1.5% Mn–0.06% Ti. The tensile strength of the specimen that was aged at 200 °C was ~1500 MPa. Transmission electron microscopy of the specimen confirmed ε-carbide precipitates. A specimen without tempering was also tested. The HE susceptibility of the specimens was evaluated by the U-bending method with a bending radius of 2~10 mm [20]. The U-bending specimen was 30 mm wide, 100 mm in length, and 1.6 mm thick. An applied stress of 900 MPa was measured by X-ray. The concentration of NH_4SCN was 0.1%. The testing was carried out at room temperature.

Preliminary and qualitative experimental results are presented here. The tempered specimens showed no cracking after immersion for 300 h, whereas the specimens without tempering displayed cracks. Thus, the ε-carbide precipitates lowered the HE susceptibility of the specimens. Hydrogen contents in the tested specimens were measured by thermal desorption analysis with gas chromatography. Figure 1 compares the hydrogen-evolution curves of the quenched and tempered specimens that were charged previously with hydrogen for 100 h at room temperature. The heating rate was 200 K/h. The dotted line represents the evolution curve of the as-quenched specimen. The peak at ~100 °C is ascribed to diffusible hydrogen atoms in the specimen. The solid curve is for the tempered specimen with ε-carbide precipitates. A large peak starts at ~300 and ends at 400 °C. This range corresponds to the decomposition of ε-carbide to cementite in the Fe–C phase diagram. Through the decomposition, the hydrogen atoms trapped in the ε-carbide are forced to the outside of the specimen. Although this experiment is qualitative in nature, the result suggests that hydrogen atoms are trapped strongly at the ε-carbide precipitates. It is believed that once trapped, they will not easily return to diffusible states, which lowers the HE susceptibility.

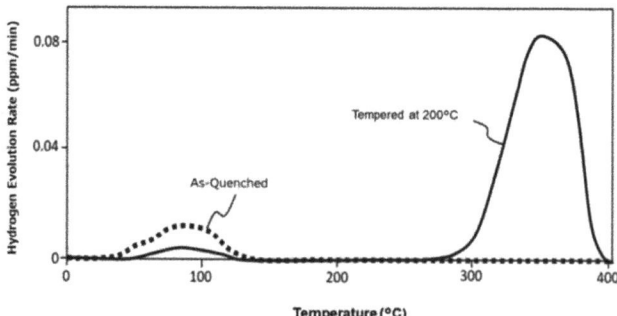

Figure 1. Hydrogen-evolution curves of steel specimens with martensitic structure. The dotted line represents the quenched specimen, and the solid line represents the quench-tempered specimen.

The lattice sites for the hydrogen atoms in ε-carbide will be interstitial and/or substitutional sites in the crystal. In the case of hydrogen-storage materials such $LaNi_5$, the absorbed hydrogen atoms occupy octahedral and tetrahedral interstitial sites. These sites facilitate easy desorption of hydrogen atoms that are required for such materials. However, for ε-carbide, which may be regarded as a non-stoichiometric compound between Fe_2C (=$Fe_{2.4}C_{1.2}$) and Fe_3C (=$Fe_{2.4}C_8$), vacant substitutional sites in the carbon-lattice are available. If the hydrogen atoms occupy all vacant substitutional sites, the chemical formula is expressed as $Fe_{2.4}CH_{0.2}$, compared with the reported value of $Fe_{2.4}CH_{0.04}$ [9]. The substitutional hydrogen atoms will be stable compared with the interstitial ones. Further studies are needed to determine the full hydrogen-absorption capacity of ε-carbide. Theoretically, ab initio calculations of the trapping energy of hydrogen atoms in the ε-carbide lattice are desirable. Experimentally, the influence of hydrogen absorption on the mechanical spectrum of carbon dipoles in ε-carbide may provide insight into the locations of hydrogen atoms in the ε-carbide lattice [21,22].

4. Heuristic Design of Steels

The literature survey and the experimental studies described above lead to a design of advanced steels with low HE susceptibility. Requirements for this kind of steel are martensitic microstructure with ε-carbide precipitates. The precipitates must be stabilized by alloying and should have a proper shape, size, and dispersion. Some carbides, nitrides, or carbonitrides that may act as the inoculants for ε-carbide have to be discussed. The factors that contribute to the yield strength of martensitic steel have been classified as solid solution strengthening, precipitation or dispersion strengthening, dislocation strengthening, and lath-boundary strengthening [23,24]. In this case, the second to fourth strengthening mechanisms are involved. The coherency and/or semi-coherency of ε-carbide to the host crystal is expected to have a close relationship with the tensile elongation and impact toughness of the steels. Coherence loss with the introduction of fresh dislocations through the punching of dislocation loops may enhance the steel elongation and toughness. This issue is left for further study.

4.1. Alloying Elements

Carbon: The carbon contents are important in martensitic microstructure and carbide precipitation to design advanced steels that are highly resistant to HE with the assistance of ε-carbide. According to an early study [25], the lowest limit for carbon content to observe ε-carbide in the tempered martensite is 0.2%. The other importance of carbon as an alloying element is shown in the relationship between M_s temperature and the chemical composition [26]:

$$M_s (°C) = 539 - 423(\%C) - 30.4(\%Mn) - 17.7(\%Ni) - 12.1(\%Cr) - 7.5(\%Mo) \quad (1)$$

where M_s is the martensitic transformation temperature. Equation (1) shows carbon has the most significant effect on M_s temperature. The M_s temperature can neither be increased nor decreased by varying the cooling rate. In contrast, the temperature at which martensitic transformation ends, M_f, is dependent on the cooling rate.

Silicon: Silicon addition to steels shifts cementite precipitation to higher temperature. This results in the stabilization of ε-carbide up to ~400 °C. The mechanism by which silicon retards the cementite precipitation is that the driving force for the precipitation is dramatically reduced when the cementite is forced to inherit the silicon in the parent phase, which is the martensite [27]. There is also evidence that the silicon can enhance the formation of ε-carbide [28]. Silicon addition of more than 0.8% has been used for martensitic steels. Excessive addition of silicon will induce remarkable hardening that is undesirable for industrial production.

Manganese: This element has a serious effect on M_s temperature as shown in Equation (1). One can control the steel M_s by manganese addition.

4.2. Inoculants for ε-Carbide

Inclusions and precipitates in steels as inoculants for phase transformations have attracted much attention for controlling the grain size of ferritic microstructure. Inclusions, such as VN, TiC, TiN, AlN, and Al_2O_3, can stimulate ferrite nucleation in austenite [29]. TiC is reported to act as the inoculant to precipitate ε-carbide in ferrite [12]. In this case, one must look for effective inoculants to precipitate ε-carbide in martensite. Nano-sized carbides, nitrides, or carbonitrides of metals are the candidate inclusions. Two procedures are probable for introduction of the inoculants: static and dynamic procedures. The starting point of the former is interphase precipitation [30–32]. In interphase precipitation, almost equally spaced rows of nanometer-sized carbides are precipitated in the ferritic microstructure during the austenite-ferrite semi-static transformation. If these precipitates survive coarsening and/or dissolution during reheating of the ferritic steel to the austenitic region for subsequent quenching to martensite, they may act as inoculants for the controlled precipitation of ε-carbide during tempering.

Alternatively, the inoculants may be introduced dynamically during the martensitic transformation. The process takes advantage of solubility difference of nitrogen or carbon between austenite and martensite. Nitrides or carbonitrides of aluminum, titanium, and vanadium are promising precipitates. For low-carbon aluminum-killed steels, AlN and MnS are supposed to nucleate heterogeneously at dislocation nodes [33]. In this case, dislocation nodes in the lath martensite are believed to act as nucleation sites. Although the cooling speed is critically important in this procedure, the heat pattern of steels during processing is simple.

5. Conclusions

Many advanced steels have tempered martensitic microstructures. Their mechanical strength is characterized by a fine sub-grain structure with a high density of free dislocations and metallic carbides and/or nitrides. However, their strength for practical use has been limited mostly to below 1400 MPa, owing to delayed fracture caused by hydrogen. A literature survey suggests that ε-carbide in the tempered martensite is effective for strengthening. An experimental survey on hydrogen absorption and hydrogen embrittlement of a tempered martensitic steel with ε-carbide precipitates has revealed its low HE susceptibility. Based on these surveys, martensitic steels with ε-carbide precipitates are proposed. The steels are expected to be highly resistant to hydrogen embrittlement and to have a high strength and toughness. Silicon is the most important alloying element to stabilize ε-carbide. Inoculants are required for controlled nucleation and dispersion of ε-carbide precipitates. Procedures to introduce inoculants in martensitic steels are discussed.

Funding: The study was not supported by any grant.

Data Availability Statement: Not applicable.

Acknowledgments: The author wishes to acknowledge all those who aided in the experiment, but were too modest to be mentioned.

Conflicts of Interest: The author has no conflict of interest to declare.

References

1. Michler, T.; Naumann, J. Microstructural aspects upon hydrogen environment embrittlement of various bcc steels. *Int. J. Hydrogen Energy* **2010**, *35*, 821–832. [CrossRef]
2. Hughes, P.C.; Lamborn, I.R.; Liebert, B.B. Delayed fracture of a low-alloy steel in corrosive environments. *J. Iron Steel Inst.* **1965**, *203*, 154–159.
3. Berg, T.G.O. Hydrogen in Carbide in Steel. In Proceedings of the 61st Annual Meeting, American Ceramic Society, Chicago, IL, USA, 17–21 May 1959; Ceramic Bulletin, Am. Ceramic Soc.: Westerville, OH, USA, 1961; Volume 40, pp. 78–80.
4. Hofer, L.J.E.; Cohn, E.M.; Peebles, W.C. The Modifications of the Carbide, Fe2C; Their Properties and Identification. *J. Am. Chem. Soc.* **1949**, *71*, 189–195. [CrossRef]
5. Jack, D.H.; Jack, K.H. Invited review: Carbides and nitrides in steel. *Mater. Sci. Eng.* **1973**, *11*, 1–27. [CrossRef]

6. Caballero, F.; Miller, M.K.; Garcia-Mateo, C. Atom Probe Tomography Analysis of Precipitation during Tempering of a Nanostructured Bainitic Steel. *Met. Mater. Trans. A* **2011**, *42*, 3660–3668. [CrossRef]
7. Song, W.; Von Appen, J.; Choi, P.-P.; Dronskowski, R.; Raabe, D.; Bleck, W. Atomic-scale investigation of ε and θ precipitates in bainite in 100Cr6 bearing steel by atom probe tomography and ab initio calculations. *Acta Mater.* **2013**, *61*, 7582–7590. [CrossRef]
8. Nagakura, S.; Oketani, S. Structure of Transition Metal Carbides. *Trans. Iron Steel Inst. Jpn.* **1968**, *8*, 265–294. [CrossRef]
9. Zhu, X.; Li, W.; Hsu, T.; Zhou, S.; Wang, L.; Jin, X. Improved resistance to hydrogen embrittlement in a high-strength steel by quenching–partitioning–tempering treatment. *Scr. Mater.* **2015**, *97*, 21–24. [CrossRef]
10. Leslie, W.C. The quench-aging of low-carbon iron and iron-manganese alloys. *Acta Met.* **1961**, *9*, 1004–1022. [CrossRef]
11. Orowan, E. The Theory of Yield without Particle Shear. In Proceedings of the Symposium on Internal Stresses in Metals and Alloys, London, UK, 15–16 October 1947; Institute of Metals: London, UK, 1948; p. 451.
12. Saito, N.; Abiko, K.; Kimura, H. Effects of Small Addition of Titanium, Vanadium and Chromium on the Kinetics of ε-carbide Precipitation in High Purity Fe–C Alloys. *Mater. Trans. JIM* **1995**, *36*, 601–609. [CrossRef]
13. Dilmore, M.; Ruhlman, J.D. Eglin Steel-A Low Alloy High Strength Composition. U.S. Patent 7537727, 26 May 2009.
14. Leister, B.M.; Dupont, J.N.; Watanabe, M.; Abrahams, R.A. Mechanical Properties and Microstructural Evolution of Simulated Heat-Affected Zones in Wrought Eglin Steel. *Met. Mater. Trans. A* **2015**, *46*, 5727–5746. [CrossRef]
15. Abrahams, R.A. Low Alloy High Performance Steel. U.S. Patent 10450621, 22 October 2019.
16. Xia, S.; Zhang, F.; Zhang, C.; Yang, Z. Mechanical Properties and Microstructures of a Novel Low-carbon High-silicon Martensitic Steel. *ISIJ Int.* **2017**, *57*, 558–563. [CrossRef]
17. Krauss, G. Tempering of Lath Martensite in Low and Medium Carbon Steels: Assessment and Challenges. *Steel Res. Int.* **2017**, *88*, 1700038. [CrossRef]
18. Thompson, S.W. A Two-Tilt Analysis of Electron Diffraction Patterns from Transition-Iron-Carbide Precipitates Formed During Tempering of 4340 Steel. *Met. Microstruct. Anal.* **2016**, *5*, 367–383. [CrossRef]
19. Takagi, S.; Toji, Y. Application of NH4SCN Aqueous Solution to Hydrogen Embrittlement Resistance Evaluation of Ultra-high Strength Steels. *ISIJ Int.* **2012**, *52*, 329–331. [CrossRef]
20. Takagi, S.; Toji, Y.; Hasegawa, K.; Tanaka, Y.; Roessler, N.; Hammer, B.; Heller, T. Hydrogen Embrittlement Evaluation Methods for Ultra-high Strength Steel Sheets for Automobiles. *Int. J. Automot. Eng.* **2010**, *1*, 7–13. [CrossRef]
21. Shimotomai, M. Coherent-incoherent transition of ε-carbide in steels found with mechanical spectroscopy. *Metall. Mater. Trans. A* **2016**, *47*, 1052–1060. [CrossRef]
22. Shimotomai, M. Study of carbon steel by mechanical spectroscopy beyond the old limitations. *Res. Rep. Metals 1:3* **2017**, *1*, 1000107.
23. Maruyama, K.; Sawada, K.; Koike, J.-I. Advances in Physical Metallurgy and Processing of Steels. Strengthening Mechanisms of Creep Resistant Tempered Martensitic Steel. *ISIJ Int.* **2001**, *41*, 641–653. [CrossRef]
24. Abe, F. Precipitate design for creep strengthening of 9% Cr tempered martensitic steel for ultra-supercritical power plants. *Sci. Technol. Adv. Mater.* **2008**, *9*, 013002. [CrossRef]
25. Speich, G.R. Tempering of low-carbon martensite. *Trans. Metall. Soc. AIME* **1969**, *245*, 2553–2564.
26. Andrews, K.W. Empirical formulae for the calculation of some transformation temperatures. *J. Iron Steel Inst.* **1965**, *203*, 721–727.
27. Kozeschnik, E.; Bhadeshia, H.K.D.H. Influence of silicon on cementite precipitation in steels. *Mater. Sci. Technol.* **2008**, *24*, 343–347. [CrossRef]
28. Jang, J.H.; Kim, I.G.; Bhadeshia, H. ε-carbide in alloy steels: First-principles assessment. *Scr. Mater.* **2010**, *63*, 121–123. [CrossRef]
29. Zhang, S.; Hattori, N.; Enomoto, M.; Tarui, T. Ferrite Nucleation at Ceramic/Austenite Interfaces. *ISIJ Int.* **1996**, *36*, 1301–1309. [CrossRef]
30. Morrison, W.B. The influence of small niobium additions on the properties of carbon-manganese steels. *J. Iron Steel Inst.* **1963**, *201*, 317–325.
31. Funakawa, Y.; Shiozaki, T.; Tomita, K.; Yamamoto, T.; Maeda, E. Development of High Strength Hot-rolled Sheet Steel Consisting of Ferrite and Nanometer-sized Carbides. *ISIJ Int.* **2004**, *44*, 1945–1951. [CrossRef]
32. Chen, M.-Y.; Yen, H.-W.; Yang, J.-R. The transition from interphase-precipitated carbides to fibrous carbides in a vanadium-containing medium-carbon steel. *Scr. Mater.* **2013**, *68*, 829–832. [CrossRef]
33. Chen, Y.; Wang, Y.; Zhao, A. Precipitation of AlN and MnS in low carbon Aluminum-killed steel. *J. Iron Steel Res. Int.* **2012**, *19*, 51–56. [CrossRef]

Article

Evolution of Microstructure during Isothermal Treatments of a Duplex-Austenitic 0.66C11.4Mn.9.9Al Low-Density Forging Steel and Effect on the Mechanical Properties

Idurre Kaltzakorta [1,*], Teresa Gutierrez [1], Roberto Elvira [2], Pello Jimbert [3] and Teresa Guraya [3]

1. TECNALIA, Basque Research and Technology Alliance (BRTA), Parque Científico y Tecnológico de Bizkaia, C/Astondoa, Edificio 700, 48160 Derio, Spain; Teresa.Gutierrez@tecnalia.com
2. Sidenor I+D, Barrio Ugarte s/n, 48970 Basauri, Spain; Roberto.Elvira@sidenor.com
3. Faculty of Engineering in Bilbao, University of the Basque Country UPV/EHU, Paseo Rafael Moreno "Pitxitxi" 3, 48003 Bilbao, Spain; Pello.Jimbert@ehu.eus (P.J.); Teresa.Guraya@ehu.eus (T.G.)
* Correspondence: Idurre.Kaltzakorta@tecnalia.com

Abstract: In the last decades, low-density steels for forging have increasing interest in the automotive industry, and good mechanical properties are required for their real application. This paper describes the results obtained for a 0.66C11.4Mn9.9Al duplex austenitic low-density steel after applying a set of isothermal treatments at different combinations of time and temperature, aimed to promote kappa carbide precipitation, and improve the mechanical properties obtained with a water quenching treatment. The effects of the different isothermal treatments on the microstructure and on the mechanical properties have been analyzed and compared to those obtained from a quenching heat treatment. We found that isothermal treatments in the range temperature between 550–750 °C promoted the profuse precipitation of coarse kappa carbides at grain boundaries, which dramatically reduced the ductility of the alloy, whereas a traditional quenching treatment resulted in a better combination of ductility and mechanical strength.

Keywords: low density steels; forging; kappa carbide; FeCMnAl

1. Introduction

To comply with the severe policies regarding CO_2 emissions, the automotive industry is focused on reducing the weight of cars without penalizing the safety of passengers (a reduction of 100 kg in the vehicle weight implies ~8.5 g less CO_2 emissions per km).

For this purpose, they are adopting different strategies such as the redesign of components (to eliminate unnecessary material), the use of high-strength steels (to reduce thickness) and the reduction of the density of the material used. In the literature, many studies have focused on the development of lightweight steels with a high strength and low density [1–3], the Fe-Mn-Al-C system being the most widely explored. Low-density steels that consider additions of Al (more than 10%), obtaining a density reduction of around 15%, are the most studied ones [3]. Depending on the chemical composition (weight %), low-density steels can be classified into three different categories [4]: austenitic steels (0.5 < %C < 2, 8 < %Al < 12, 15 < %Mn < 30), duplex steels (0.1 < %C < 0.7, 3 < %Al < 10, 5 < %Mn < 30) and ferritic steels (%C < 0.03, 5 < %Al < 8, %Mn < 8); Mn and C are the austenite former while Al is the strong ferrite former. Therefore, the mechanical properties of the Fe–Mn–Al–C system are dependent on the deformation characteristics of the constituent phase(s) [5].

A promising improvement for these low-density alloys comes from the proper precipitation of kappa carbides (k-carbide). The k-carbide is a carbide with an L1′$_2$-type structure with a perfect formula of $(Fe,Mn)_3AlC$, and the lattice parameters vary with the C and Mn content [6]. The addition of aluminium accelerates the precipitation of ordered perovskite structure carbide, i.e., k-carbide [1,7–9]. Even if in the beginning, k-carbide was thought to be harmful to ductility [10–12], it has recently been documented that k-carbide can enhance strength and ductility at the same time by optimizing its morphology, distribution, fraction and size [13,14]. For the enhancement of mechanical properties, the intragranular nanosized k-carbides are usually desirable, whereas the intergranular k-carbide must be avoided [13]. In a ferritic Fe-3.0Mn-3.0Al-0.3C (wt.%) alloy, the same author described that lamellar-type k-carbides were formed as a result of a eutectoid reaction during isothermal annealing between 500 °C and 600 °C associated with nucleation and growth. For austenitic low-density alloys with a high-Mn content, in aging treatments between 500 °C and 600 °C, rectangular-shaped k-carbides were reported to precipitate out, mainly from the austenitic matrix via spinodal decomposition during aging [13–16].

In a previous work [17], we carried out a preliminary study of several low-density steels formed by ferritic, austenitic and duplex microstructures. For this investigation, a 0.66C11.4Mn9.9Al duplex austenitic low-density steel was selected to be submitted to a set of isothermal treatments at different combinations of time and temperature, aimed to promote kappa carbide precipitation in order to study the influence of this k-carbide precipitation on the final mechanical properties and to define an appropriate isothermal heat treatment for obtaining a good combination of strength and ductility for forged components. First, the kinetics of k-carbide precipitation under different isothermal conditions was studied. Samples were quenched from 1150 °C and isothermally heat-treated by varying the treatment temperature in the range of 550–750 °C and the holding time in the range of 0.5–5 h. Second, tensile tests were carried out to study the effect of the k-carbide precipitation on the final mechanical properties of the treated materials, and the results were compared to those obtained for the same material after water quenching.

2. Materials and Methods

The selected duplex low-density steel with the theoretical composition of 0.65C12Mn10Al (wt.%) was melted in a vacuum levitation furnace using iron pieces (99.97+%), graphite flakes (99.9%), manganese pieces (99.9%) and pure aluminium pieces (99%+) as raw materials. The advantage of this equipment is that it works in a protective atmosphere, which eliminates risks of contamination and therefore yields a pure and reproducible cast material that also allows for remelting. To ensure the homogeneous distribution of the composition throughout the ingot, two refusions were performed, and ingots of about 1 kg were cast. In Figure 1a the cylindrical cast ingot is shown, whereas Figure 1b corresponds to the bar forged from the same ingot.

Figure 1. (**a**) Ingot cast in the levitation vacuum furnace and (**b**) bar forged from the ingot.

Two analytical techniques were used to determine the actual composition of the cast alloy. The C content was analyzed by Infra-Red (IR) detection after a sample combustion using an IR LECO CS-400 (LECO- St. Joseph, MI, USA). The Al and Mn contents were determined by plasma emission spectrometry (ICP-OES) in a THERMO-ICAP 7400 DUO equipment (Thermofisher Scientific, Waltham, Massachusetts, USA). The sample was dissolved in a mixture of nitric and clorhydric acids (1:3). Then, the solution was filtered, and the residue was collected. The filtrate was made up to the mark with Milli-Q water.

The residue was calcined at 550 °C, melted at 950 °C in lithium metaborate and dissolved in an acid medium for further analysis. The final composition is the sum of the contents obtained by analyzing the acid-soluble and the acid-insoluble parts. Four samples were extracted from different areas of the outer and the inner parts of the ingot from the upper and lower sides, in order to analyze the homogeneity of the ingot, obtaining a similar composition (standard deviation values of 0.005 for the carbon, 0.05 for the aluminium and 0.09 for the manganese were obtained).

The ingots were forged using a mechanical hammer. A ThermoCalc (Thermocalc®Sofware version 2020a, database TCFe6)phase diagram of the cast alloy was used to determine the forging conditions based on the temperature range, where the highest fraction of austenite, appropriate for hot working, was expected (Figure 2). During the forging process, to keep the material in the temperature range where austenite was the dominant phase, and considering that the ingots cool down very fast during manipulation, the ingots were heated up to 1150 °C, and when the temperature dropped below 900 °C they were heated again until the entire bar was forged. A Land Ametek Cyclops L Series Digital Pyrometer (Land Instruments International Ltd, Dronfield, UK) was used in all cases to control the ingot temperature. The average reduction that was applied was in the range of 50–65%, from an initial 30-mm diameter to about 11–15 mm at the end of the process (Figure 1b).

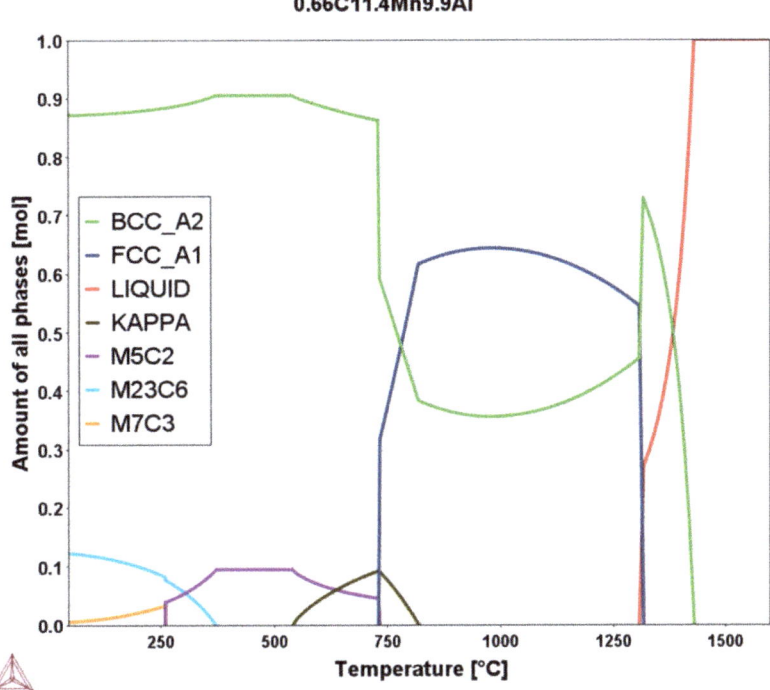

Figure 2. Phase diagram of the analyzed composition calculated with ThermoCalc®software (version 2020a, TCFe6 Database).

The density of the cast alloy was measured by calculating the density of the sample by measuring the mass of the sample both in air and immerged in a liquid of known density using a Radwag AS 120.X2 PLUS analytical balance (Radwag, Radom, Poland) with an accuracy of 0.01 mg. The density of the sample ρ_s (in g·cm^{-3}) was calculated using Equation (1):

$$\rho_s = \frac{m_a}{m_a - m_l} \cdot \rho_l \tag{1}$$

where m_l is the mass of the sample in the liquid, m_a is the mass of the sample in air (g), and ρ_l is the density of the liquid (g·cm^{-3}). In all cases, the liquid water and its density were considered to be 1 g·cm^{-3} regardless of the temperature.

For the microstructural characterization, the samples were sectioned, mechanically grinded and polished to up to 0.05 microns γ-Al$_2$O$_3$, and etched with 5% Nital solution. Microstructures were studied in Leica DM 400 (LEICA, Wetzlar, Germany) and Zeiss AxioCamMRc5 (ZEISS, Jena, Germany) optical microscopes and a FEI Quanta 450 (FEI, Hillsboro, OR, USA) field-emission scanning electron microscope (FEG-SEM) operating at 20 kV. The FEG-SEM was equipped with an energy dispersive spectrometer (EDS) and EBSD detector. Samples were also analyzed by X-ray diffraction (XRD) in a Rigaku DMAX-RB X-ray diffractometer (Rigaku Europe SE, Neu-Isenburg, Germany) with a Cu target operating at 40 kV and 150 mA.

Dilatometry studies were carried out using a Dilatometer BAHR 805L (TAInstruments, New Castle, DE, USA). In a vacuum chamber, the sample was located between quartz rods and heated inside an induction coil. The temperature was controlled by an RhPt thermocouple, and the sample dilatation was measured by a linear variable displacement transducer (LVDT). Sample cooling was carried out by blowing an argon or helium flow, depending on the required cooling rate. Heat treatments were performed in a Carbolite Wire Wound Tube Furnace TZF 12/65/550 (Nabertherm GmbH, Lilienthal, Germany) with temperature and atmosphere control. Finally, an Instron Machine, Model 5500R (INSTRON; Barcelona, Spain), was used to carry out the tensile tests.

3. Results and Discussion

The actual chemical composition of the steel was a 11.4 wt.% Mn, 9.9 wt.% Al, 0.66 wt.% C and Fe balance. The measured density of 6.86 g/cm^3 was in good agreement with the theoretical value of 6.80 g/cm^3, calculated with ThermoCalc software. The cast low-density steel resulted in being 13% lighter than pure iron.

3.1. Microstructural Evolution in Heat Treatments

3.1.1. Starting Microstructure

As the material was forged at 1150 °C, the microstructure at this temperature was defined as the initial condition to study the microstructural changes and, particularly, the k-carbide precipitation during different isothermal heat treatments. To analyze these changes, several samples of the alloy were heated up to 1150 °C at a heating rate of 5 °C/min, kept at that temperature for an extra 30 min and then water-quenched. According to the ThermoCalc diagram (Figure 2), the expected duplex microstructure would be formed by 38% ferrite and 62% austenite. Figure 3 shows the microstructure of the quenched samples. The optical microscope image shows a homogeneous duplex microstructure (Figure 3a) formed by austenite (darker regions) and ferrite (lighter regions). The volume fraction of each phase was calculated by an automatic image analysis performed on low magnification images according to ASTM E562-08 [18]. The calculated results were 37% ferrite and 63% austenite. Both phases were also confirmed as ferrite and austenite by SEM-EBSD (Figure 3b,c). The phase proportion measured on the EBSD map was a 65% area for austenite and a 35% area for ferrite. Although the EBSD results correspond to the area % of a reduced area and the ThermoCalc results correspond to the volume %, both are in good agreement with the volume fraction of each phase measured by point counting. X-ray diffraction (XRD) was used to confirm whether k-carbides were formed during quenching. The diffractogram shown in Figure 3d confirms the presence of the FCC and BCC structures by the peaks at (2θ) values of 42.5° and 44.3°, respectively. There was no evidence of the k-carbide peak, which was expected to be at 40.6°.

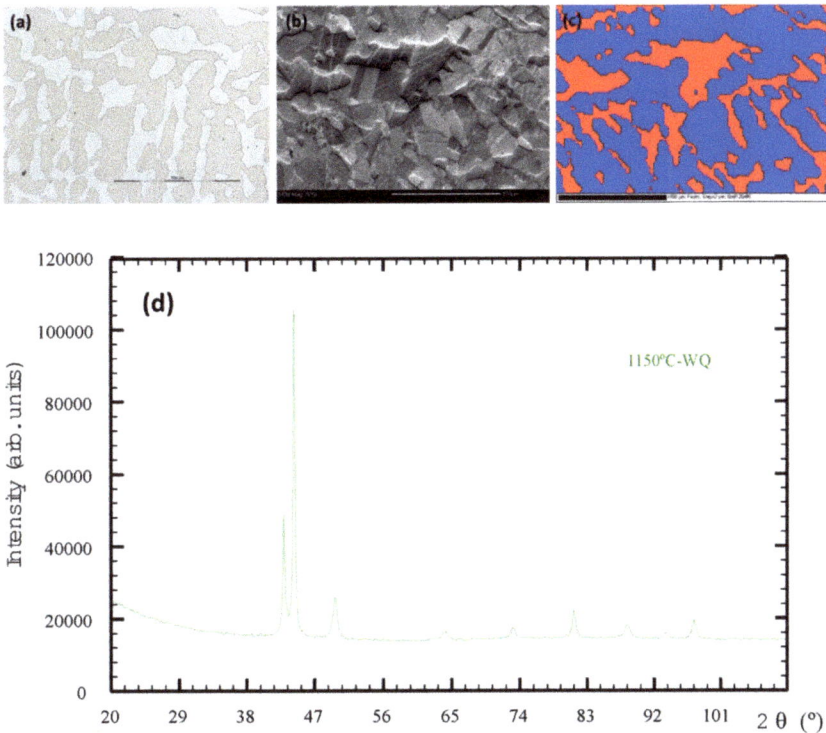

Figure 3. Microstructure images obtained by (**a**) optical microscope, (**b**) SEM image and (**c**) its corresponding phase EBSD mapping (ferrite in red and austenite in blue) of the sample that was water-quenched from 1150 °C, and (**d**) its XRD diffractogram.

3.1.2. Phase Transformation and K-Carbide Precipitation

A dilatometry test was carried out on a sample quenched from 1150 °C in order to observe the transformations that took place during heating and cooling. Heating from room temperature to 1000 °C was conducted at 0.05 °C/s, and reverse cooling to room temperature was conducted at 0.01 °C/s. Figure 4 shows the dilatometry curves and their derivative, corresponding to the heating (Figure 4a) and the cooling (Figure 4b) transformations. In the heating curve, the peak observed at 500 °C can be assigned to the beginning of the k-carbide precipitation. According to the phase diagram (Figure 2), these precipitates react with ferrite in the range temperature of 740–820 °C and transform into austenite by a eutectoid reaction [19–22]. This transformation ends close to 900 °C. Similarly, the reverse transformations can be assigned during the cooling process. At the beginning of the curve, the slope decreases progressively, which is related to the gradual transformation of austenite to ferrite. The peak above 900 °C can be attributed to the beginning of carbide precipitation during cooling, since at this temperature the decomposition of austenite to ferrite and k-carbide occurs through the eutectoid reaction. To study the precipitation kinetics of the k-carbides under isothermal conditions (those corresponding to an annealing treatment at a constant temperature), a permanence dilatometry test was carried out at 750 °C for 100 h. Figure 4c shows the dilatometry curve and its derivative, and Figure 4d shows a detail of the first two hours of heating. Although it is difficult to differentiate the part corresponding to the test regulation itself and homogenization of the temperature in the specimen from the part corresponding to the phase transformation, the precipitation of k-carbides seems, apparently, to take place during the first hour of maintenance.

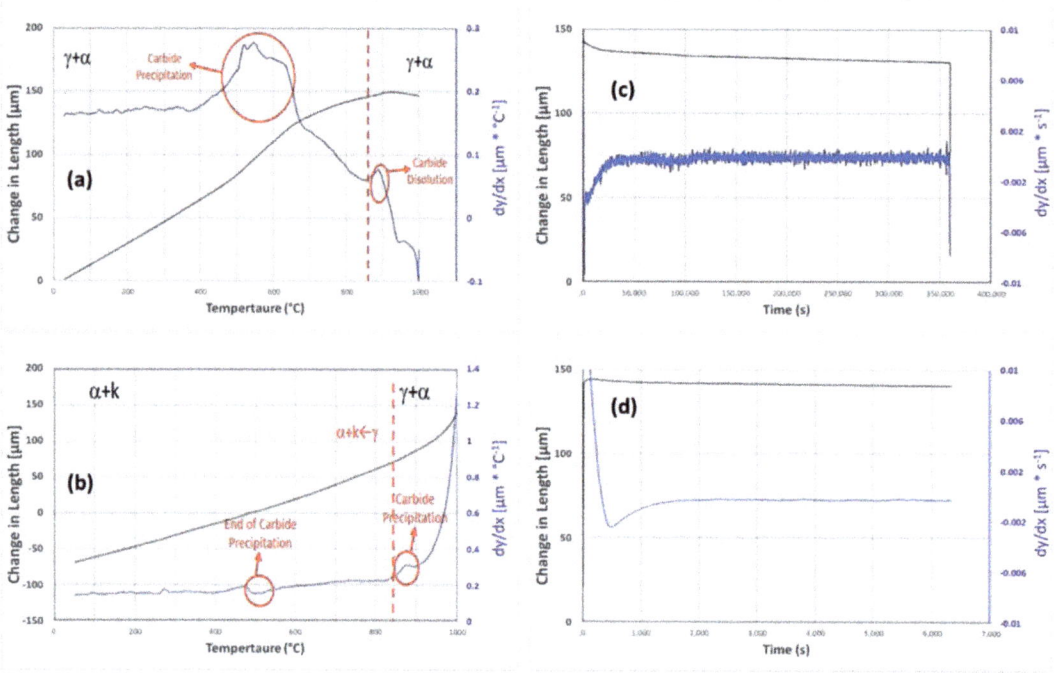

Figure 4. Dilatometry tests. (a) The curve (black) and its derivative (blue) corresponding to the heating process up to 1000 °C; (b) curves for the reverse cooling process; (c) curve (black) and its derivative (blue) corresponding to an isothermal treatment (750 °C-100 h); (d) detail of the beginning of the isothermal treatment. The symbols γ, α and k refer to austenite, ferrite and k-carbide phases, respectively.

3.2. K-Carbide Precipitation Kinetic Study

Based on the results obtained from the dilatometry test carried out at 750 °C, the effect of the holding time on the k-carbide precipitation was studied. Heat treatments were performed in a furnace. The samples were heated up to 750 °C while varying the holding times to 0.5, 1, 2 and 5 h and were finally water-quenched. Figure 5 shows the optical microscope and SEM micrographs obtained for each treatment with their corresponding XRD diffractograms. At 750 °C, the precipitation of the k-carbides begins at around 30 min. The precipitates mainly formed at grain boundaries, growing from those locations into the austenite phase. The lamellar structure suggests that the eutectoid transformation reported by other authors [19–22] is occurring. After a 1 h treatment, optical microscope micrographs confirm that the transformation is significant, and it is slightly detected by XRD. Regarding the accuracy of the XRD test under experimental conditions, it can be stated that k-carbide precipitation is lower than 0.5% for a heat treatment shorter than 1 h. As the holding time rises up to 2 h or more, the austenite peak decreases and k-carbide peak increases. After two hours, the progress of the transformation is high, while in Figure 6 the SEM analysis confirms that small austenite islands remain in the microstructure, in agreement with the corresponding diffractogram. After 5 h at 750 °C, the transformation is mostly completed, the peak corresponding to austenite has completely disappeared from the diffractogram, and SEM confirms that only residual grains of austenite remain in the material.

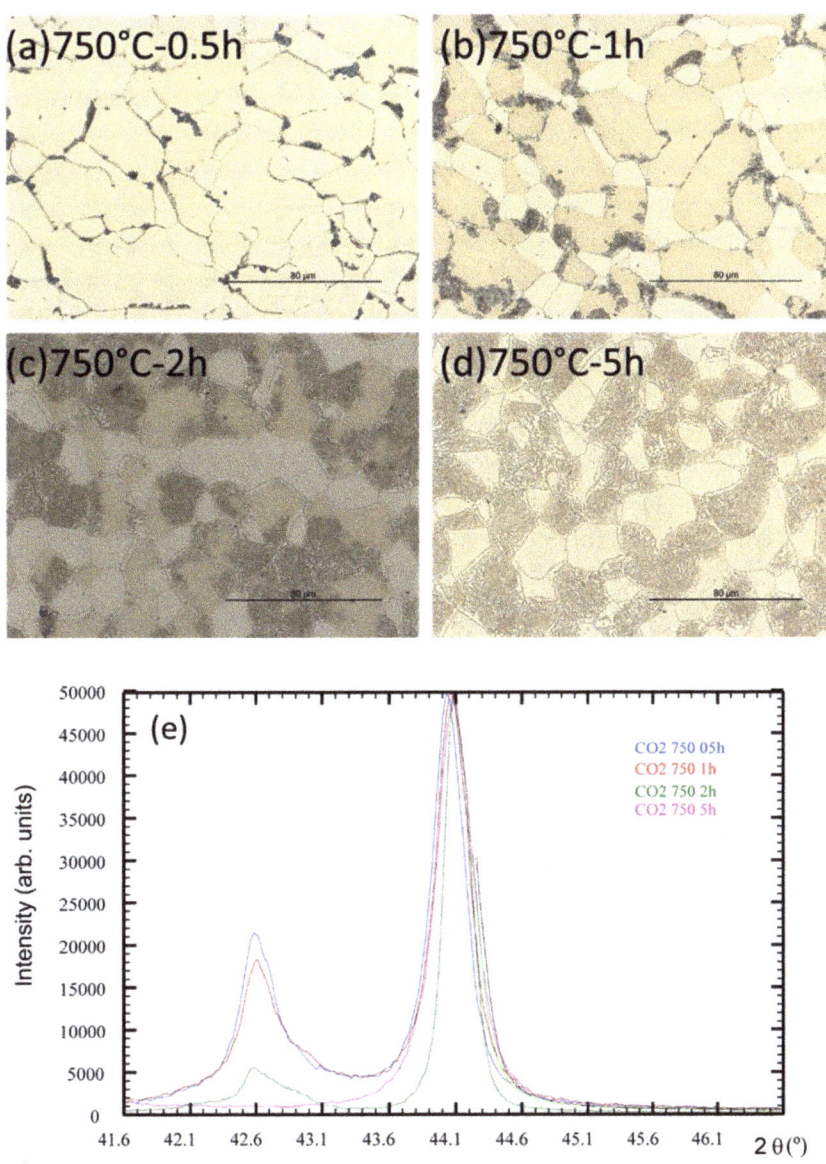

Figure 5. Optical microscope micrographs of the alloy after heat treatments performed at 750 °C for (**a**) 0.5 h, (**b**) 1 h, (**c**) 2 h and (**d**) 5 h. (**e**) Their corresponding diffractograms.

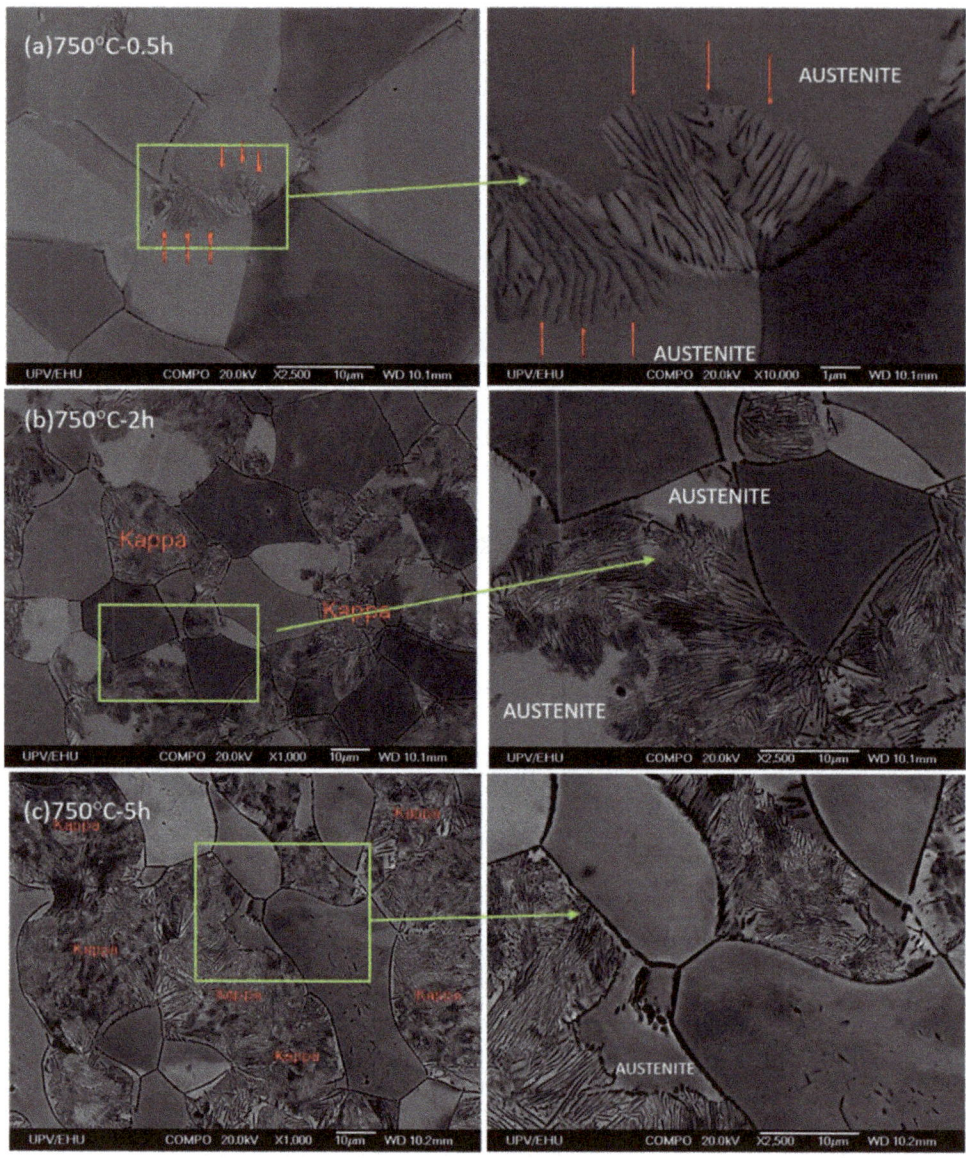

Figure 6. SEM micrographs of the alloy after heat treatments performed at 750 °C for (**a**) 0.5 h, (**b**) 2 h and (**c**) 5 h.

In order to study the influence of temperature in the first stages of k-carbide precipitation, additional treatments were performed at 700 °C and 650 °C with a holding time of 2 h. Temperatures were selected inside the range predicted by Thermocalc Software for the existence of k-carbides. Figure 7 shows the diffractograms and optical microscope images for the material after these treatments. In Figure 8's SEM images, tiny precipitates at the austenite-ferrite grain boundaries can be seen growing into the austenite phase at 650 °C. These precipitates are slightly larger at 700 °C and remain undetectable in XRD.

Taking into account the obtained results, a longer holding time of 5 h was considered for those temperatures, that is to say for 650 °C and 700 °C. All diffractograms corresponding to this holding time, shown in Figure 9, present a k-carbide peak at 40.5° and a progressive reduction of the austenite phase peak. The progress of the eutectoid transformation is depicted in the optical microscope images, where a reduction of the amount of this phase progresses as the temperature rises from 650 °C to 700 °C and 750 °C.

Table 1 summarizes whether k-carbides were identified by the analytic techniques applied for each isothermal treatment (under the testing conditions, the XRD limit detection is about 0.5% vol). Moreover, the most remarkable findings obtained from this study are:

- The precipitation of k-carbides begins at the grain boundaries, mostly austenite-ferrite, and progresses into austenite grains.
- K-carbides form from austenite decomposition into ferrite and k-carbides, resulting in a recognizable eutectoid lamellar structure. In these microstructures, primary and eutectoid ferrite are clearly distinguishable.
- At 750 °C, the precipitation of k-carbides begins after 30 min and ends at about 5 h, when approximately all the austenite has transformed into ferrite and k-carbide.
- For the same holding time, as the temperature decreases, the k-carbide precipitation kinetic slows down.

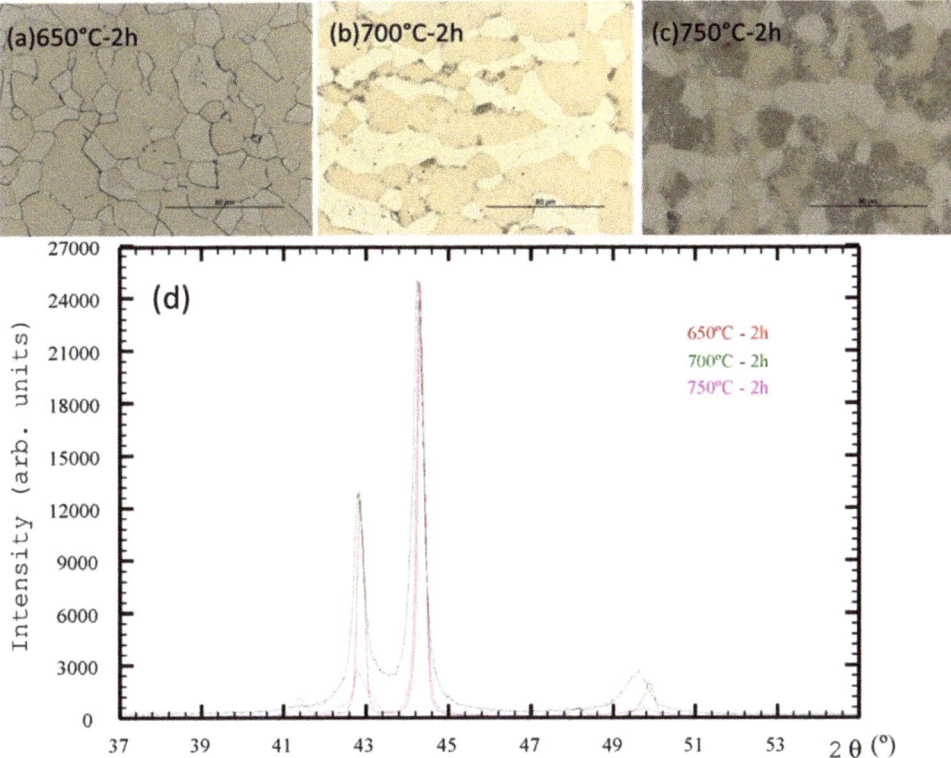

Figure 7. Optical microscope micrographs of the alloy after heat treatments performed for 2 h at (**a**) 650 °C, (**b**) 700 °C and (**c**) 750 °C. In (**d**), their corresponding diffractograms are shown.

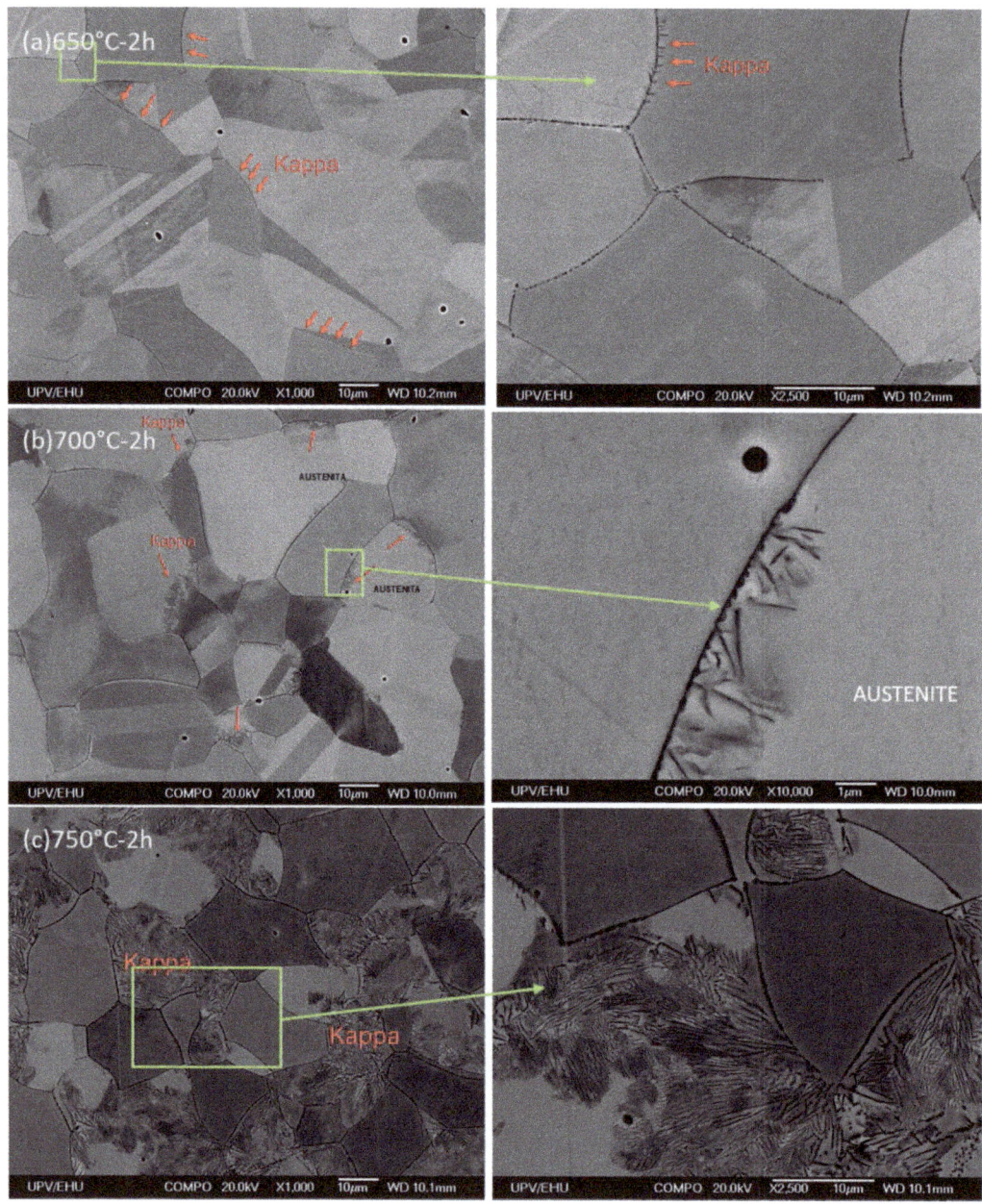

Figure 8. SEM micrographs of the alloy after 2 h of heat treatment at (**a**) 650 °C, (**b**) 700 °C and (**c**) 750 °C.

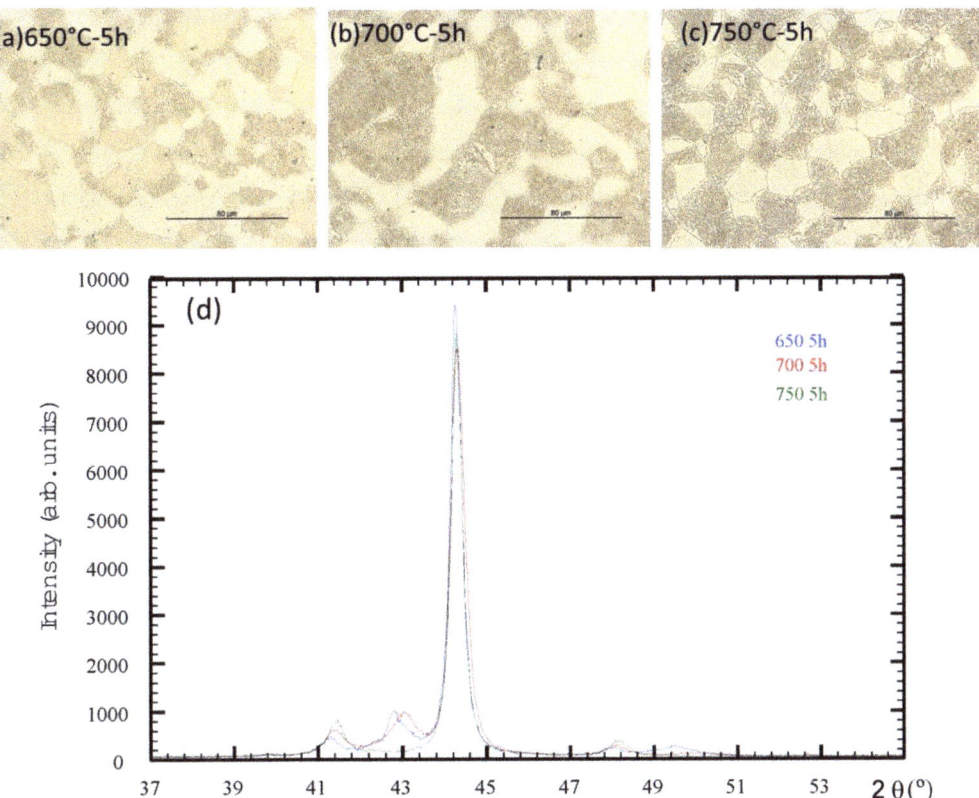

Figure 9. Optical micrographs of the alloy after 5 h of heat treatment at (**a**) 650 °C, (**b**) 700 °C and (**c**) 750 °C. In (**d**), their corresponding diffractograms are shown.

Table 1. Applied isothermal heat treatments and results from the detection of Kappa carbides using XRD, Optical Microscope and SEM.

QT	HEAT TREATMENT		SAMPLE CODE	Kappa XRD	Kappa (OM, SEM)
	Temperature (°C)	Time (h)			
1150 °C	750 °C	0.5	750 °C-0.5 h	NO	YES
		1	750 °C-1 h	YES	YES
		2	750 °C-2 h	YES	YES
		5	750 °C-5 h	YES	YES
	700 °C	2	700 °C-2 h	NO	YES
		5	700 °C-5 h	YES	YES
	650 °C	2	650 °C-2 h	NO	YES
		5	650 °C-5 h	YES	YES

3.3. Study of the Influence of K-Carbides on the Mechanichal Properties

Regarding the microstructures obtained from the performed heat treatments, the following ones were selected to produce samples for mechanical testing:

- Ferrite and austenite formed during water quenching from 1150 °C.
- Ferrite and austenite with an initial precipitation of k-carbides formed at 750 °C for 1 h.
- Ferrite and austenite with a partial transformation of austenite into lamellar ferrite plus k-carbide, formed at 750 °C for 2 h and at 650 and 700 °C for 5 h.
- Ferrite and lamellar ferrite plus k-carbide, formed from the complete transformation of austenite performed at 750 °C for 5 h.

Minicylindrical tensile specimens of samples treated in the mentioned conditions, with a calibrated length of 30 mm, were machined and tested following the UNE-EN ISO 6892-1B:2010 standard. Figure 10 shows the tensile curves obtained for each treatment, and Table 2 summarizes the values of the mechanical properties obtained.

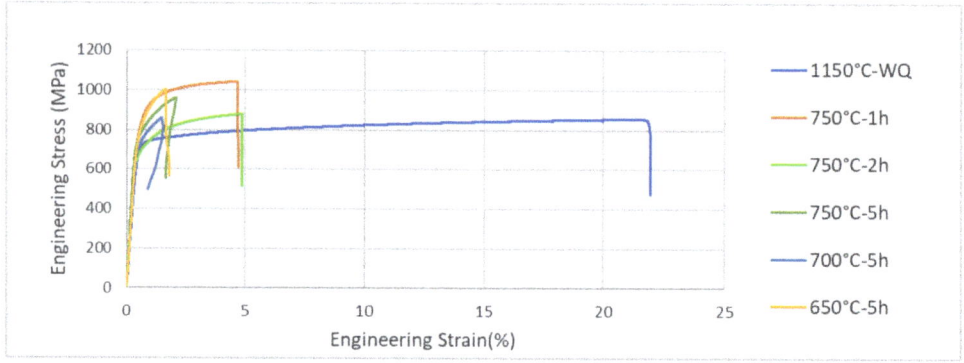

Figure 10. Tensile curves obtained for each heat treatment.

Table 2. Mechanical properties obtained for the different treatments.

	HEAT TREATMENT			MECHANICAL PROPERTIES			
QT	ISOTHERMAL		SAMPLE CODE	YS (MPa)	UTS (MPa)	El (%) A30	TSxEl (GPa)
	T(°C)	Time(h)					
1150 °C	QUENCHING		1150 °C -WQ	720	855	22.1	18.9
1150 °C	750 °C	1	750 °C-1 h	838	1041	4.7	4.9
1150 °C		2	750 °C-2 h	709	878	4.88	4.28
1150 °C		5	750 °C-5 h	777	959	2.05	1.96
1150 °C	700 °C	5	700 °C-5 h	762	943	3.5	3.3
1150 °C	650 °C	5	650 °C-5 h	803	1001	1.5	1.5

From the mechanical test results, it can be concluded that the quenched sample has a greater elongation than those submitted to isothermal heat treatment and analyzed in this work. It also has the best Resistance x Elongation product value. Applying the isothermal treatments defined in this work and based on the results obtained from the ThermoCalc simulations and dilatometry tests, k-carbide precipitation is promoted and the presence of k-carbides noticeably increases the resistance of the material when compared to the quenching treatment. However, the precipitation of coarse k-carbides in the grain boundaries leads to a very poor ductility: the higher the degree of transformation, the lower the deformation to fracture. It has been reported [3] that k-carbides contribute to the enhancement of elongation when their size, shape and distribution are optimally controlled during hot working and heat treatments. From our results, it was observed that when k-carbides grew over a nanoscale size and in the grain boundaries, their contribution to mechanical properties maintained high levels of resistance, but the elongation to fracture decreased dramatically when compared to the quenched material.

Finally, to investigate the influence of the k-carbides and final microstructures on the fracture modes, two of the tested samples were selected in order to analyze the type of fracture: On the one hand, the sample that provided the greatest elongation and the best balance of properties (1150 °C-WQ) and, on the other hand, the sample that provided the greatest resistance but a very low elongation (750 °C-1 h). Figure 11 shows the fractographies obtained by SEM of the two tested samples that were selected. In the sample that was quenched from 1150 °C, some areas of ductile breakage at the grain limit were observed, while in the sample subjected to the isothermal treatment, a crack propagating along the grain boundaries was observed, showing an intergranular fracture mode, which could be attributed to the precipitation of brittle k-carbides close to the grain limits [23]. These precipitates deteriorate the cohesion of the grain boundaries, leading to an intergranular fracture [24] and resulting in an extremely low ductility.

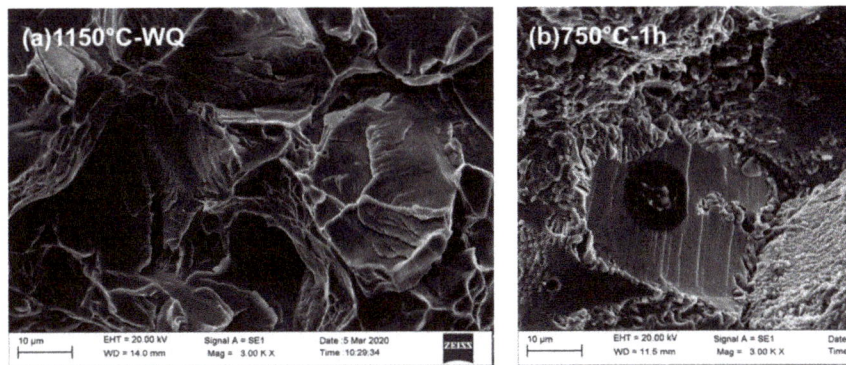

Figure 11. Fractography images obtained after mechanical tests of the samples that were (**a**) water-quenched from 1150 °C and (**b**) treated at 750 °C for one hour.

4. Conclusions

In this study, it has been observed that for austenitic duplex steels, such as the 0.66C11.4Mn9.9Al composition that was analyzed in this investigation, heat treatments that promote the precipitation of coarse kappa carbides at the grain limit can apparently increase the strength of the material when compared to a quenching treatment but cause a very detrimental effect on its ductility. For a good combination of resistance and ductility, more heat treatments must be analyzed. These heat treatments can be: isothermal heat treatments at lower temperatures or with shorter times, in order to control the size, morphology and distribution of the kappa carbide precipitates; or different temperature quenching/normalization treatments that can improve the ductility or resistance of steel while avoiding the precipitation of k-carbide.

The great potential of these low-density steels to forge components justifies their study, and further investigations must be carried out.

Author Contributions: I.K., T.G. (Teresa Gutierrez) and T.G. (Teresa Guraya) wrote the paper. All the authors conceived and designed the experiments; R.E. performed the dilatometric tests; P.J. performed the optical and electron microscope experiments. All authors discussed and reviewed the paper. All authors have read and agreed to the published version of the manuscript.

Funding: This research has been carried out with a financial grant of the Basque Government under its ELKARTEK Research Program, KK-2016/00029-ABADE project and KK-2018/00016-COFADEN project.

Data Availability Statement: The data are not publicly available.

Acknowledgments: The authors also would like to thank the technical and human support provided by SGIker of UPV/EHU.

Conflicts of Interest: The authors declare no conflict of interest.

References

1. Kim, H.; Suh, D.-W.; Kim, N.J. Fe-Al-Mn-C lightweight structural alloys: A review on the microstructures and mechanical properties. *Sci. Technol. Adv. Mater.* **2013**, *14*, 1–11. [CrossRef] [PubMed]
2. Chen, S.; Rana, R.; Haldar, A.; Ray, R.K. Current state of Fe-Mn-Al-C low density steels. *Prog. Mater. Sci.* **2017**, *89*, 345–391. [CrossRef]
3. Frommeyer, G.; Brüx, U. Microstructures and mechanical properties of high strength Fe-Mn-Al-C lightweight triplex steels. *Steel Res. Int.* **2006**, *77*, 627–633. [CrossRef]
4. Rana, R.; Liu, C.; Ray, R.K. Evolution of microstructure and mechanical properties during thermomechanical processing of a low-density multiphase steel for automotive application. *Acta Mater.* **2014**, *75*, 227–245. [CrossRef]
5. Haa, M.C.; Koob, J.-M.; Leeb, J.-K.; Hwanga, S.W.; Parka, K.-T. Tensile deformation of a low density Fe–27Mn–12Al–0.8C duplex steel in association with ordered phases at ambient temperature. *Mater. Sci. Eng. A* **2013**, *586*, 276–283. [CrossRef]
6. Cheng, P.; Li, X.; Yi, H. The k-Carbides in Low-Density Fe-Mn-Al-C Steels: A Review on Their Structure, Precipitation and Deformation Mechanism. *Metals* **2020**, *10*, 1021. [CrossRef]
7. Chin, K.; Lee, H.; Kwak, J.; Kang, J.; Lee, B. Thermodynamic calculation on the stability of (Fe,Mn)3AlC carbide in high aluminum steels. *J. Alloys Compd.* **2010**, *505*, 217–223. [CrossRef]
8. Jiménez, J.A.; Frommeyer, G. The ternary iron aluminum carbides. *J. Alloys Compd.* **2011**, *509*, 2729–2733. [CrossRef]
9. Lu, W.J.; Zhang, X.F.; Qin, R.S. Structure and properties of k-carbides in duplex lightweight steels. *Ironmak. Steelmak.* **2015**, *42*, 626–631. [CrossRef]
10. Grässel, O.; Frommeyer, G. Effect of martensitic phase transformation and deformation twinning on mechanical properties of Fe–Mn–Si–Al steels. *Mater. Sci. Technol.* **1998**, *14*, 1213–1217.
11. Frommeyer, G.; Jiménez, J.A. Structural superplasticity at higher strain rates of hypereutectoid Fe-5.5Al-1Sn-1Cr-1.3C steel. *Metall. Mater. Trans. A* **2005**, *36*, 295–300. [CrossRef]
12. Kim, K.-H.; Lee, J.-S.; Lee, D.-L. Effect of silicon on the spheroidization of cementite in hypereutectoid high carbon chromium bearing steels. *Met. Mater. Int.* **2010**, *16*, 871–876. [CrossRef]
13. Raabe, D.; Springer, H.; Gutierrez-Urrutia, I.; Roters, F.; Bausch, M.; Seol, J.-B.; Koyama, M.; Choi, P.-P.; Tsuzaki, K. Alloy Design, Combinatorial Synthesis, and Microstructure–Property Relations for Low-Density Fe-Mn-Al-C Austenitic Steels. *JOM* **2014**, *66*, 1845–1856. [CrossRef]
14. Seol, J.B.; Jung, J.E.; Jang, Y.W.; Park, C.G. Influence of carbon content on the microstructure, martensitic transformation and mechanical properties in austenite/ε-martensite dual-phase Fe–Mn–C steels. *Acta Mater.* **2013**, *61*, 558–578. [CrossRef]
15. Gutierrez-Urrutia, I.; Raabe, D. Influence of Al content and precipitation state on the mechanical behaviour of austenitic high-Mn low-density steels. *Scr. Mater.* **2013**, *68*, 343–347. [CrossRef]
16. Choi, K.; Seo, C.-H.; Lee, H.; Kim, S.K.; Kwak, J.H.; Chin, K.G.; Park, K.-T.; Kim, N.J. Effect of aging on the microstructure and deformation behaviour of austenite base lightweight Fe–28Mn–9Al–0.8C steel. *Scr. Mater.* **2010**, *63*, 1028–1031. [CrossRef]
17. Kaltzakorta, I.; Gutierrez, T.; Elvira, R.; Guraya, T.; Jimbert, P. Low density steel for forging. *Mater. Sci. Forum* **2018**, *941*, 287–291. [CrossRef]
18. ASTM E562-08. *Standard Test Method for Determining Volume Fraction by Systematic Manual Point Count*; ASTM International: West Conshohocken, PA, USA, 2008. [CrossRef]
19. Cheng, W.C. Phase transformations of an Fe-0.85C-17.9Mn-7.1Al austenitic steel after quenching and annealing. *JOM* **2014**, *66*, 1809–1820. [CrossRef]
20. Song, H.; Yoo, J.; Kim, S.H.; Sohn, S.S.; Koo, M.; Kim, N.J.; Lee, S. Novel ultra-high-strength Cu-containing medium-Mn duplex lightweight steels. *Acta Mater.* **2017**, *135*, 215–225. [CrossRef]
21. Jeong, J.; Lee, C.-Y.; Park, I.-J.; Lee, Y.-K. Isothermal precipitation behaviour of κ-carbide in the Fe–9Mn–6Al–0.15C lightweight steel with a multiphase microstructure. *J. Alloys Compd.* **2013**, *574*, 299–304. [CrossRef]
22. Zhao, C.; Song, R.; Zhang, L.; Yang, F.; Kang, T. Effect of annealing temperature on the microstructure and tensile properties of Fe-10Mn-10Al-0.7C low-density steel. *Mater. Des.* **2016**, *91*, 348–360. [CrossRef]
23. Meng, S.; Sugiyama, S.; Yanagimoto, J. Effects of heat treatment on microstructure and mechanical properties of Cr-V-Mo steel processed by recrystallization and partial melting method. *J. Mater. Process. Technol.* **2014**, *214*, 87–96. [CrossRef]
24. Raabe, D.; Herbig, M.; Sandlöbes, S.; Li, Y.; Tytko, D.; Kuzmina, M.; Ponge, D.; Choi, P.-P. Grain boundary segregation engineering in metallic alloys: A pathway to the design of interfaces. *Curr. Opin. Solid State Mater. Sci.* **2014**, *18*, 253–261. [CrossRef]

Article

Correlation between Microstructures and Ductility Parameters of Cold Drawn Hyper-Eutectoid Steel Wires with Different Drawing Strains and Post-Deformation Annealing Conditions

Jin Young Jung [1,2], Kang Suk An [1], Pyeong Yeol Park [2] and Won Jong Nam [1,*]

[1] School of Advanced Materials Engineering, Kookmin University, Seoul 02707, Korea; jyjung@kiswire.com (J.Y.J.); ags0826@kookmin.ac.kr (K.S.A.)
[2] KISWIRE R&D Center, Pohang 37872, Gyeongbuk, Korea; pypark@kiswire.com
* Correspondence: wjnam@kookmin.ac.kr; Tel.: +82-2-910-4649

Abstract: The relationship between microstructures and ductility parameters, including reduction of area, elongation to failure, occurrence of delamination, and number of turns to failure in torsion, in hypereutectoid pearlitic steel wires was investigated. The transformed steel wires at 620 °C were successively dry-drawn to drawing strains from 0.40 to 2.38. To examine the effects of hot-dip galvanizing conditions, post-deformation annealing was performed on cold drawn steel wires (ε = 0.99, 1.59, and 2.38) with a different heating time of 30–3600 s at 500 °C in a salt bath. In cold drawn wires, elongation to failure dropped due to the formation of dislocation substructures, decreased slowly due to the increase of dislocation density, and saturated with drawing strain. During annealing, elongation to failure increased due to recovery, and saturated with annealing time. The variation of elongation to failure in cold drawn and annealed steel wires would depend on the distribution of dislocations in lamellar ferrite. The orientation of lamellar cementite and the shape of cementite particles would become an effective factor controlling number of turns to failure in torsion of cold drawn and annealed steel wires. The orientation and shape of lamellar cementite would become microstructural features controlling reduction of area of cold drawn and annealed steel wires. The density of dislocations contributed to reduction of area to some extent.

Keywords: pearlitic steel wire; elongation to failure; torsion; reduction of area; annealing

1. Introduction

The increase of the deformation limit in the steel wire industry has an advantage of obtaining a high strength level and eliminating the lead patenting process. The recent improvement of strength in steel wires has achieved a high strength above 5 GPa in steel cords for automobile tires and above 2 GPa in cable wires for suspension bridge. In cold drawn steel wires, high strength has been achieved by the refinement of interlamellar spacing [1–3], the increase of drawing strain [4–6], the increase of carbon content [7–9], and the addition of alloying elements [10–12]. However, the increase of strength generally accompanies the deterioration of ductility in cold drawn steel wires. For the development of high strength steel wire, the important subject becomes how to increase the strength of steel wires without a significant loss of ductility. The manufacturing process of wire products includes bending, stranding, stretching, and coiling, etc. Thus, ductility parameters required for steel wire products are the reduction of area, elongation to failure, occurrence of delamination, and number of turns to failure in torsion. Occurrence of delamination acts as an indicator of brittle fracture, splitting longitudinally along the wire axis during torsion. On the other hand, the number of turns to failure reflects the gradual variation of torsional ductility with manufacturing conditions of steel wires [13,14]. Thus, both occurrence of delamination (DEL) and number of turns to failure (NT) could become good parameters to evaluate torsional ductility of steel wire products.

In cold drawn steel wires, the increase of drawing strain decreases elongation to failure (EL) and increases tensile strength (TS) [15,16], while the reduction of area (RA) and number of turns to failure in torsion (NT) show the sequential variation; increasing, showing the maximum peak, and decreasing continuously [17,18]. These mechanical behaviors are closely related to microstructural evolution occurred during wire drawing. The main features of microstructural evolution during wire drawing are a progressive realignment of lamellar cementite along the drawing axis [19–21], a reduction of interlamellar spacing, a thinning of lamellar cementite and ferrite, a fracture, and dissolution of lamellar cementite [22–24]. Recently, the decomposition of cementite and the subsequent distribution of carbon atoms in heavily drawn wires were directly observed using a high resolution transmission electron microscopy (HRTEM) and an atom probe tomography (APT). While the relationship between microstructural evolution during wire drawing and mechanical properties of TS and RA in cold drawn steel wires has been widely investigated [24–26], the main microstructural features controlling other ductility parameters, which are related to the total amount of deformation in the manufacturing process, remain unclear until now.

Meanwhile, the manufacturing process of wire products includes post-deformation annealing, such as bluing for wires for tire bead, hot-dip galvanizing for suspension bridge cables, or stress-relief annealing for coil springs. Since the degree of age hardening or age softening depends on drawing and annealing conditions, post-deformation annealing conditions alter strength, and ductility of steel wire products through microstructural evolution during annealing. Age hardening proceeds with the break-up and decomposition of lamellar cementite, the diffusion of carbon atoms to dislocations, and the pinning of dislocations by carbon atoms in lamellar ferrite [27,28]. The mechanism of age softening is described as the break-up and spheroidization of lamellar cementite, the growth of cementite particles, the reprecipitation of cementite particles, and the recovery or recrystallization of ferrite [28–30]. Thus, post-deformation annealing conditions and drawing conditions contribute significantly to strength and ductility of steel wire products. Several works about the evaluations of TS, DEL, and NT with different annealing conditions were reported [31–33]. However, those works merely focused on microstructures and mechanical properties with the variations of annealing conditions. Thus, to improve ductility of cold drawn and annealed steel wires, the variations of ductility with drawing strain and annealing conditions need to be explained in terms of microstructural features.

In the present work, the effects of microstructural features on ductility with the variations of drawing strain and annealing time in hypereutectoid steel wires cold drawn and annealed at 500 °C. Particularly, the present work focused on the relationship between microstructural evolution during wire drawing and subsequent annealing processes and ductility parameters of EL, RA, DEL, and NT in hypereutectoid pearlitic steel wires.

2. Materials and Methods

The chemical composition of steel wire rods used in this work is Fe-0.92%C-1.3%Si-0.5%Mn-0.3%Cr (wt %). Hot rolled and Stelmor-cooled wire rods with a diameter of 13 mm were cold-drawn to 4.9 mm in diameter. To investigate the effects of the drawing strain and annealing time on tensile strength and ductility, the 4.9 mm-diameter wires were austenitized at 900 °C for 3 min, followed by quenching in a salt bath at 620 °C for 3 min. The transformed wires were pickled and successively dry-drawn to wires with various final diameters from 4.02 (ε = 0.40) to 1.49 mm (ε = 2.38) as shown in Table 1. Drawing speed was set to 3 m/min to avoid dynamic strain aging, and the average reduction per pass was about 18%. Additionally, to investigate the effects of hot-dip galvanizing conditions on mechanical properties, post-deformation annealing was performed on cold drawn steel wires (ε = 0.99, 1.59, and 2.38) with different heating time of 30 s, 2 min, 20 min, and 1 h at a temperature of 500 °C in a salt bath.

Table 1. Diameter and drawing strain of a cold drawn steel wire for each drawing step.

Drawing step	0	1	2	3	4	5	6	7	8	9	10	11	12
Diameter (mm)	4.9	4.44	4.02	3.64	3.29	2.98	2.70	2.45	2.21	2.01	1.82	1.65	1.49
Total reduction (%)	-	17.9	32.7	44.8	54.9	63.0	69.6	75.0	79.7	83.2	86.2	88.7	90.8
Drawing strain (ε)	-	0.20	0.40	0.59	0.80	0.99	1.19	1.39	1.59	1.78	1.98	2.18	2.38

Tensile strength of steel wires was determined by tensile tests at room temperature with a constant speed of 20 mm/min. Elongation to failure was measured with an extensometer.

To examine torsional ductility, torsion tests were carried out as free-end twist tests at a rotational speed of 30 r.p.m. The torsion test machine (Hongduk Eng., Busan, Korea) (Figure 1) equipped with torque sensors. The length of cold drawn steel wires was chosen as 200 mm. Tension was applied to the specimen by using a weight of 1% of the maximum tensile load of the wire. During torsion testing, the occurrence of delamination was evaluated by assessing NT or examining the shape of fracture surfaces.

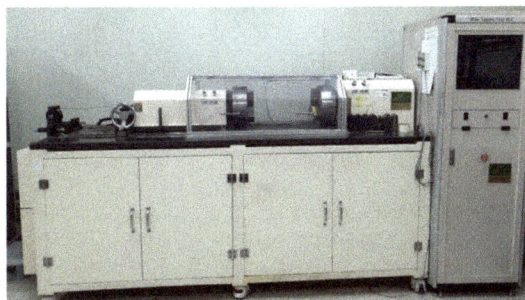

Figure 1. Torsion test equipment.

For detailed understanding of microstructural evolution during annealing, a scanning electron microscope (SEM, JEOL Ltd., Tokyo, Japan), and a transmission electron microscope (TEM, JEOL Ltd., Tokyo, Japan) were used. Thin foils parallel to the longitudinal cross section of the wire were prepared by utilizing a jet polishing technique in a mixture of 90% acetic acid and 10% perchloric acid at room temperature. In order to measure the block size of a steel wire transformed at 620 °C, the longitudinal section of a steel wire was grinded with silicon carbide papers and electropolished. The inverse pole figure (IPF) color maps of steel wires were examined by using electron backscatter diffraction (EBSD, JEOL Ltd., Tokyo, Japan). Microstructure and IPF map of an isothermal transformed steel wire are shown in Figure 2. The average lamellar spacing of pearlite was 72.6 nm, the average size of block, consisting of colonies with the misorientation less than 15°, was 11.6 µm.

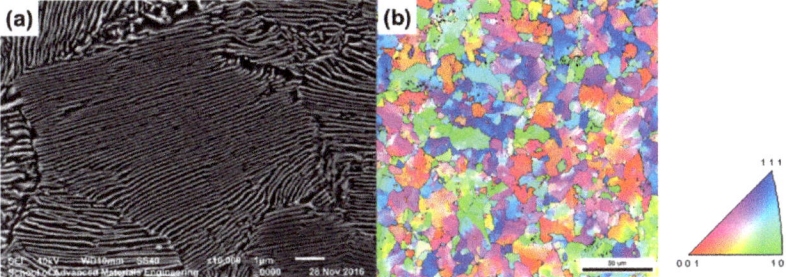

Figure 2. SEM micrograph (a) and inverse pole figure (IPF) map (b) of a hypereutectoid steel wire transformed at 620 °C.

3. Results

3.1. Tensile Strength

In Figure 3a, a drawing strain of 2.38 increased TS continuously from 1335 (as-transformed) to 2287 MPa, due to work hardening of lamellar ferrite and solid solution hardening of carbon atoms dissolved in lamellar ferrite, which was caused by the occurrence of dynamic strain aging. According to Zhang et al. [34], TS of cold drawn pearlitic steel wires can be expressed with the following strengthening mechanisms: (1) boundary strengthening, which represents the refinement of interlamellar spacing; (2) strain hardening of ferrite; and (3) solid solution hardening due to the increased amount of carbon atoms dissolved in lamellar ferrite. The process of dynamic strain aging consists of the fracture and decomposition of lamellar cementite, and the interaction of dissolved carbon atoms with dislocations in lamellar ferrite during deformation [23,25,27]. The rapid increase of TS above a drawing strain of 1.5 in Figure 3a was attributed to the increased solid solution hardening of carbon atoms at high strain and the efficient work hardening of completely realigned lamellae along the wire axis, compared with the less work hardening during the progressive realignment of randomly oriented lamellae at low strain. The increase of TS during wire drawing was attributed to strain hardening of lamellar ferrite and solid solution hardening of dissolved carbon atoms by dynamic strain aging.

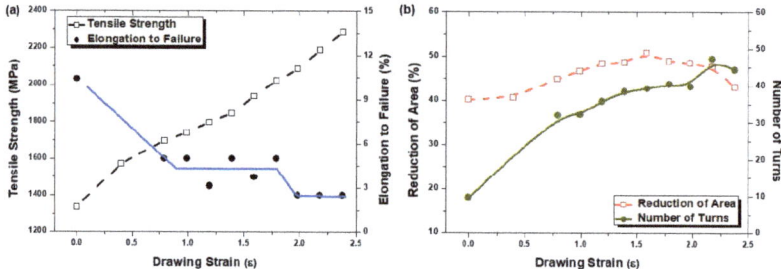

Figure 3. The variations of (**a**) tensile strength (TS) and elongation to failure (EL), and (**b**) reduction of area (RA) and numbers of turns to failure in torsion (NT) with a drawing strain in cold drawn steel wires.

TS of cold drawn and annealed steel wires reflects the degree of age hardening or age softening during post-deformation annealing. While age hardening and age softening occur simultaneously at low annealing temperature or for a short annealing time, the increases of annealing temperature and time make age softening become the main operating mechanism during annealing. Figure 4 shows the effects of drawing strain and annealing time on TS of steel wires cold drawn and annealed at 500 °C. The different drawing strains in cold drawn steel wires provided two different typed TS curves. For steel wires received strains of 0.99 and 1.59, the variation of TS with annealing time showed the sequential behaviors; increasing, showing the peak, and decreasing continuously after the peak. For cold drawn steel wires with $\varepsilon = 1.59$, TS increased from 1938 to 1951 MPa (30 s annealing), and then decreased to 1632 MPa (1 h annealing). The increment of TS for a short annealing time of 30 s was due to age hardening. Age hardening during annealing consists of two stages; the segregation of carbon atoms to dislocations in ferrite and a partial decomposition of lamellar cementite [28]. The partial dissolution of lamellar cementite, the diffusion of carbon atoms into lamellar ferrite and the interaction with dislocations would be responsible to the increase of TS during annealing. Thus, the increase of TS for 30 s annealing (13 MPa) would be attributed to the occurrence of age hardening, which corresponds to the solid solution hardening. Takahashi et al. [9] showed that the occurrence of the maximum TS during aging would be closely related to the decomposition of lamellar cementite in cold drawn hypereutectoid steel wires, using the field ion micrograph (FIM). However, as annealing time increased, lamellar cementite started to spheroidize. Additionally, the occurrence of recovery in ferrite led to the decrease of TS. Accordingly, TS decreased continuously with the increased contribution of age softening with annealing time.

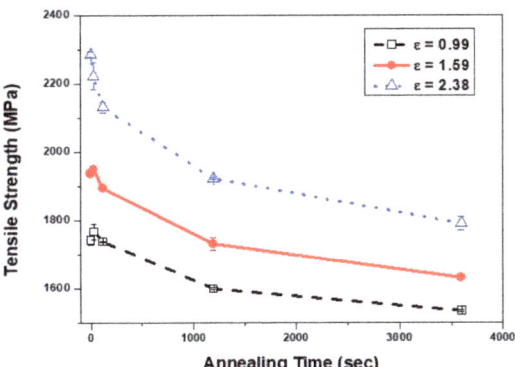

Figure 4. The variation of TS with annealing time at 500 °C in cold drawn steel wires received different drawing strains.

Meanwhile, for steel wires with a drawing strain above 2.38, TS decreased continuously from 2287 (as-drawn) to 1791 MPa (1 h annealing). The continuous decrease of TS with annealing time in Figure 3 indicates that age hardening does not occur during annealing for steel wires cold drawn to $\varepsilon = 2.38$. Li et al. [25] have reported that the carbon concentration in lamellar ferrite saturates at the critical strain and above the critical strain the decomposition of lamellar cementite does not proceed in their work. Thus, it is expected that there exists a limit of the carbon content dissolved in lamellar ferrite, which corresponds to the specific drawing strain or TS that no more age hardening occurs during post-deformation annealing. During annealing, the severe fracture of lamellar cementite and the high concentration of carbon atoms in lamellar ferrite in cold drawn steel wires would encourage the age softening process; spheroidizing and reprecipitation of cementite particles, and recovery of ferrite, rather than age hardening even for a short annealing time.

It is interesting to note that the degree of age softening during annealing in Figure 4 was significantly influenced by the amount of drawing strains in cold drawn steel wires. A steel wire with $\varepsilon = 2.38$ showed the larger decrease of TS, 496 MPa, than that with $\varepsilon = 0.99$, 208 MPa, for 1 h annealing. The more damage and fracture in lamellar cementite and the higher dislocation density in cold drawn steel wires received the higher drawing strain that would accelerate the age softening process during annealing. Thus, annealing of a steel wire that received a high drawing strain showed the larger decrease of TS than that with a low drawing strain.

During post-deformation annealing, TS of steel wires with a high drawing strain decreased continuously with annealing time, since age softening, such as the fracture and spheroidization of lamellar cementite, the growth of cementite particles, and the recovery of lamellar ferrite, became the major process to control TS. Meanwhile, steel wires with low drawing strain showed the sequential variation of TS; increasing, showing the peak, and decreasing continuously with annealing time. The increment of TS for a short annealing time was due to the occurrence of age hardening, which was related to the decomposition of lamellar cementite and the interaction of dislocations with carbon atoms dissolved in lamellar ferrite.

3.2. Elongation to Failure

In Figure 3a, EL (10.4%) of a transformed steel wire dropped to 5.6% at a drawing strain of 0.79, decreased gradually with drawing strain, and reached a saturation value of 2.5% for a drawing strain above 2.0 [17,35]. It is interesting to note that the rapid drop of EL occurs even at a low strain of 0.79. According to Zhang et al. [34,36], the dislocation structure in lamellar ferrite transformed from threading dislocations, tangles at low and medium strains, to dislocation cells at high strain. In pearlitic steels, the bundle of dislocations ending at ferrite/cementite boundaries act like dislocation cell boundaries during plastic deformation, due to the existence of lamellar cementite. The presence of dislocation tangles in Figure 5 indicated that the formation of dislocation tangles, i.e., the formation of dislocation substructures, would become one of the main causes for the EL drop at low strain. He et al [37] reported that the formation of dislocation tangles was observed at a drawing strain of 0.8 in cold drawn steel wires.

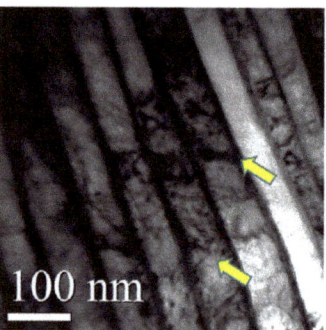

Figure 5. TEM micrographs, showing dislocation tangles in a steel wire drawn with $\varepsilon = 1.13$.

In Figure 6, post-deformation annealing raised EL continuously with annealing time. EL of cold drawn steel wires increased rapidly by 2.5% ($\varepsilon = 0.99$) and 3.1% ($\varepsilon = 1.59$) for a short annealing time of 30 s, although TS increased due to age hardening at the same condition. This implies that the effect of age softening, such as the recovery of ferrite and the spheroidization of lamellar cementite, would have the stronger effect on EL than the age hardening effect. Meanwhile, the increasing rate of EL slowed down or EL saturated at the 8–9% level for an annealing time above 2 min.

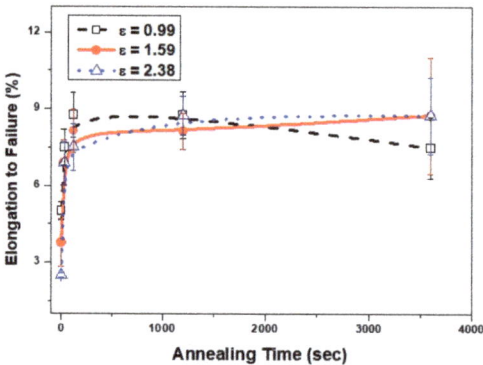

Figure 6. The plot of EL as a function of annealing time at 500 °C in cold drawn steel wires received different drawing strains.

Annealing at a high temperature of 500 °C for 2 min enhanced the spheroidization of lamellar cementite (Figure 7a) and the formation of subgrains (Figure 7b) in cold drawn steel wires with $\varepsilon = 2.38$. Therefore, it is expected that the initial increasing rate of EL at 2 min annealing in Figure 6 would be related to the spheroidization of lamellar cementite or the formation of subgrains as a recovery process. Li et al. [33] reported that the formation of subgrains from dislocation cells was observed during annealing at a low temperature of 250 °C in heavily drawn steel wires.

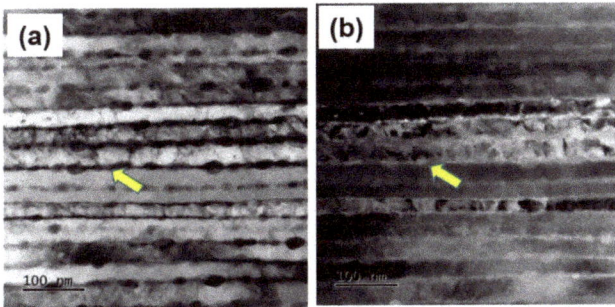

Figure 7. TEM micrographs, showing (**a**) the presence of subgrains in a steel wire drawn with ε = 2.38 and annealed at 500 °C for 2 min, and (**b**) the dark field image of (**b**).

3.3. Torsional Ductility

Torsional ductility is evaluated as the occurrence of delamination (DEL) or number of turns to failure (NT) in cold drawn and annealed steel wires. NT reflects the gradual variation of torsional ductility with manufacturing conditions of steel wires, while DEL is the qualitative indication of bad torsional ductility [14,34]. The variation of NT with a drawing strain in Figure 3b showed the similar trend to that of RA; increasing, showing the peak, and decreasing continuously, although the peak strain of 2.18 was different from that of 1.59 in RA.

In Figure 8, NT showed the steady decrease with annealing time, except for the drop of NT region. DEL, which is closely related to the interaction of dissolved carbon atoms with dislocations in lamellar ferrite, would cause the drop of NT in torsion. In Figure 8, steel wires with the higher drawing strain showed the larger range of the DEL region and the larger decrease of NT during annealing. The higher drawing strain induced more damage in lamellar cementite during wire drawing. The damaged cementite easily dissolved in ferrite during annealing. The increased amount of dissolved carbon atoms interacted with dislocations more frequently. Thus, the higher drawing strain resulted in the larger drop of NT and more frequent DEL during annealing [11,25]. Additionally, the disappearance of DEL with 2 min annealing at 500 °C in steel wires received a strain of 1.59 indicates that the increase of annealing time would decrease the amount of dissolved carbon atoms, due to the reprecipitation of cementite particles [9]. NT of steel wires drawn with ε = 0.99 decreased continuously from 30.6 turns (as-drawn) to 14.5 turns (1 h annealing) without the occurrence of DEL. Except for DEL, NT of cold drawn steel wires decreased continuously with annealing time at 500 °C.

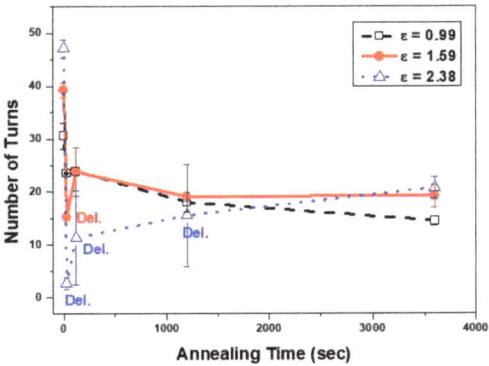

Figure 8. The plot of NT as a function of annealing time at 500 °C in hypereutectoid steel wires received different drawing strains.

3.4. Reduction of Area

Figure 3b shows the plot of reduction of area (RA) in cold drawn wires as a function of drawing strain. RA of a transformed steel wire (40.2%) increased up to 50.9% ($\varepsilon = 1.59$) and decreased gradually to 41.2% ($\varepsilon = 2.38$) with increasing strain. The initial increase of RA was due to the realignment of randomly oriented lamellar cementite along the wire axis. Thus, the maximum peak of RA at $\varepsilon = 1.59$ corresponded to the completion of realignment of lamellar cementite. The continuous decrease of RA after the peak resulted from work hardening of lamellar ferrite and thinning and/or fragmenting lamellar cementite at high strain.

Generally, annealing of cold deformed steels improves ductility of EL, RA, and torsional ductility at the expense of strength in cold deformed steels, since the occurrence of recovery or recrystallization during annealing encourages the softening of matrix. However, the variations of RA with annealing time in Figure 9 showed the different behavior from that of EL. For a short annealing time of 30 s, RA of cold drawn wires ($\varepsilon = 1.59$), 48%, significantly dropped to 34%. Meanwhile, the increase of annealing time from 30 s to 1 h resulted in only the 3.5% increase of RA.

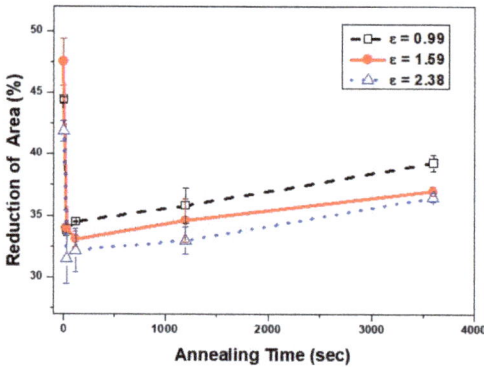

Figure 9. The variation of RA with an annealing time at 500 °C in hypereutectoid steel wires received different drawing strains.

4. Discussion

Ductility parameters of cold drawn steel wires must reflect the variations of microstructural features during wire drawing. Microstructural evolution during wire drawing includes the reduced thickness of lamellar ferrite and lamellar cementite, the work hardened lamellar ferrite, the progressive reorientation of lamellae along the wire axis, the fracture of lamellar cementite, and the increased content of carbon atoms dissolved in lamellar ferrite. Depending on the post-deformation annealing conditions, the occurrence of age hardening or age softening would alter ductility parameters of steel wires. Microstructural evolution of age hardening consists of the break-up and decomposition of lamellar cementite, the diffusion of carbon atoms to dislocations, and dislocation pinning by carbon atoms. Thus, the occurrence of age hardening is closely related to the amount of carbon atoms dissolved in lamellar ferrite. Meanwhile, age softening occurs through the break-up and spheroidization of lamellar cementite, the reprecipitation of cementite particles, and the occurrence of recovery or recrystallization of lamellar ferrite.

4.1. Elongation to Failure

To find the relationship between EL and microstructural evolution during wire drawing, the effects of microstructural features on EL were examined. The drop of EL at a drawing strain of 0.79 in Figure 3a could not be explained with the reductions of ferrite thickness and cementite thickness, the reorientation of lamellar cementite, and the gradual increase of dislocation density during wire drawing. Meanwhile, the increase of dislocation density accompanies with the formation of dislocation substructures, such as dislocation tangles and dislocation cells. The presence of dislocation tangles in Figure 5 indicated that the formation of dislocation substructures would become one of the main causes for the EL drop at low strain.

The variation of EL in Figure 6 shows which microstructural feature would have an effect on EL of cold drawn and annealed steel wires. The rapid increase of EL for 30 s annealing would be related to the formation of subgrains as the recovery process or the spheroidization of lamellar cementite (Figure 7). Additionally, the presence of yield plateaus in true stress–true strain curves of cold drawn and annealed steel wires (Figure 10) confirmed the occurrence of recovery in lamellar ferrite during annealing at 500 °C for 2 min, while the continuous yielding occurred in cold drawn steel wires.

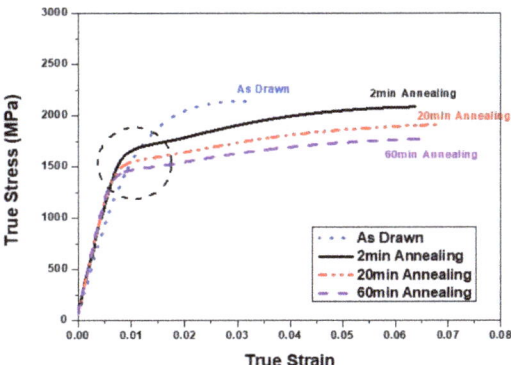

Figure 10. True stress–true strain curves of steel wires drawn with a strain of 1.59 and annealed at 500 °C for 2 min to 1 h.

In Figure 11, an annealing at 500 °C for 1 h induced the fragmentation of cementite particles in most area of cold drawn steel wires with $\varepsilon = 2.38$, while the fragmentation of cementite particles occurred in less than half area of cold drawn steel wires with $\varepsilon = 0.99$. This meant that the severer damage in lamellar cementite accelerated the spheroidization process during annealing. In spite of microstructural difference in Figure 11, both steel wires showed the similar EL as 8% and 9% for an annealing time of 1 h, respectively. This implies that the increased mean free path due to the spheroidization and growth of cementite particles would not have a significant influence on EL.

Additionally, the saturation of EL at 8–9% level for annealing time above 2 min in Figure 6 also indicates that the further progress of recovery during annealing would not change EL. According to Xiang et al, [38], when annealing temperature increased from 250 to 325 °C, EL of cold drawn steel wires ($\varepsilon = 2.34$) increased rapidly, accompanying the slight variation of dislocation density. This indicated that EL did not depend on the dislocation density significantly.

From the above, it was found that EL of cold drawn wires dropped due to the formation of dislocation tangles or cells, decreased slowly due to work hardening, and saturated with the drawing strain. During post-deformation annealing EL initially increased rapidly due to the transformation from dislocation cells to subgrains as a recovery process, and saturated at 8–9% with annealing time.

Thus, the variation of EL in cold drawn and annealed steel wires depends on the formation of dislocation substructures in lamellar ferrite. The formation of dislocation substructures showed the stronger effect on the variation of EL than dislocation density in cold drawn and/or annealed steel wires.

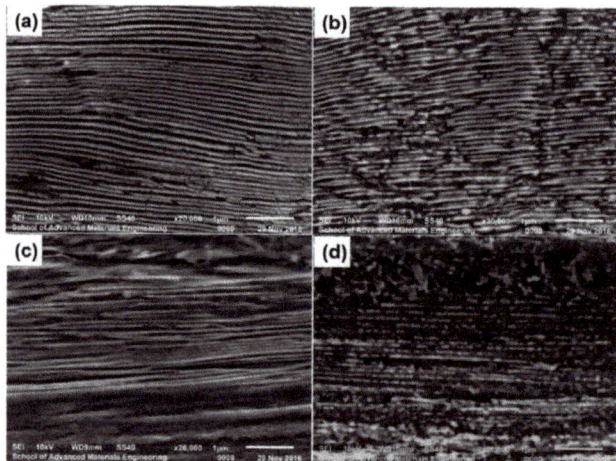

Figure 11. SEM micrographs of steel wires received strains of (**a**) 0.99 and (**b**) annealed at 500 °C for 1 h, (**c**) received strains of 2.38, (**d**) and annealed at 500 °C for 1 h.

4.2. Torsional Ductility

The similar sequential behaviors between NT and RA with drawing strain (Figure 3b); increasing, showing the maximum peak, and decreasing continuously, suggests that microstructural features affecting RA would possibly control NT of cold drawn steel wires. The initial increase of RA comes from the realignment of randomly oriented lamellar cementite, and the maximum peak of RA at $\varepsilon = 1.59$ corresponds to the completion of the realignment of lamellar cementite along the wire axis. The subsequent decrease of RA, after the maximum peak, results from work hardening of lamellar ferrite and thinning and/or fragmenting lamellar cementite at high strain. From the above, it is obvious that the increase of RA is attributed to the realignment of lamellar cementite at low strain, and the decrease of RA after the peak comes from the break-up of lamellar cementite and work hardened lamellar ferrite.

Meanwhile, the peak strain of 2.18 in NT was different from that of 1.59 in RA. This difference of peak strains between RA and NT implies that microstructural features controlling RA of cold drawn steel wires would not have the similar degree of influence on NT in cold drawn steel wires. From Figure 3b, it is expected that the realignment of lamellar cementite increases NT with drawing strain, and the decrease of NT would depend on the fracturing of lamellar cementite, and dynamic strain aging in cold drawn steel wires [39].

Figure 8 shows the steady decrease of NT with annealing time, except for the NT drop due to DEL. This steady decrease of NT with annealing time would be related to the age softening process, including the decrease of dissolved carbon atoms due to the reprecipitation of cementite particles, the recovery, and the spheroidization and growth of cementite particles. The larger NT for cold drawn steel wires than NT of post-deformation annealed wires in Figure 8, indicates that NT was not significantly affected by dislocation density in ferrite. The reduction of carbon atoms dissolved in ferrite with annealing time also did not change NT significantly. However, it is obvious that DEL, which causes the NT drop during torsion, depends on the amount of carbon atoms in ferrite. Meanwhile, the fracture of lamellar cementite at high drawing strain and the spheroidization and growth of lamellar cementite during annealing decreased NT. This meant that the destruction of lamellar structure, i.e., the increase of mean free path for mobile dislocations, would decrease NT, since lamellar cementite, aligned parallel to the wire axis, blocked the movement of mobile dislocations during torsion [39].

Figure 12 shows a schematic diagram of different mean free path (MPF) due to annealing of cold drawn steel wires. The fracture of lamellar cementite at high drawing strain and the spheroidization and growth of cementite particles increased MPF for the dislocation movement during torsion. Thus, the fragmented cementite particles reduced NT in cold drawn and annealed steel wires.

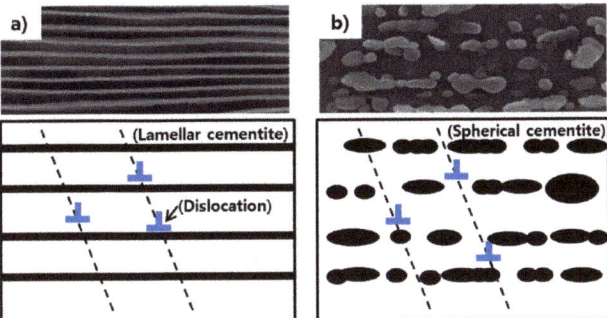

Figure 12. A schematic diagram of mean free path (MPF) difference due to annealing of cold drawn steel wires; (**a**) as-drawn with a strain of 1.95 and (**b**) annealed at 500 °C for 1 h [39].

Thus, the orientation of lamellar cementite and the shape of cementite particles would become a major factor in controlling NT of cold drawn and annealed steel wires. Additionally, the amount of dissolved carbon atoms would have an influence through DEL, which drops NT rapidly during torsion.

4.3. Reduction of Area

It is well known that RA of cold drawn steel wires increases with the realignment of randomly oriented lamellar cementite parallel to the wire axis, and decreases with the fracture of lamellar cementite and work hardening of lamellar ferrite. However, the variations of RA with annealing time in Figure 9 showed the different behavior from RA with drawing strain (Figure 3b). When the annealing is carried out at low temperature with a short annealing time, both age hardening and age softening occur in cold drawn steel wires. With increasing annealing temperature or time, age softening dominantly controls mechanical properties of cold drawn steel wires. The significant drop of RA was observed in steel wires received 30 s annealing at 500 °C, and then RA increased slowly with increasing annealing time. For 30 s annealing, all cold drawn steel wires showed the significant RA drop, whether cold drawn and annealed steel wires showed DEL during torsion or not in Figure 8. This meant that the occurrence of age hardening (the amount of carbon atoms dissolved in ferrite) did not have any influence on the drop of RA. It is interesting to note that RA of cold drawn steel wires is larger than RA of post-deformation annealed wires in Figure 9. This implies that the occurrence of recovery would not contribute significantly to the increase of RA in cold drawn and annealed steel wires.

Considering the dependence of RA on the orientation of lamellar cementite to the wire axis and the fragmentation of lamellar cementite in cold drawn steel wires, the shape change (the degree of spheroidization) of lamellar cementite could become one of the main causes for the rapid RA drop. Figure 12 shows the gradual spheroidization with annealing time in steel wires cold drawn and annealed at 500 °C. A short annealing time of 2 min was enough to start the fracture of lamellar cementite in steel wires with $\varepsilon = 2.38$ (Figures 7 and 13). For steel wires annealed for 20 min (Figure 13c), most lamellar cementite fractured into cementite particles. As annealing time increased to 1 h, the spheroidization of cementite particles was almost completed and the size of cementite particles varied from 20 to 200 nm, although cementite particles kept the lamellar typed array (Figure 13d) [29].

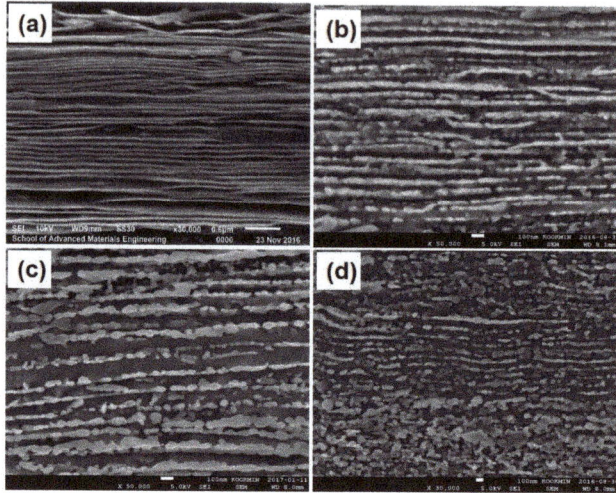

Figure 13. SEM micrographs, showing the microstructural evolution, (**a**) as-drawn with ε = 2.38, and annealed at 500 °C for annealing time of (**b**) 2 min, (**c**) 20 min, and (**d**) 1 h.

Meanwhile, the slow increasing rate of RA after the drop in Figure 9 would come from the softening behavior of recovery. Among cold drawn wires, the RA of a steel wire with ε = 1.59 showed the largest RA due to the completion of realignment of lamellar cementite. However, after annealing longer than 2 min RA of a steel wire with ε = 0.99 was larger than that with ε = 2.38. This means that the larger drawing strain would accelerate the spheroidization and growth of cementite particles during annealing and result in the lower RA in steel wires.

Therefore, the orientation and shape of lamellar cementite would become microstructural features controlling RA of cold drawn and annealed steel wires. The occurrence of recovery during annealing caused the slight increase of RA with annealing time.

Mechanical properties of steel wires reflect the variations of microstructural features during wire drawing and annealing. Figure 14 shows a schematic diagram, describing the evolution of pearlitic microstructure during wire drawing and subsequent annealing.

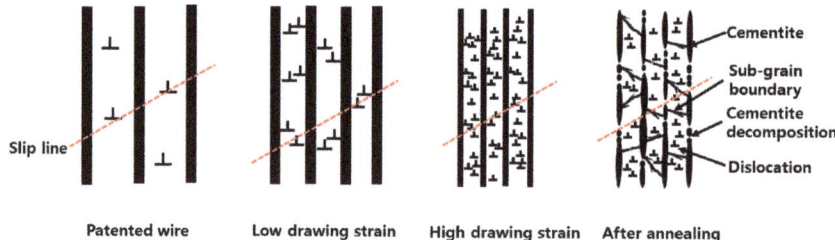

Figure 14. A schematic diagram, describing the evolution of pearlitic microstructure during wire drawing and subsequent annealing.

5. Conclusions

The effects of microstructural features on ductility with the variations of drawing strain and annealing time in hypereutectoid steel wires were investigated. Especially, the relationship between microstructural evolution during wire drawing and subsequent annealing and ductility parameters of reduction of area, elongation to failure, occurrence of delamination, and number of turns to failure in torsion in hypereutectoid pearlitic steel wires was discussed.

(1) The increase of tensile strength (TS) during wire drawing was attributed to work hardening of lamellar ferrite and solid solution hardening of dissolved carbon atoms by dynamic strain aging. During post-deformation annealing, TS of steel wires with high drawing strain decreased continuously with annealing time, since age softening became the major process to control TS. Meanwhile, steel wires with low drawing strain showed the sequential variation of TS; increasing, showing the peak, and decreasing with annealing time. The increment of TS for a short annealing time was due to the occurrence of age hardening.

(2) The variation of elongation to failure (EL) in cold drawn and/or annealed steel wires depends on the formation of dislocation substructures in lamellar ferrite. The formation of dislocation tangles or cells would become one of main causes for the EL drop at low strain. The rapid increase of EL during annealing came from the transformation from dislocation cells to subgrains as a recovery process. The formation of dislocation substructures showed the stronger effect on the variation of EL than dislocation density in cold drawn and/or annealed steel wires.

(3) Occurrence of delamination (DEL) caused a significant drop of number of turns to failure in torsion (NT). Since DEL depends on the amount of carbon atoms dissolved in ferrite, steel wires with the higher drawing strains showed the larger range of DEL region and the larger decrease of NT during annealing. The higher drawing strain induced the more damage in lamellar cementite and resulted in the increased amount of dissolved carbon atoms in ferrite during annealing. Thus, the higher drawing strain resulted in the larger drop of NT and more frequent DEL during annealing.

(4) Number of turns to failure (NT) increased with the realignment of lamellar cementite and decreased with fracturing of lamellar cementite and dynamic strain aging in cold drawn steel wires. During post-deformation annealing, NT of steel wires decreased with annealing time, except for DEL. The orientation of lamellar cementite and the shape of cementite particles would become an effective factor controlling NT of cold drawn and annealed steel wires.

(5) With drawing strain, reduction of area (RA) increased due to the realignment along the wire axis, showed the peak, and decreased gradually due to work hardening of ferrite and fragmenting lamellar cementite at high strain. During post-deformation annealing, RA of cold drawn wires significantly dropped and increased slowly with annealing time. The orientation and shape of lamellar cementite would become a dominant microstructural feature in controlling RA of cold drawn and annealed steel wires. The occurrence of recovery during annealing also contributed to RA to some extent.

Author Contributions: Conceptualization, writing draft, J.Y.J.; investigation, preparation, data interpretation, K.S.A.; supervision, writing review, P.Y.P.; study design, writing review and editing, funding acquisition, W.J.N. All authors have read and agreed to the published version of the manuscript.

Funding: This research was supported by KISWIRE.

Data Availability Statement: Not applicable.

Conflicts of Interest: The authors declare no conflict of interest.

References

1. Borchers, C.; Kirchheim, R. Cold-drawn Pearlitic Steel Wires. *Prog. Mater. Sci.* **2016**, *82*, 405–444. [CrossRef]
2. Nam, W.J.; Bae, C.M.; Oh, S.J.; Kwon, S.J. Effect of Interlamellar Spacing on Cementite Dissolution during Wire Drawing of Pearlitic Steel Wires. *Scr. Mater.* **2000**, *42*, 457–463. [CrossRef]
3. Toribio, J.; Ovejero, E. Effect of Cumulative Cold Drawing on the Pearlite Intrlamellar Spacing in Eutectoid Steel. *Scr. Mater.* **1998**, *39*, 323–328. [CrossRef]
4. Li, Y.J.; Choi, P.; Goto, S.; Borchers, C.; Raabe, D.; Kirchheim, R. Evolution of Strength and Microstructure during Annealing of Heavily Cold-drawn 6.3 GPa Hypereutectoid Pearlitic Steel Wire. *Acta Mater.* **2012**, *60*, 4005–4016. [CrossRef]
5. Lu, X. Correlation between Microstructural Evolution and Mechanical Properties of 2000 MPa Cold-Drawn Pearlitic Steel Wires during Galvanizing Simulated Annealing. *Metals* **2019**, *9*, 326. [CrossRef]

6. Tashiro, H.; Tarui, T. State of the Art for High Tensile Strength Steel Cord. *Nippon Steel Tech. Rep.* **2003**, *88*, 87–91.
7. Tarui, T.; Maruyama, N.; Takahashi, J.; Nishida, S.; Tashiro, H. Microstructure Control and Strengthening of High-carbon Steel Wires. *Nippon Steel Tech. Rep.* **2005**, *91*, 56–61.
8. Sakamoto, M.; Teshima, T.; Nakamura, K. Wire Rod for High Tensile Strength Steel Cords. *Nippon Steel Tech. Rep.* **2019**, *122*, 129–136.
9. Takahashi, J.; Kosaka, M.; Kawakami, K.; Tarui, T. Change in Carbon State by Low-temperature aging in Heavily Drawn Pearlitic Steel Wires. *Acta Mater.* **2012**, *60*, 387–395. [CrossRef]
10. Song, H.R.; Kang, E.G.; Bae, C.M.; Lee, C.Y.; Lee, D.L.; Nam, W.J. The Effect of a Cr Addition and Transformation Temperature on the Mechanical Properties of Cold Drawn Hyper-Eutectoid Steel Wires. *Met. Mater. Int.* **2006**, *12*, 239–243. [CrossRef]
11. Tarui, T.; Takahashi, T.; Ohashi, S.; Uemori, R. Effect of Silicon on the Age Softening of High Carbon Steel. *Iron Steel Maker* **1994**, *21*, 25–30.
12. Joung, S.; Nam, W. Effects of Alloying Elements, Si and Cr, on Aging and Delamination Behaviors in Cold-Drawn and Subsequently Annealed Hyper-eutectoid Steel Wires. *Met. Mater. Int.* **2019**, *25*, 34–44. [CrossRef]
13. Zhao, T.Z.; Zhang, S.H.; Zhang, G.L.; Song, H.W.; Cheng, M. Hardening and Softening Mechanisms of Pearlitic Steel Wire under Torsion. *Mater. Des.* **2014**, *59*, 397–405. [CrossRef]
14. Zhou, L.; Fang, F.; Wang, L.; Hu, X.; Xie, Z.; Jiang, J. Torsion Performance of Pearlitic Steel Wires: Effects of Morphology and Crystallinity of Cementite. *Mater. Sci. Eng. A* **2019**, *743*, 425–435. [CrossRef]
15. Toribio, J.; Ayaso, F.J.; González, B.; Matos, J.C.; Vergara, D.; Lorenzo, M. Tensile Fracture behavior of Progressively-Drawn Pearlitic Steels. *Metals* **2016**, *6*, 114. [CrossRef]
16. Shiota, Y.; Tomota, Y.; Moriai, A.; Kamiyama, T. Structure and Mechanical Behavior of Heavily Drawn Pearlite and Martensite in a High Carbon Steel. *Met. Mater. Int.* **2005**, *11*, 371–376. [CrossRef]
17. Zelin, M. Microstructure Evolution in Pearlitic Steels during Wire Drawing. *Acta Mater.* **2002**, *50*, 4431–4447. [CrossRef]
18. Joung, S.W.; Kang, U.G.; Hong, S.P.; Kim, Y.W.; Nam, W.J. Aging Behavior and Delamination in Cold Drawn and Post-deformation Annealed Hyper-eutectoid Steel Wires. *Mater. Sci. Eng. A* **2013**, *586*, 171–177. [CrossRef]
19. Nam, W.J.; Bae, C.M. Void Initiation and Microstructural Changes during Wire Drawing of Pearlitic Steels. *Mater. Sci. Eng. A* **1995**, *203*, 278–285. [CrossRef]
20. Toribio, J.; Ovejero, E. Microstructure Orientation in a Pearlitic Steel Subjected to Progressive Plastic Deformation. *J. Mater. Sci. Lett.* **1998**, *17*, 1045–1048. [CrossRef]
21. Toribio, J.; Ovejero, E. Microstructure Evolution in a Pearlitic Steel Subjected to Progressive Plastic Deformation. *Mater. Sci. Eng. A* **1997**, *234–236*, 579–582. [CrossRef]
22. Read, H.G.; Reynolds, W.T., Jr.; Hono, K.; Tarui, T. AFPIM and TEM Studies of Drawn Pearlitic Wire. *Scr. Mater.* **1997**, *37*, 1221–1230. [CrossRef]
23. Takahashi, J.; Tarui, T.; Kawakami, K. Three-dimensional Atom Probe Analysis of Heavily Drawn Steel Wires by Probing Perpendicular to the Pearlitic Lamellae. *Ultramicroscopy* **2009**, *109*, 193–199. [CrossRef] [PubMed]
24. Li, Y.J.; Choi, P.; Borchers, C.; Chen, Y.Z.; Goto, S.; Raabe, D.; Kirchheim, R. Atom probe tomography characterization of heavily cold drawn pearlitic steel wire. *Ultramicroscopy* **2011**, *111*, 628–632. [CrossRef] [PubMed]
25. Li, Y.J.; Choi, P.; Borchers, C.; Westerkamp, S.; Goto, S.; Raabe, D.; Kirchheim, R. Atomic-scale Mechanisms of Deformation-induced Cementite Decomposition in Pearlite. *Acta Mater.* **2011**, *59*, 3965–3977. [CrossRef]
26. Maruyama, N.; Tarui, T.; Tashiro, H. Atom Probe Study on the Ductility of Drawn Pearlitic Steels. *Scr. Mater.* **2002**, *46*, 599–603. [CrossRef]
27. Hammerle, J.R.; de Almeida, L.H.; Monteiro, S.N. Lower Temperatures Mechanism of Strain Aging in Carbon Steels for Drawn Wires. *Scr. Mater.* **2004**, *50*, 1289–1292. [CrossRef]
28. Watte, P.; Humbeeck, J.V.; Aernoudt, E.; Lefever, I. Strain ageing in heavily drawn eutectoid steel wires. *Scr. Mater.* **1996**, *34*, 89–95. [CrossRef]
29. Hinchliffe, C.E.; Smith, G.D.W. Strain Aging of Pearlitic Steel Wire during Post-drawing Heat Treatments. *Mater. Sci. Technol.* **2001**, *17*, 148–154. [CrossRef]
30. Lee, J.W.; Lee, J.C.; Lee, Y.S.; Park, K.T.; Nam, W.J. Effects of Post-deformation Annealing Conditions on the Behavior of Lamellar Cementite and the Occurrence of Delamination in Cold-drawn Steel Wires. *J. Mater. Process. Technol.* **2009**, *209*, 5300–5304. [CrossRef]
31. Fang, F.; Hu, J.; Chen, S.H.; Xie, Z.H.; Jiang, J.Q. Revealing Microstructural and Mechanical Characteristics of Cold-drawn Pearlitic Steel Wires undergoing Simulated Galvanization Treatment. *Mater. Sci. Eng. A* **2012**, *547*, 51–54. [CrossRef]
32. Zhou, L.C.; Fang, F.; Wang, L.; Chen, H.Q.; Xie, Z.H.; Jiang, J.Q. Torsion Delamination and Recrystallized Cementite of Heavy Drawing Pearlitic Wires after Low Temperature Annealing. *Mater. Sci. Eng. A* **2018**, *713*, 52–60. [CrossRef]
33. Xiang, L.; Liang, L.W.; Wang, Y.J.; Chen, Y.; Wang, H.Y.; Dai, L.H. One-step Annealing Optimizes Strength-ductility Tradeoff in Pearlitic Steel Wires. *Mater. Sci. Eng. A* **2019**, *757*, 1–13. [CrossRef]
34. Zhang, X.; Godfrey, A.; Huang, X.; Hansen, N.; Liu, Q. Microstructure and strengthening Mechanisms in cold-drawn pearlitic steel wire. *Acta Mater.* **2011**, *59*, 3422–3430. [CrossRef]

35. Gondo, S.; Tanemura, R.; Suzuki, S.; Kajino, S.; Asakawa, M.; Takemoto, K.; Tashima, K. Microstructures and Mechanical Properties of Fiber Textures forming Mesoscale Structure of Drawn Fine High Carbon Steel Wire. *Mater. Sci. Eng. A* **2019**, *747*, 255–264. [CrossRef]
36. Zhang, X.; Hansen, N.; Godfrey, A.; Huang, X. Dislocation-based Plasticity and Strengthening Mechanisms in sub-20 nm Lamellar Structures in Pearlitic Steel Wire. *Acta Mater.* **2016**, *114*, 176–183. [CrossRef]
37. He, Y.; Xiang, S.; Shi, W.; Liu, J.; Ji, X.; Yu, W. Effect of Microstructure Evolution on Anisotropic Fracture Behavior of Cold Drawing Pearlitic Steels. *Mater. Sci. Eng. A* **2017**, *683*, 153–163. [CrossRef]
38. Li, Y.J.; Kosta, A.; Choi, P.; Goto, S.; Ponge, D.; Kirchheim, R.; Raabe, D. Mechanisms of Subgrain Coarsening and Its Effect on the Mechanical Properties of Carbon-supersaturated Nanocrystalline Hypereutectoid Steel. *Acta Mater.* **2015**, *84*, 110–123. [CrossRef]
39. Jung, J.Y.; An, K.S.; Park, P.Y.; Nam, W.J. Effects of Wire Drawing and Annealing Conditions on Torsional Ductility of Cold Drawn and Annealed Hyper-Eutectoid Steel Wires. *Metals* **2020**, *10*, 1043. [CrossRef]

Article

Effect of Intercritical Annealing and Austempering on the Microstructure and Mechanical Properties of a High Silicon Manganese Steel

Mattia Franceschi [1,*], Luca Pezzato [1], Claudio Gennari [1], Alberto Fabrizi [2], Marina Polyakova [3], Dmitry Konstantinov [3], Katya Brunelli [1] and Manuele Dabalà [1]

1. Department of Industrial Engineering, University of Padua, Via Marzolo 9, 35131 Padova, Italy; luca.pezzato@unipd.it (L.P.); claudio.gennari@unipd.it (C.G.); katya.brunelli@unipd.it (K.B.); manuele.dabala@unipd.it (M.D.)
2. Department of Management and Engineering, University of Padova, Stradella San Nicola 3, 36100 Vicenza, Italy; alberto.fabrizi@unipd.it
3. Department of Mechanical Engineering and Metallurgical Technologies, Nosov Magnitogorsk State Technical University, pr. Lenina, 38, 455000 Magnitogorsk, Russia; m.polyakova-64@mail.ru (M.P.); const_dimon@mail.ru (D.K.)
* Correspondence: mattia.franceschi@studenti.unipd.it; Tel.: +39-0498-275-503

Received: 29 September 2020; Accepted: 27 October 2020; Published: 29 October 2020

Abstract: High Silicon Austempered steels (AHSS) are materials of great interest due to their excellent combination of high strength, ductility, toughness, and limited costs. These steel grades are characterized by a microstructure consisting of ferrite and bainite, accompanied by a high quantity retained austenite (RA). The aim of this study is to analyze the effect of an innovative heat treatment, consisting of intercritical annealing at 780 °C and austempering at 400 °C for 30 minutes, on the microstructure and mechanical properties of a novel high silicon steel (0.43C-3.26Si-2.72Mn wt.%). The microstructure was characterized by optical and electron microscopy and XRD analysis. Hardness and tensile tests were performed. A multiphase ferritic-martensitic microstructure was obtained. A hardness of 426 HV and a tensile strength of 1650 MPa were measured, with an elongation of 4.5%. The results were compared with those ones obtained with annealing and Q&T treatments.

Keywords: austempering; high silicon steel; retained austenite; mechanical properties

1. Introduction

Nowadays, one of the most important objectives of steel producers and researchers in metallurgy is to bring to the market materials with improved properties and performance, high strength-to-weight ratios, and low costs. To achieve these objectives, high-alloyed steels, such aluminum and titanium alloys are not preferred due to the high cost of their raw material; on the contrary, the use of high strength steels (AHSS) is strongly recommended. Great attention is devoted to the study of multiphase steels due to their interesting mechanical characteristics, and both TRIP (TRansformation Induced Plasticity) steels, which belong to the second generation of Advanced High Strength Steels and High Silicon Austempered Steels are considered as more promising.

High silicon austempered steels are attractive grades for their particular combination of mechanical properties and ausfererritic microstructure, which is a mixture of ferrite and high carbon enriched austenite [1,2]. This particular microstructure leads to better mechanical performance, in terms of strength, hardness, and impact toughness in comparison with austempered ductile irons [1].

A silicon weight percentage higher than 1% prevents cementite formation [1–5] and favors austenite carbon enrichment during austempering [6], permitting its retention at room temperature.

According to Zhu et al. [7] silicon also retards static and dynamic recrystallization and retained austenite grain growth, enhancing its stability down to room temperature. Matsumura et al. [4,8] demonstrated that Si slows down bainitic transformation kinetic, widening its stability field.

Furthermore, significant silicon addition raises the critical transformation temperatures (Ac1 and Ac3) in the Fe-C carbon phase diagram [4]. Also. manganese promotes austenite retention, being an austenite stabilizer [4]. Moreover, it prevents pearlite formation, reduces the martensite start temperature (Ms), and delays bainitic transformation. A beneficial effect in term of mechanical properties, due to solid solution strengthening, can also be observed.

Hence, the key to obtain an ausferritic microstructure is to achieve retention and stabilization of a significant amount of retained austenite at room temperature. Retained austenite stability depends on several factors: its carbon content, shape, size, crystallographic orientation, temperature, and the state of the applied stresses [9–12].

A particular heat treatment called austempering, consisting of several steps, should be performed to enable austenite carbon enrichment and consequently its stabilization. During the first step, which is the so called Intercritical Annealing (IA), the material is heated in the dual phase region ($\alpha + \gamma$) between Ac1 and Ac3 [4]. Within this regime a mixture of ferrite and austenite is formed, with different weight fractions depending on the temperature. Austenite dissolves most of the carbon due to ferrite's low carbon solubility but this is not enough to retain austenite at room temperature and further carbon partitioning is required [4]. During the second step, known as the cooling step, additional austenite carbon enrichment takes place due to austenite transformation into ferrite. Most of the carbon is rejected by the newly formed ferrite diffusing into austenite, increasing its stability. The third important part of this treatment is an isothermal soak at temperature for the so called isothermal bainitic transformation (IBT). Part of the remaining volume fraction of austenite transforms in free carbide bainite with low carbon concentration and all the carbon is distributed in the remaining austenite. Once bainitic transformation is completed, the material should be cooled to room temperature retaining a consistent volume fraction of the austenite.

Several works have been devoted to the research of effects of austenitization at intercritical temperatures. According to Yi et al. [13] during IA, austenite nucleates along ferritic grain boundaries (GBs), where carbon is rapidly supplied to austenitic islands. Its growth proceeds rapidly at the beginning and then slows when GBs are saturated with austenite islands. Kang and coworkers demonstrated that the final volume fraction of retained austenite is enhanced by an increase in manganese content [14].

The temperature and holding time of IA have strong influence on the microstructure and mechanical properties of the steels. According to Samajdar et al. [4,15], an increase in annealing temperature reduces the carbon concentration in austenite but increases the martensite start temperature, leading to a decrease in the austenite's stability. This phenomenon was also confirmed by Erişir et al. [16], who also observed austenitic grain growth at higher temperatures. Furthermore, complete recrystallization could be achieved at higher IA temperature, as verified in [4,17]. Emadoddin et al [18] observed that, as the annealing temperature increases, the final volume fraction of residual austenite and its carbon content also increase.

Concerning the soaking time effect on the microstructure during IA, the long dwell time is accompanied by austenite grain growth, which reduces its stability.

Isothermal holding at IBT temperature is the most important step of the heat treatment [4]. If the soaking is performed at a temperature close and/or corresponding to the nose of the bainitic transformation, the time to complete the bainitic transformation is reduced. Moreover, the longer the holding time, higher the bainite volume fraction and the carbon partitioned in austenite, and this produces an increase in the austenite's stability.

Another relevant heat treatment consists in Quenching and Partitioning (Q&P). The treatment differs substantially from Quenching and Tempering (Q&T), which is a common treatment in industrial applications [19,20]. During Q&P, once steel is austenitized, it is quenched at a temperature between martensite start temperature (Ms) and Mf (martensite finish temperature) and then soaked at the so-called partition temperature (PT). At PT carbon diffuses from martensite to retained austenite favoring its retention at room temperature and enhancing strength and ductility [21].

Moreover martensite, formed after quenching, speeds up the kinetic of bainitic transformation and the process of austenite stabilization [22,23].

Oliveira et al [21], analyzing the behavior of a MnSi steel after austempering and Q&P, observed that in the first case a microstructure consisting in a mixture of plate and granular bainite, martensite, and retained austenite is present. After Q&P a martensite matrix with bainitic islands and austenite films/blocks is instead formed.

Novelty of the Work

In this paper, a new dual phase high silicon steel with a novel composition was investigated. The effect of intercritical annealing and austempering on microstructure and mechanical behavior was studied The samples were characterized by optical microscopy (OM, LEICA DMRE, Leica Microsystems S.r.l., Milan, Italy), scanning electron microscopy (SEM, LEICA™ Cambridge Stereoscan LEO 440, Leica Microsystems S.r.l., Milan, Italy, EBSD, AMETEK BV, Tilburg, The Netherlands), transmission electron microscopy (TEM, JEOL JEM 200CX, Jeol Ltd, Tokyo, Japan), X-ray diffraction (XRD, Bruker D8 Advance, Karlsruhe, German) techniques, and mechanical tests. The results were compared with those obtained after annealing and Q&T treatments.

2. Materials and Methods

The material used in this study was a high silicon manganese steel, produced by Magnitogorsk Nosov State Technical University of Magnitogorsk (Russia). The chemical composition is presented in Table 1.

Table 1. Chemical composition of the investigated alloy (wt.%).

C	Si	Mn	P	S	Cr	Ni	Cu	Mo	Ti	V	Al
0.43	3.26	2.72	0.010	0.0082	0.043	0.074	0.060	0.022	0.0010	0.0051	0.105

The steel was prepared melting high purity raw material in an induction furnace. Liquid metal was cast, and the produced ingots were forged, and water cooled to room temperature

2.1. Heat Treatments

Heat treatment setup was composed by a Carbolite tubular electrical furnace and Nabetherm 3000 muffle electrical furnace. Tubular furnace was used for austenitization during quenching and for tempering the samples. It was also used for the intercritical annealing step. The other furnace was employed for austempering treatment. An air-cooling system was realized for the cooling step from the biphasic region to the Isothermal Bainitic transformation temperature. The final cooling to room temperature, from the IBT temperature, was performed in water. The samples' temperatures were recorded during the heat treatments with a K-thermocouple.

Heat treatments were designed using JmatPro software. A theoretical phase diagram was built to study the biphasic austenite-ferrite region, to evaluate the volume fraction of the two microstructural constituents and to determine austenite composition as a function of temperature (Figure 1). The obtained information permitted us to choose the temperatures to perform the IA step, highlighted by the red line in Figure 1b. Theoretical CCT and TTT curves (Figure 2) were built to define required cooling rates in the transition between annealing and the second austempering phase. This operation was necessary to avoid ferrite and perlite precipitation during treating cycles. Finally, holding times in the bainitic region were set thanks to the relative start and end transformation curves.

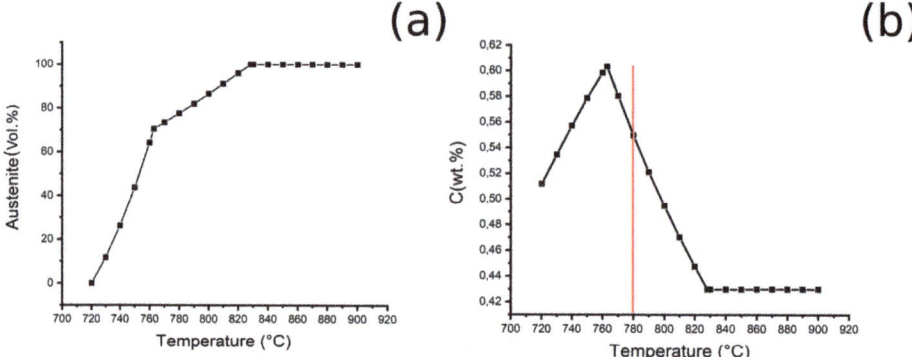

Figure 1. Volume percentage of austenite (**a**) and carbon content (**b**) as a function of temperature during austenitization. The red line in (**b**) indicates intercritical annealing condition.

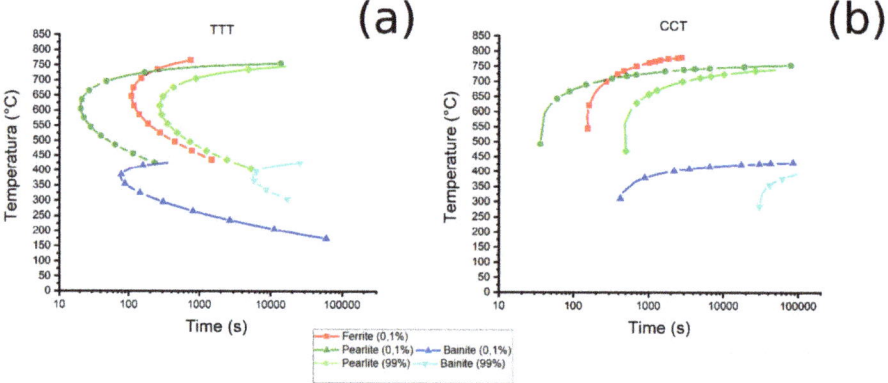

Figure 2. (**a**) TTT, (**b**) CCT curves calculated with JMat Pro software for heat treatment design.

We decided to perform the intercritical annealing with the parameters reported in Table 2.

Table 2. Intercritical annealing parameters.

A_{c1} (°C)	A_{c3} (°C)	Intercritical Annealing Temperature (°C)	Ferrite (wt.%)	Austenite (wt.%)	C_γ (wt.%)
~763	~839	~780	~23	~76	0.55

For each treatment described below, three samples were prepared.
To summarize, the subsequent heat treatments were performed:

i. *Annealing*: heating at 870 °C at 1 °C/s, 10 min holding time and furnace cooling (0.15 °C/s).
ii. *Quenching and tempering (Q&T)*: heating at 900 °C at 1 °C/s, 15 min dwell time, and water quenching (cooling rate: 40 °C/s); tempering at 600 °C for 30 min and air cooling (5 °C/s) (Figure 3a).
iii. *Intercritical Annealing and Austempering (IA&A)*: Pre-quenching treatment from 870 °C (15 min) and water cooling. Heating at 780 °C for 30 min at 0.8 °C/s, air cooling at 10 °C/s to 400 °C and holding for 30 min followed by water cooling to room temperature at 40 °C/s (Figure 3b).

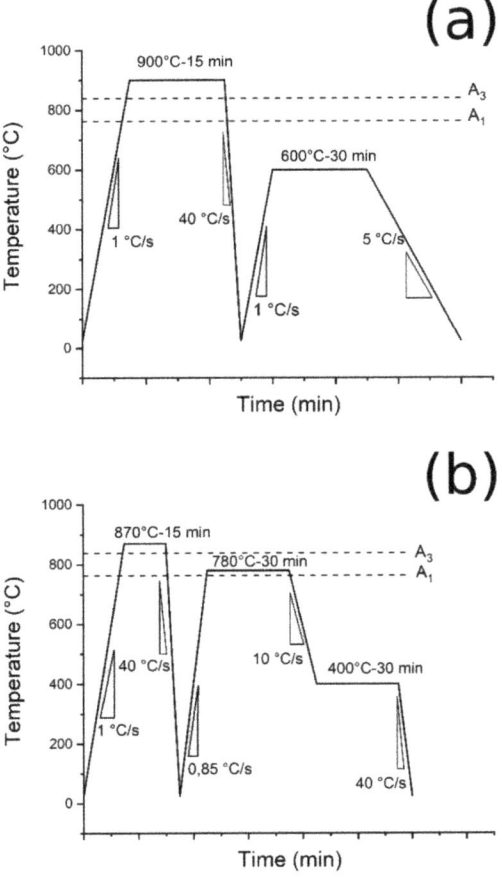

Figure 3. Heat treatment processing scheduled for the high silicon steel: (**a**) Quenching and Partitioning (Q&P), (**b**) Austempering.

The heat treatments were performed starting from a martensitic microstructure, because martensite allows for the faster recrystallization processes of ferrite during intercritical annealing and higher performance after the treatment, according to other works [4,24,25]. Furthermore, Kim et al. [4,25] demonstrated that fine microstructure, resulting from a rapid cooling, favors austenite formation and its carbon enrichment when material is re-heated.

2.2. Microstructural Study

The microstructure of the samples was analyzed along the cross section of the specimens. To perform the analysis an optical microscope, LEICA DMRE (Leica Microsystems S.r.l., Milan, Italy), and a scanning electron microscopy (SEM) LEICA™ Cambridge Stereoscan LEO 440 (Leica Microsystems S.r.l., Milan, Italy) were used. The preparation of the samples was carried out following the standard technique. The specimens were cut with SiC discs lubricated with a mixture of water and oil, mounted in phenolic resin, grinded with SiC papers (320, 500, 800, 1200 grit), polished with clothes and 6 μm and 1 μm polycrystalline diamond suspensions. EBSD analysis occurred after the samples were further polished with 200 nm and 40 nm silica colloidal particles suspension. In order to reveal the microstructure of the samples they were etched with Nital 2.

Phase identification and phase quantification were carried out through X-ray diffraction with a Bruker D8 Advance (XRD, Bruker D8 Advance, Karlsruhe, German), operating at 40 kV and 40 mA and a Cu radiation tube (Kβ radiation was filtered by mean of nickel filter on the tube side). The investigated angular range was between 40° and 105°, steps scan of 0.025° and counting time of 3 s. The obtained patterns were analyzed using High Score Plus software in order to identify the constituent phases. Volume fraction calculation of the phases was performed through Rietveld analysis on the same software.

Electron Backscattered diffraction (EBSD) analysis was performed after the heat treatments in order to study in detail the obtained microstructure. In particular, the identification and quantification of the phases were carried out. Moreover, phase distribution, orientation, and the presence of textures were investigated. For EBSD investigations, we used a FEI QUANTA 205 FEG SEM (Thermo Fisher Scientific, Hillsboro, OR, USA), equipped with AMETEK EBSD (AMETEK BV, Tilburg, The Netherlands) system and OIM Analysis™, operating at 20 kV. The analyses were performed with the following parameters: Scan step: 200 nm, area: ~150 × 200 µm^2; confidence Index >5% and Scan step: 100 nm; area: ~35 × 40 µm^2, confidence Index >5%.

Transmission electron microscopy (TEM) was also performed to complete the characterization of RA on the samples using a JEOL JEM 200CX (Jeol Ltd, Tokyo, Japan) operating at 160 kV. The preparation of the thin foils was realized by mechanical grinding until thickness of 70 µm, followed by mechanical punching to obtain 3 mm diameter specimens. The final polishing and etching were performed electrochemically using a twin-jet polisher STRUERS TENUPOL-3 (Struers S.A.S., Milan, Italy), with 95% acetic acid (CH_3COOH) and 5% perchloric acid ($HClO_4$) solution, at 45 V and room temperature [26,27].

2.3. Mechanical Tests

Vickers micro-hardness, with a Leitz™ DURIMET (Leica Microsystem S.r.l., Milan, Italy) hardness tester, were conducted on each sample performing three indentations, with a 300 g load.

The treated samples were also subjected to tensile tests according to ASTM A370 $\varepsilon^{19\text{-}1}$. The tests were carried out on dog-bone samples, at strain rate of 5×10^{-3} s^{-1} with a MTS tensile test machine (MTS System Corporation, Eden Prairie, MN, USA), using a maximum force of 50 kN. The displacement was measured through the crosshead movement and the force by the machine load cell.

3. Results

3.1. Microstructure

The microstructure of the as-received material is shown in Figure 4a–c. The as-received structure shows a predominant martensitic matrix with some allotriomorphic ferritic colonies (F_A) formed at the prior austenitic grain boundaries during quenching. From the SEM micrograph (Figure 4b,c) it can be further observed the presence of idiomorphic ferrite (F_I), nucleated inside the original austenite grains (GB), present before the quenching.

Figure 5a–c reports the evolution of the microstructure caused by the annealing treatment at 870 °C for 10 min. From OM and the SEM micrograph (Figure 5a,b) it is possible to observe the presence of pearlitic islands (P). In the red box, shown in Figure 5c at higher magnification, eutectoidic islands with different lamellae orientation, surrounded by ferritic (F) grains, can be distinguished. This is the typical microstructure of a medium carbon steel after annealing treatment [28].

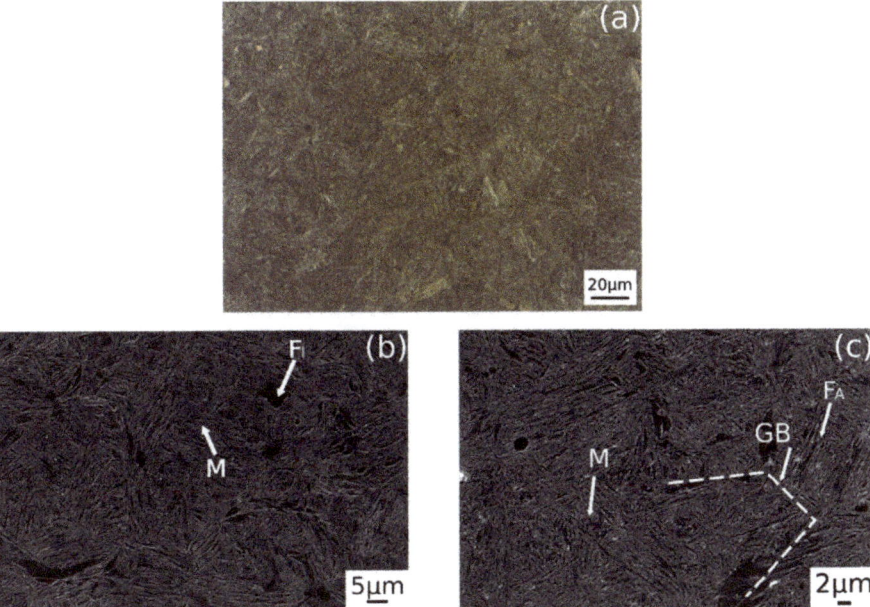

Figure 4. Optical (**a**) and SEM (**b,c**) micrographs showing the microstructure of the material in as-received state.

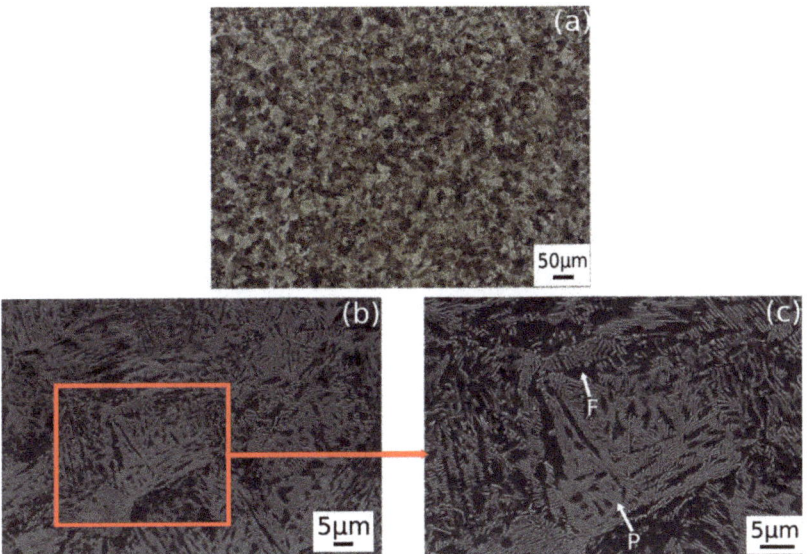

Figure 5. Optical (**a**) and SEM (**b,c**) micrographs showing the microstructure after annealing at 870 °C.

Figure 6 refers to the microstructure of the material after water quenching, before tempering. As expected, a complete martensitic microstructure (M) was obtained. SEM images (Figure 6b,c) reveal the presence of allotriomorphic ferrite, indicated in the image with F_A at the prior austenite grain boundaries (GB) [29,30]. In Figure 7c idiomorphic ferrite (F_I) at the center of the grain is shown.

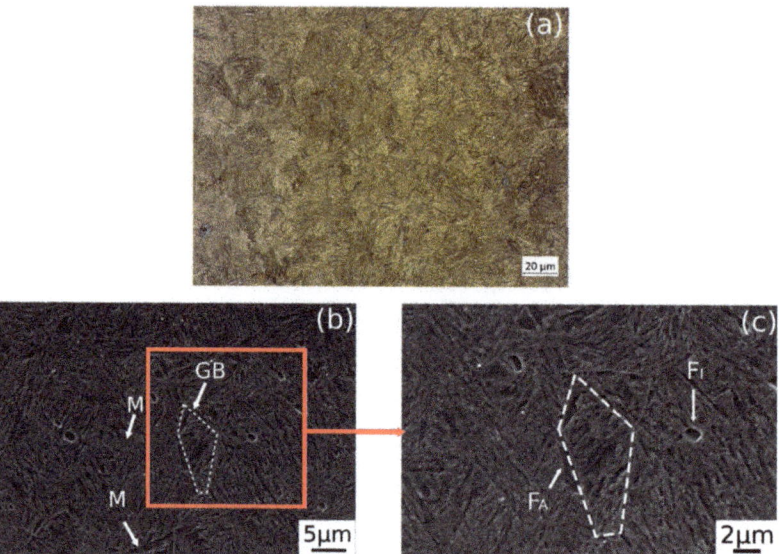

Figure 6. Optical (**a**) and SEM (**b**,**c**) micrographs after water quenching from 900°C (15 min holding).

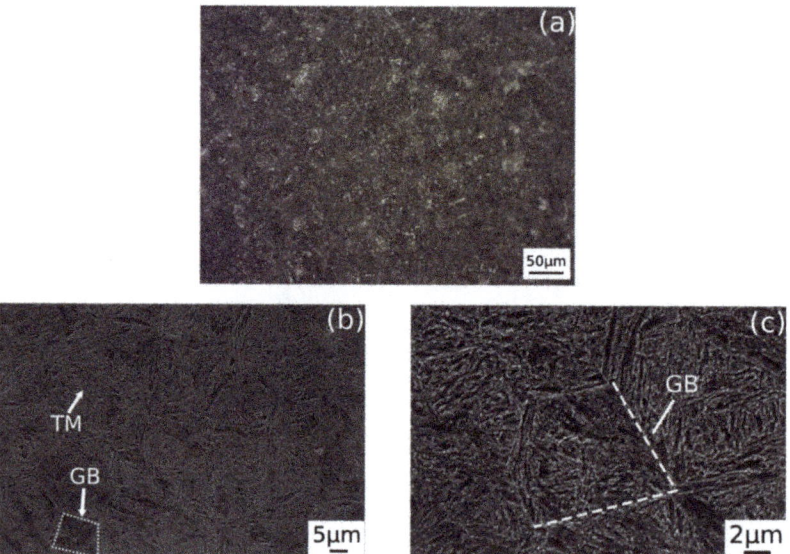

Figure 7. Microstructure of material after tempering at 600 °C and air cooled. (**a**) Optical micrograph, (**b**) and (**c**) SEM micrograph.

After tempering treatment, the material shows a microstructure consisting of tempered martensite (TM), as reported in Figure 7. SEM analysis (Figure 7b,c) allows us observe the presence of tempered martensite and trace of prior austenite grain boundary (GB).

After Quenching, Intercritical Annealing, and Austempering treatments, the samples show a different microstructure compared to that of the previous ones. It is possible to observe (Figure 8a,b) a dual phase microstructure, characterized by the presence of ferrite surrounded by a martensitic matrix and retained austenite. Ferrite is formed during the intercritical annealing at 780 °C. The ferrite presents in microstructure, visible in Figure 8, is a combination of different types and morphologies. A certain volume fraction of ferrite is formed during the intercritical annealing in the biphasic region. It is also possible to observe allotriomorphic and idiomorphic ferrite, as result of the partial transformation of austenite during the first cooling phase. The formation of these kinds of ferrite contributes strongly to the carbon enrichment of austenite. The low carbon solubility of ferrite forces carbon partitioning in austenite, increasing its stability and the possibility to retain it at room temperature. Martensite derives from austenite formed during heating in biphasic regime; austenite partially transforms into ferrite during the first heating step and into martensite during the last cooling phase. Bainite is not present during the austempering phase of the treatment and it is not visible in the microstructure, even if expected from the cooling transformation curve. This phenomenon could be related to the evolution of the curves and transformation temperatures caused by carbon partitioning during the heat treatment steps. It can be supposed that carbon partitioning leads to a decrease of Bs.

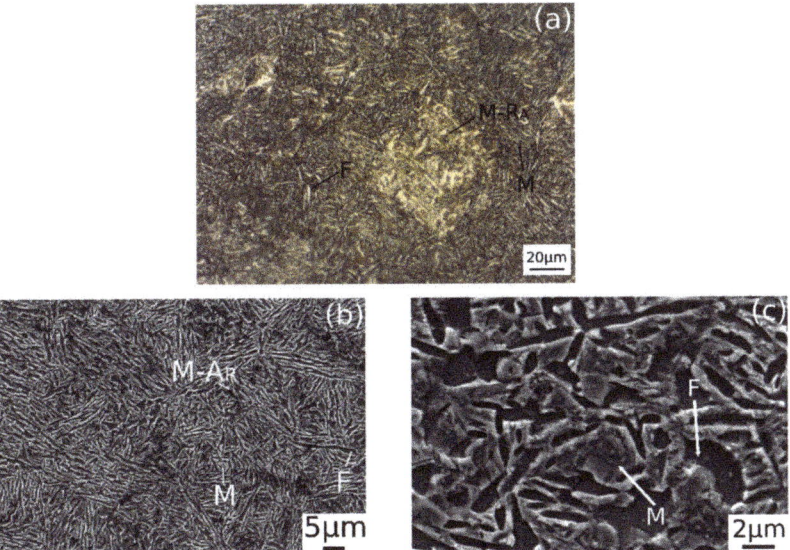

Figure 8. Microstructure of material after austempering. (**a**) Optical micrograph, (**b**) and (**c**) SEM micrograph.

FEG-SEM micrograph (Figure 9), taken in backscattered electron mode, allows us to see in detail the microstructure of the specimen subjected to austempering treatment. In the red box Figure 9b, it is clearly visible lath martensite in light grey and ferrite in black. This type of martensite morphology was confirmed in a research work by Ahmad and colleagues, which demonstrated that with intercritical annealing a lath morphology is acquired [31].

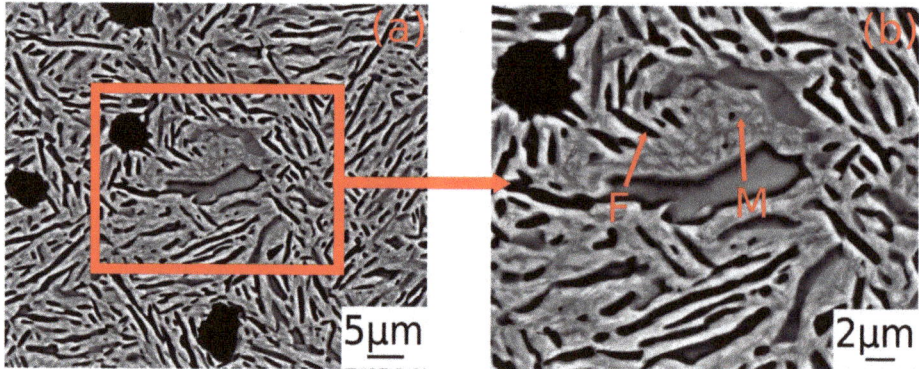

Figure 9. (**a**) Backscattered electrons FEG-SEM image of specimen subjected to Austempering (**b**) lath martensite detail.

3.2. X-ray Diffraction

One of the main goals of this thermal cycle is to achieve a multiphase material with a sufficiently high carbon content in the retained austenite which promote its retention at room temperature and leads a significant volume fraction of RA.

Figure 10 reports the X-ray diffraction patterns of the specimens subjected to annealing, Q&P, and Austempering heat treatment. X-ray pattern of annealed sample (Figure 10a) shows BCC iron peaks in the investigated angle range. Moreover, as 2θ increases it is possible to resolve Kα1 and Kα2 line peaks (indicated with the arrows). This could be related with the presence of coarse grains, which allows a better peak definition and high pattern resolution.

A slight difference in the peak position in the X-ray patterns can be observed, which could be attributed to the different dimension of the cell, caused by the different heating cycles.

Asymmetrical broadening of ferrite diffraction peaks towards low 2θ angles can be observed in the austempered samples (Figure 10b). Those peaks are related to the presence of tetragonal martensite. The presence of a displacement between ferrite and martensite peaks is due to the high carbon content of martensite and the strong distortion of the cell [32].

The presence of martensite in samples can be related to room temperature cooling that causes austenite transformation. High carbon austenite, which is not transformed in bainite during austempering, transforms in martensite when cooled down.

The X-ray diffraction technique allowed us to confirm the presence of retained austenite in the austempered samples, with a volume fraction similar to the one found in several TRIP steels and austempered high silicon steels [4]. The results of the phase quantification, reported in Table 3, demonstrate that the treatment cycle favors the process of carbon partitioning, allowing austenite retention at room temperature.

Table 3. Rietveld analysis results for phase quantification for the austempered sample.

Retained Austenite (%)	Martensite (%)	Ferrite (%)
14.9	40.5	44.6

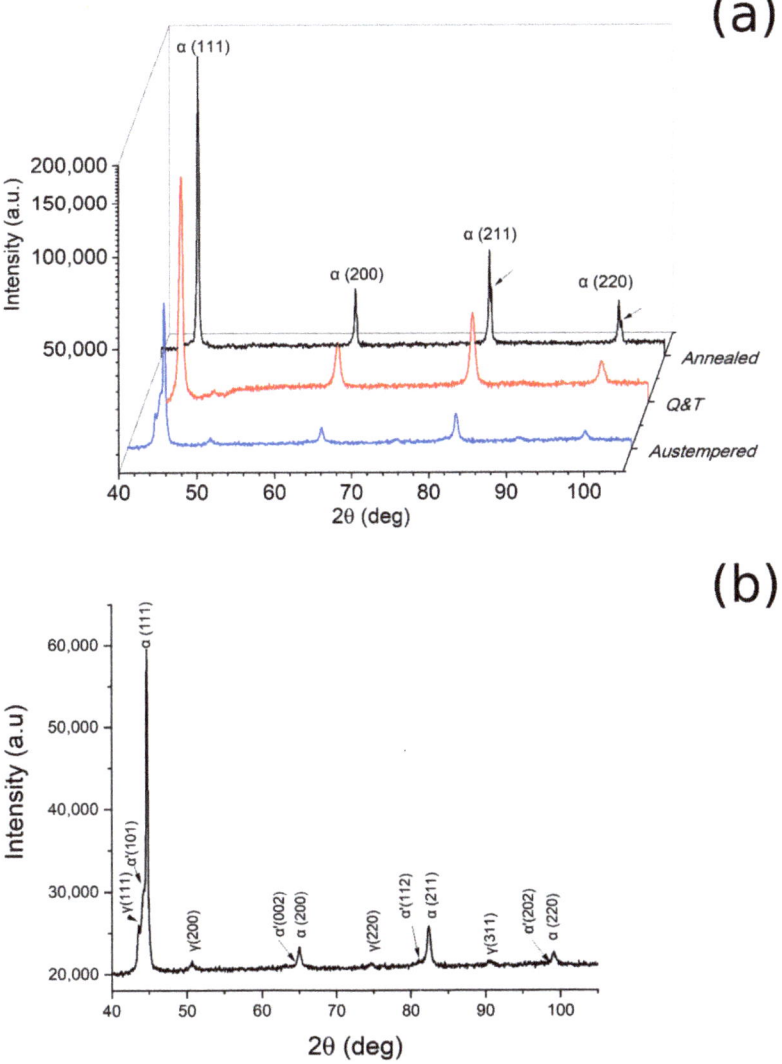

Figure 10. X-ray diffraction pattern of high silicon steel after each heat treatment (**a**), X-ray pattern of austempered steel in detail (**b**).

3.3. Electron BackScattered Diffraction

In order to better analyze the microstructural features of the austempered samples, EBSD analysis was performed. The main goal of the investigation was to understand distribution retained austenite, and its morphology and size.

No preferential orientation was observed in the material subjected to the austempering treatment (Figure 11). The absence of texture and preferred grain orientation can be explained considering that: a) the material was not subjected to deformation or forming process during heat treatment, b) a pretreatment, which produces recrystallization, was performed.

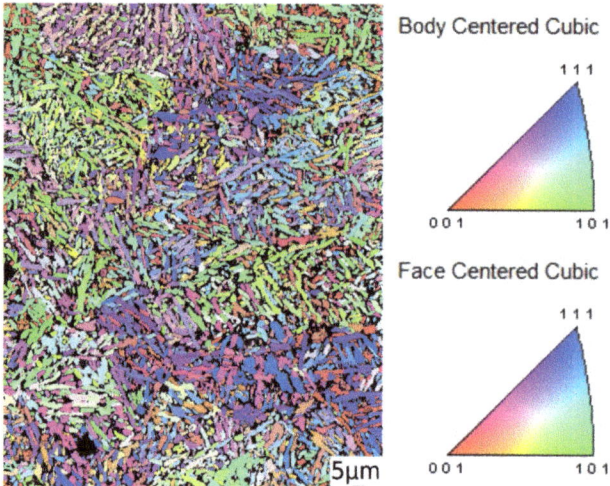

Figure 11. Electron Backscattered diffraction (EBSD) micrograph and related inverse pole figure of the austempered sample.

Figure 12 shows the detailed EBSD analysis on smaller sample areas. It can be noted also in this case that each phase has randomly oriented grains (Figure 12a). In Figure 12b, a phase distribution map is shown: red zones represent FCC iron, green ones refer to tetragonal martensite and retained austenite (FCC iron) is colored in yellow. A matrix of ferrite can be distinguished, with uniformly distributed austenitic and martensitic islands. Austenite is located at ferrites grain boundaries close to the martensitic islands, with a grain size of 1–5 μm. Such dimension of the FCC iron refer to high stable austenite with high carbon content. These islands should exhibit a strain induced martensitic transformation during plastic deformation at high strains [4,9,10,33–39].

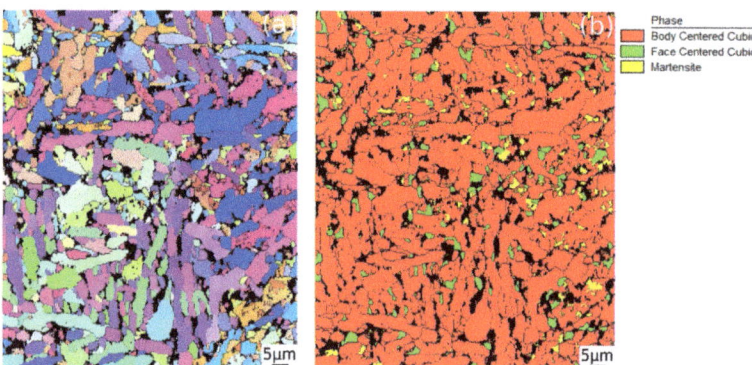

Figure 12. Detailed inverse pole (**a**) figure and phase identification of the material subjected to austempering treatment (**b**).

3.4. Transmission Electron Microscopy

A further microstructural investigation was carried out by TEM on the austempered samples (Figure 13a,b) in order to deeply analyze the morphology of the martensitic–austenitic regions formed during the heat treatment and shown with the other techniques.

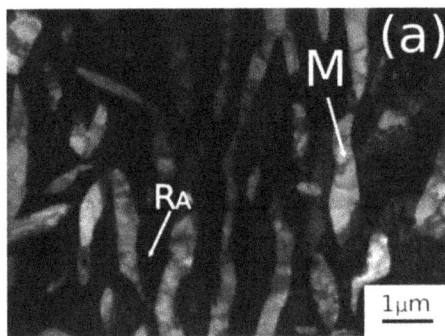 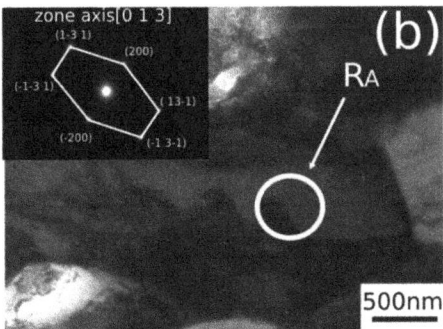

Figure 13. TEM micrographs (**a**) and (**b**) and electron diffraction patterns.

Figure 13a shows the presence of martensitic zones (light grey islands) and films of retained austenite (black islands). A detail of a retained austenite film between martensitic region is shown in Figure 13b. SAED was performed in the correspondence of the white circle zone, and the resulting diffraction pattern, indexed according to [40], confirm the presence of austenite. Therefore, it is possible to state that the treatment allows the austenite retention, due to the carbon partitioning.

EBSD and TEM analysis show that retained austenite has two morphologies: films between martensitic islands and blocks. The obtained retained austenite is characterized by different carbon content and different stability, in particular RA located near the martensitic islands is richer in carbon and more stable, according to [41]. Blocky austenite, which is poor in carbon, transforms earlier than lath austenite during the deforming process, as evidenced by [9].

3.5. Mechanical Properties

3.5.1. Microhardness Test

Microhardness results are reported in Table 4. It is possible to observe that the material, when annealed, has lower hardness compared to the samples subjected to the other treatments. Moreover, material subjected to austempering treatment has higher hardness than the sample treated by Q&T. The lower hardness of the Q&T sample could be explained considering the effect of the tempering treatment on martensite. The subsequent heating of martensite, below austenitization temperature, favors stress release, carbon diffusion, and carbide precipitation.

Table 4. Microhardness on material before tensile test.

Treatment	Average ($HV_{0.3}$)	St. Deviation
Annealed	316	22
Q&T	364	29
Austempering treatment	426	19

3.5.2. Tensile Tests

Mechanical properties were evaluated in terms of yield strength (YS), ultimate tensile strength (UTS), and fracture strain. The specimens, after annealing treatment and Q&P, showed the highest uniform elongation (Figure 14 and Table 5), while after austempering treatment the sample exhibited poor elongation, without evident yielding and plastic deformation. In detail, in the austempered sample, a fracture strain of approximately 4.5% and UTS of 1650 MPa were recorded. The brittle nature of the material could be explained considering the amount of martensite and its high dislocation density which limits further plastic deformation and the low strength of the martensite-ferrite cohesion.

A similar behavior was observed by Ahmad and colleagues [31]. The presence of a slight plastic deformation can be attributed to the presence of a considerable amount of ferrite.

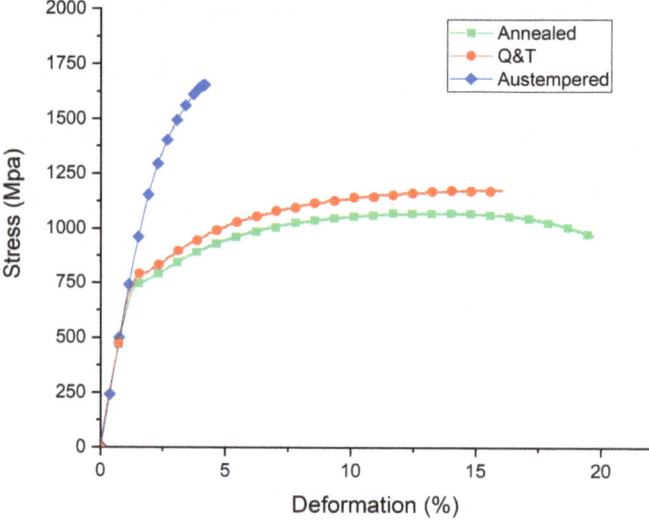

Figure 14. Engineering stress-strain curves. Blue curve with blue circle is related to austempered samples, red curve with diamonds referred to the Q&P treatment and green curve with squares representing the annealed samples.

Table 5. Results of tensile tests.

Material	Yield Strength (MPa)	Tensile Strength (MPa)	Fracture Strain (%)
Annealed	730	1130 ± 5	20 ± 2
Q&T	760	1200 ± 5	16.5 ± 2
Austempering treatment	1250	1650 ± 5	4.5 ± 0.5

The fracture surface of specimens after annealing (Figure 15) shows the typical features of a ductile fracture, confirming the mechanical behavior evidenced in the stress-strain curves. It also is possible to observe in Figure 15b the fracture surfaces of pearlitic grains [42].

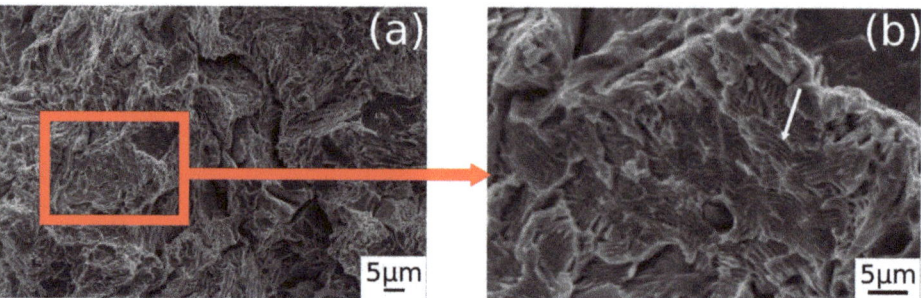

Figure 15. (**a**) SEM fractography of the sample in annealed condition (**b**) detail of a pearlitic fractured grain. White arrow in (**b**) indicates fracture surface of a pearlitic grain.

45. Lee, Y.K. Empirical formula of isothermal bainite start temperature of steels. *J. Mater. Sci. Lett.* **2002**, *21*, 1253–1255. [CrossRef]
46. Krauss, G. *Steels: Processing, Structure, and Performance*; ASM International: : New York, NY, USA, 2005; ISBN 0871708175.

Publisher's Note: MDPI stays neutral with regard to jurisdictional claims in published maps and institutional affiliations.

 © 2020 by the authors. Licensee MDPI, Basel, Switzerland. This article is an open access article distributed under the terms and conditions of the Creative Commons Attribution (CC BY) license (http://creativecommons.org/licenses/by/4.0/).

Article

Heat Treatment Effect on Microstructure Evolution in a 7% Cr Steel for Forging

Andrea Di Schino [1,*], Matteo Gaggiotti [1] and Claudio Testani [2]

1. Engineering Department, University of Perugia, Via G. Duranti 93, 06125 Perugia, Italy; matteo.gaggiotti@studenti.unipg.it
2. CALEF-ENEA CR Casaccia, Via Anguillarese 301, Santa Maria di Galeria, 00123 Rome, Italy; claudio.testani@consorziocalef.it
* Correspondence: andrea.dischino@unipg.it

Received: 25 May 2020; Accepted: 16 June 2020; Published: 17 June 2020

Abstract: Well-defined heat-treatment guidelines are required to achieve the target mechanical properties in high-chromium steels for forgings. Moreover, for this class of materials, the microstructure evolution during heat treatment is not clearly understood. Thus, it is particularly important to assess the steel microstructure evolution during heat treatment, in order to promote the best microstructure. This will ascertain the safe use for long-term service. In this paper, different heat treatments are considered, and their effect on a 7% Cr steel for forging is reported. Results show that, following the high intrinsic steel hardenability, significative differences were not found versus the cooling-step treatment, although prior austenite grain size was significantly different. Moreover, retained austenite (RA) content is lower in double-tempered specimens after heat treatments at higher temperatures.

Keywords: forged steels; microstructure; high-Cr steel

---

1. Introduction

In the energy industry, every steel component is asked to have a long service duration (often longer than 30 years). The possibility to produce forged steel components characterized by high mechanical properties and competitive costs has gained more and more importance in the last years.

Medium-carbon steels have been widely used for primary energy-plant-production applications due to their excellent performances. This often needs to face with strong requirements in terms of severe service conditions (high temperature, high pressures, corrosion issues, nuclear irradiation, etc.) [1–4].

In the case of energy-sector applications (especially in oil and gas), corrosion resistance issues are often coupled with demanding mechanical requirements: in such cases, carbon steel is often cladded with stainless steels or even nickel superalloys [5–7].

Concerning mechanical properties, it is well-known that they strongly depend on grain size and steel alloying (through solution and precipitation hardening mechanisms), as discussed in the literature. Morito et al., reported the effect of microstructure refining [8]; Fan et al., discussed microstructural and precipitation effect on high-chromium steels microstructure [9]; Akbarzadeh et al., studied the hot workability of high-carbon and high-chromium steel [10]. The microstructure evolution in medium-carbon steel is well described in Reference [11], and the phase transformation evolution in high-strength low-alloyed steels is reported in Reference [12]. Some authors reported that limited amounts of retained austenite have a beneficial effect on mechanical properties [13]. It is reported that grain-size refinement should increase the yield strength up to three times [14–16]. Moreover, such a strengthening mechanism can allow the best balance of strength/toughness combination [17–19].

Solution-hardening mechanisms, if properly activated by the addition of elements, allows us to increase the wear, fatigue and high-temperature behavior [20–23].

As far as microstructure refinement is concerned, in recent years, there has been an intense focus on plastic-deformation processes, such us forging technologies [24–27]. In fact, it is reported that forging is a method for grain-size refinement via plastic-deformation accumulation. This enhances the material strength and toughness. The raw materials' internal cracks and defects can be reduced during forging, as well as mean grain size, thus allowing for a more uniform grain-size distribution [28,29]. Chaudhari et al. [30–33] reported that the grain size is clearly refined after steel multidirectional forging. This leads to the formation of fine austenite grains following the activation of continuous dynamic recrystallization and dynamic recovery. They showed that fine grains' formation is beneficial in terms of tensile strength and hardness enhancement. Gosh et al. [34] outlined that the material mechanical properties' best combination is found in the case of pure ferrite microstructure.

In high-Cr-bearing steels micro-alloying strengthening phenomena are reported to be effective to refine grains during thermomechanical processing. Kvackaj et al., analyzed austenite deformation behavior during thermomechanical processing of Nb–Ti micro-alloyed steel. They studied thermomechanical cycles aimed to favor the formation of (i) recrystallized austenite, (ii) un-recrystallized austenite, and (iii) ferrite-pearlite microstructures [35]. De Ardo reported micro-alloyed forged steels, emphasizing the possibility that pearlite-ferrite microstructures (formed during air cooling) can exhibit final strengths and fatigue resistances similar to those of the more expensive heat-treated steel [36]. Opiela reported the effect of thermomechanical processing on the microstructure and mechanical properties of Nb-Ti-V micro-alloyed steel [37]. Di Schino reported forging simulation and microstructure evolution of large-scale ingot in micro-alloyed tool steel [38]. Zitelli et al., recently discussed vanadium micro-alloyed steels for large-scale forged ingots, with the aim to balance production costs with the industrial demand of forged steels with high tensile properties. They demonstrated, at a laboratory scale, that ASTM A694 F70 forged-steel grade requirements can be fulfilled by a vanadium micro-alloyed steel with an addition of 0.15% V (plus proper heat treatment), consisting of a ferrite-pearlite microstructure [39]. Mancini et al., reported defect reduction and quality optimization by modeling plastic deformation and metallurgical evolution in ferritic stainless steels [40]. Moreover, substitutional strengthening mechanism in 7–12% Cr carbon alloyed steels has been widely studied and proved to have an excellent mechanical properties combination [41–44]. Among them, 7% Cr steels are nowadays proposed by many steelmakers for being quite promising in terms of properties/cost combination instead of the more common 9–12% steels.

Commonly, after casting, the steel ingots are deformed (forging) and heat treated. Such heat treatment, called preheating treatment or after-forging heat treatment, is aimed to develop the target microstructure for the desired mechanical properties. Finally, the heat-treatment sequence for high Cr steels (also called quality heat treatment) includes austenitization, followed by quenching and high-temperature tempering (Q & T). The microstructure of the final products typically consists of tempered martensite, undissolved carbonitrides, and/or re-precipitated carbonitrides [45]. The austenitizing temperature and soaking time determines the partitioning of carbon and alloying elements between the austenite and carbonitrides. In this processing step, a temperature increase leads to an increase of the carbonitrides dissolution degree with a positive effect on the mechanical properties [46–48]. After austenitization, cooling to below the martensite start temperature (M_s) leads to martensite formation. In its as-quenched martensitic condition, the steel is hard and brittle and may contain "pockets" of retained austenite (RA). The quenching step is followed by a high-temperature tempering step (in the range of 500–650 °C), to reduce brittleness and increase ductility and toughness by reducing residual stresses. All the parameters involved in a particular heat-treatment process strongly affect the steel final microstructure, e.g., austenitization temperature, holding time, quench medium, cooling rate, tempering temperature, and holding time. Well-defined heat-treatment guidelines are therefore required to assist the manufacturer in the selection of the correct heat treatment for achieving the target properties.

Ultimately, the microstructure evolution during the heat treatment of high-Cr steels is still not clearly understood. Thus, the understanding of microstructure evolution during heat treatment in wt.% Cr steel became a key factor for a safe long-term-service steel-microstructure design.

In this paper, the heat-treatment effect on a 7% Cr steel for forging is reported. Different heat treatments are considered. The resulting microstructure evolution is analyzed.

2. Materials and Methods

The steel nominal chemical composition is reported in Table 1.

Table 1. Main chemical composition of the considered materials (mass, %).

C	Cr	Mo	Mn	Ni	Si	V	Fe
0.42	7.0	0.70	0.65	<0.20	0.50	0.10	Balance

Specimens were withdrawn, starting from a sample-ingot for Continuous Cooling Temperature (CCT) diagrams determination. CCT diagrams were derived based on the dilatometry experiments carried out on 10 mm in length, 6 mm in diameter specimens, with controlled cooling rate. Microstructure analysis and hardness measurements of heat-treated specimens were performed by means of Light Microscopy (LM), after 4% Nital etching and Brinell indenter (HB). Prior austenite grain size was measured by image analysis software after samples etching in 10 g of CrO_3, 50 g of NaOH, 1.5 g of picric acid, and 100 mL of distilled water solution, at room temperature, for 20 s. After etching, specimens were cleaned and dried.

The following heat treatments' cycles were the performed by means of dilatometer (Table 2 and Figure 1).

Heat treatments were designed with the aim to obtain different microstructures in terms of retained austenite content. Austenite transformation during the different steps of a laboratory manufacturing route that is reproducing an industrially relevant process was considered. In particular, the following four microstructural families were targeted:

(1) Quenching and partitioning (specimen Q & P): Such heat treatment is here investigated based on very promising results reported on its effect on 9% Cr steels [49].
(2) Partially transforming austenite during first quenching at 280 °C (specimens *n*. PT-γ-1 to PT-γ-4).
(3) Partially transforming austenite during secondary cooling at 220 °C (specimens *n*. PT-γ-5 to PT-γ-8).
(4) Quenching (specimens Q, 100% martensite), quenching and tempering (specimens Q & T).
(5) Specimens PT-γ-1 and PT-γ-5 did not undergo any tempering; specimens PT-γ-2 and PT-γ-6 just underwent first tempering treatment at 533 °C; specimens PT-γ-3 and PT-γ-7 were subjected to second tempering at 528 °C; specimens PT-γ-4 and PT-γ-6 were subjected to second tempering at 505 °C.

In order to investigate the possible presence of retained austenite, X-Ray Diffraction (XRD) measurements were performed on heat-treated specimens. XRD analysis was carried out, using a Siemens D500 diffractometer (Siemens, Munich, Germany) equipped with Co-tube radiation and a graphite-monochromator on the detector side, with step size (2θ) of 0.05° and counting time of 5 s/step. The effect of possible texture was eliminated by a rotating fixture on the goniometer. Since low RA contents are expected in this class of steels, facing with XRD technique sensitivity, particular attention has been devoted to spectra postprocessing analysis. In particular, in order to avoid peaks overlapping, and thus to extract information also from "hidden peaks", spectra were processed by a tool specifically developed for peaks deconvolution (Figure 2). Three specimens were analyzed for each heat-treatment condition.

Table 2. Heat treatments' sequence.

Nr	Austenitization		Interrupted Quenching		Heating		Cooling down to
	T (°C)	t (s)	T (°C)	t (s)	T (°C)	t (s)	
Q & P	980	10,800	150	3600	450	3600	12 °C/h up to 220 °C
PT-γ-1							12 °C/h up to 220 °C
PT-γ-2			280				12 °C/h up to 220 °C
PT-γ-3							12 °C/h up to 220 °C
PT-γ-4							12 °C/h up to 220 °C
PT-γ-5							12 °C/h up to 220 °C
PT-γ-6			400				12 °C/h up to 220 °C
PT-γ-7							12 °C/h up to 220 °C
PT-γ-8							12 °C/h up to 220 °C

Nr	First Tempering		Cooling down to		Second Tempering		Cooling down to
	T (°C)	t (s)	T (°C)	t (s)	T (°C)	t (s)	
Q & P	505 °C × 57,600 s		300 °C/3600 s		505 °C × 57,600 s		Room temperature
PT-γ-1	No tempering				No tempering		
PT-γ-2	533 °C × 57,600 s				No tempering		
PT-γ-3	533 °C × 57,600 s				528 °C × 57,600 s		
PT-γ-4	505 °C × 57,600 s				505 °C × 57,600 s		
PT-γ-5	No tempering				No tempering		
PT-γ-6	533 °C × 57,600 s				No tempering		
PT-γ-7	533 °C × 57,600 s				528 °C × 57,600 s		
PT-γ-8	505 °C × 57,600 s				505 °C × 57,600 s		

Nr	Austenitization	Quenched to	First Tempering	Second Tempering
Q	980 °C × 3600 s	RT		
Q & T	980 °C × 3600 s	RT	533 °C × 57,600 s	528 °C × 57,600 s

Electron Backscattered Diffraction (EBSD) measurements were performed in selected cases, with the aim to detect the presence and position of retained austenite islands, by means of a JEOL field emission gun scanning electron microscope (FEG-SEM) (JEOL Ltd., Tokyo, Japan), using a 0.12 μm scanning step size. Retained austenite was revealed by building up phase maps, taking into account both face-centered cube (fcc) and body-centered cube (bcc) phases: automatic image analysis of such maps allowed us to determine retained austenite volume fraction. The precipitation state was analyzed by transmission electron microscope (TEM) on extraction replica specimens. The observations were performed with a JEOL 200CX transmission electron microscopy (JEOL Ltd., Tokyo, Japan). The hardness dependence upon heat treatments was assessed by means of Brinell hardness measurements.

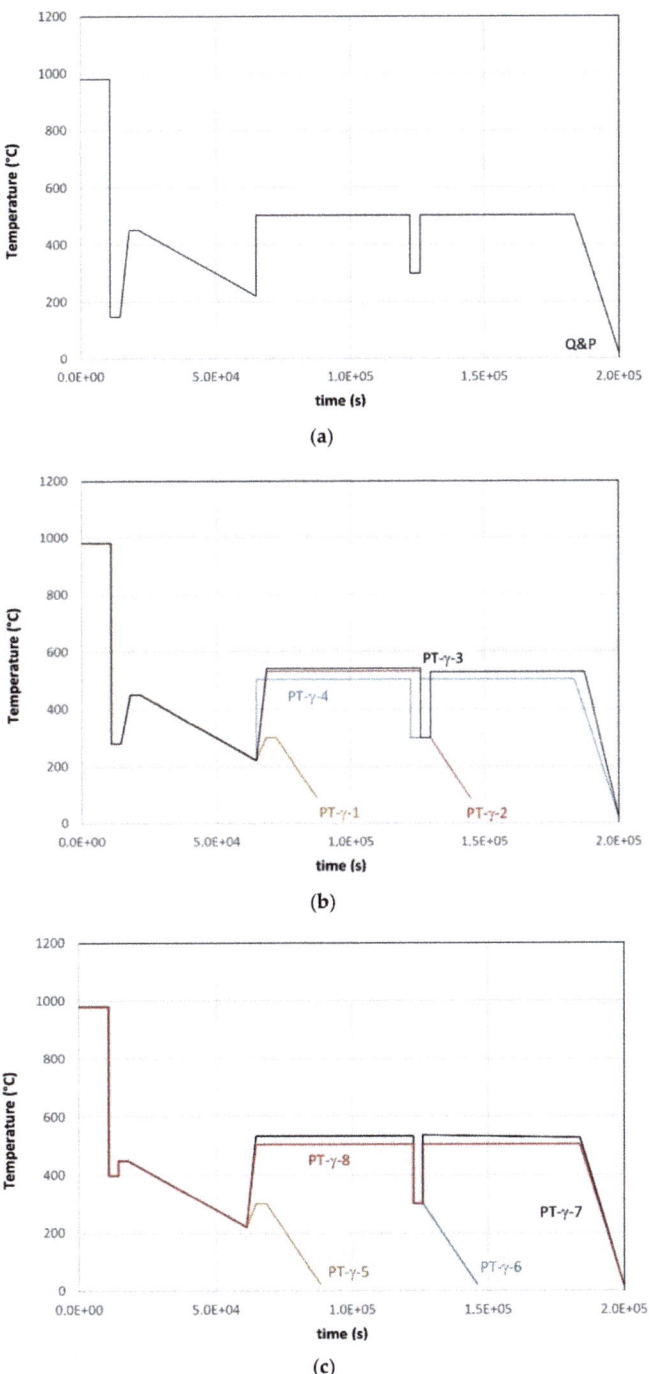

Figure 1. (**a**) Heat-treatment cycle as performed on Q & P specimen; (**b**) heat-treatment cycle as performed on specimens from PT-γ-1 to PT-γ-4; (**c**) heat-treatment cycle as performed on specimens from PT-γ-5 to PT-γ-8.

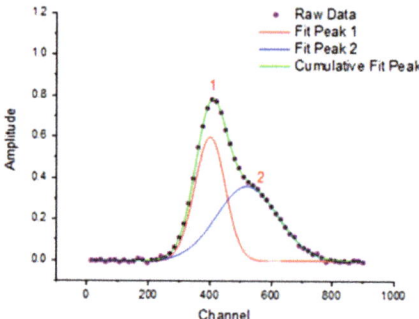

Figure 2. Example of XRD peak deconvolution.

3. Results and Discussion

3.1. Continuous Cooling Transformation

CCT diagrams were determined after austenitization at two different temperatures: 980 and 1200 °C. The first temperature is typical of that commonly adopted during quality heat treatments, and the second one of the final forging-process passes (before cooling). A 30 min soaking time was considered. Dilatometric curves acquired in order to build up the steel CCT are reported in Figures 3 and 4, respectively. The green cooling curve reported in Figure 4 represents the first cooling undergone by all specimens reported in Table 2. This assures that, in the case of complete quenching, a fully martensitic microstructure was achieved (specimens Q and Q & T). On the other hand, this also assures that, in the case of partial quenching, a mixed martensite–austenite microstructure is obtained (specimens Q & P and PT-γ-1 to PT-γ-8).

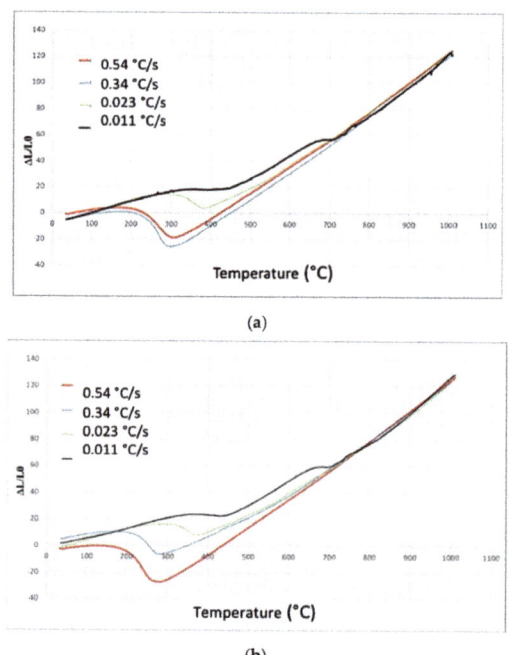

Figure 3. Dilatometric curves of the considered steel: (**a**) austenitization at 980 °C; (**b**) austenitization at 1200 °C.

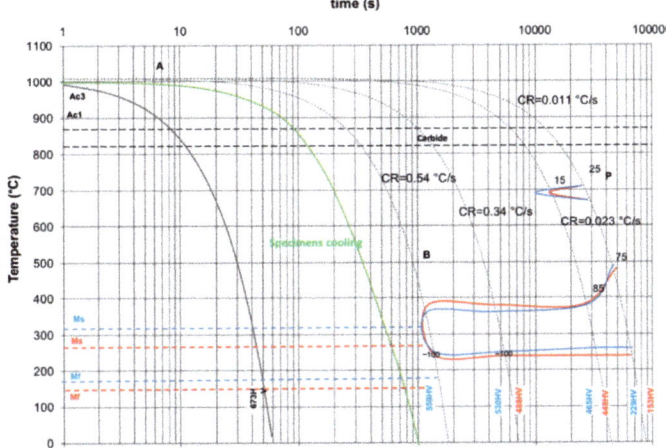

Figure 4. CCT diagrams of the considered steel (austenitization at 980 °C, blue curve; austenitization at 1200 °C, red curve). Green cooling curve: first cooling undergone by all specimens reported in Table 2.

The results show that, even if prior austenite grain size were quite different, significative differences are not found in the cooling behavior (Figure 5). In particular 10 and 75 µm grain sizes were measured in specimens after austenitization at 980 and 1200 °C, respectively. These results were expected following the high intrinsic steel hardenability due to the high C and Cr content.

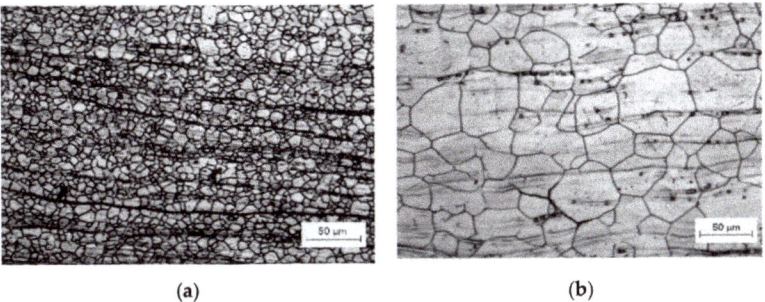

Figure 5. Austenitic grain size of two specimens after austenitization at 980 °C (**a**) and 1200 °C (**b**).

3.2. Retained Austenite Evolution

Selected XRD spectra of specimens reported in Table 2 are shown in Figure 6. Even if retained austenite values are quite low, it is clear from the spectra reported in Figure 6 that it is possible to discriminate between specimens characterized by RA presence (e.g., Figure 6, specimen PT-γ-2) and specimens where XRD is not able to detect any RA (<2.0%) (e.g., Figure 6, specimen PT-γ-3). Retained austenite contents, as measured by means of XRD spectra analysis, are reported in Table 3. XRD measurements were confirmed, in selected cases, by phase maps analysis, as obtained by means of EBSD (Figure 7). In particular, 1.7% retained austenite content was measured in specimen PT-γ-2 by EBSD (to be compared to 2.5% as measured by XRD) and 1.2% in specimen PT-γ-3 (to be compared to <2.0% as measured by XRD). The low discrepancy obtained by EBSD with respect to XRD is justified by the two different adopted techniques [50]. The agreement between RA contents, as measured by the two different methods, makes the results reported in Table 3 consistent, even if their absolute values are quite low (in the proximity of XRD-technique sensitivity).

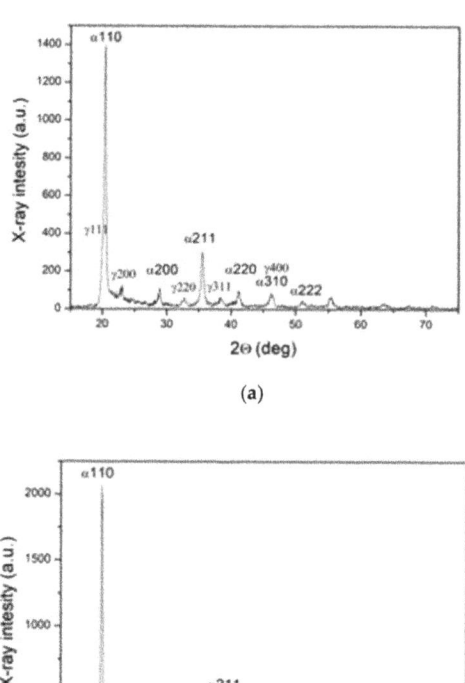

Figure 6. XRD spectra of selected heat-treated specimens: (**a**) specimen PT-γ-2 and (**b**) specimen PT-γ-3.

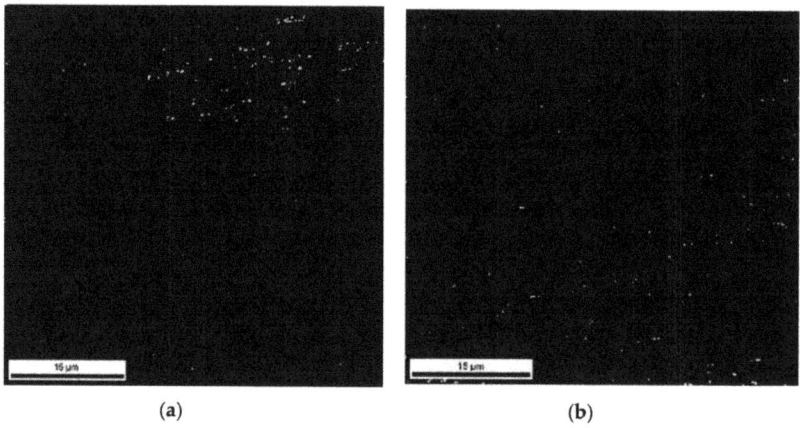

Figure 7. Phase maps by EBSD (**a**) specimen PT-γ-2 and (**b**) specimen PT-γ-3. Green areas correspond to retained austenite.

Table 3. Retained austenite content as a function of performed heat treatments (three measures for each specimen are reported).

Specimen n.	RA (%)
Q & P	2.4; 2.4; 2.3
PT-γ-1	2.3; 2.2; 2.3
PT-γ-2	2.5; 2.6; 2.5
PT-γ-3	<2.0; 2.1; <2.0
PT-γ-4	2.1; 2.2; <2.0
PT-γ-5	2.6; 2.5; 2.5
PT-γ-6	2.2; 2.1; 2.3
PT-γ-7	<2.0; <2.0; 2.0
PT-γ-8	2.7; 3.0; 2.4
Q	2.1; 2.1; 2.0
Q & T	<2.0; 2.1; <2.0

Specimens n. PT-γ-3, PT-γ-7 and Q & T, double tempered at higher temperatures (>528 °C), show RA contents lower than 2.0%. All the other variants (reported in Table 3), with none or at a lower second tempering temperature, show higher RA contents. It is worth outlining that such results are independent from the microstructures obtained after cooling, according to Table 2. The other specimens show RA content ranging from 2.4% to 2.7%: such specimens were subjected to double tempering, but at a lower temperature (505 °C). The above considerations suggest that the double tempering heat treatment at higher temperatures (>528 °C) allows us to reactivate retained austenite that, during following cooling down to room temperature, decomposes in fine carbides

In fact, a very fine precipitation is detected in specimens PT-γ-3 and PT-γ-7 (average precipitation diameter d_p = 14.6 nm and d_p = 16.4 nm, respectively), as shown in Figure 8. Such precipitates were not detected in the other specimens. Finest precipitates (d_p < 10 nm) were identified to be chromium carbides vanadium rich. Vanadium presence was not detected in coarser precipitates. The crystallographic lattice parameters' measurements obtained by diffraction patterns analysis allowed us to identify precipitates ranging d_p < 5 nm as MC, precipitates ranging 5–10 nm as M_2C. The largest precipitates were classified as $M_{23}C_6$ (ranging 70–200 nm) and M_3C in the case of (d_p > 200 nm). The Brinell hardness HB dependence on retained austenite content is reported in Figure 9. Results show that higher hardness values are found in the case of higher-retained-austenite-content specimens (hence lower tempering temperature). Results suggest therefore that the presence of retained austenite, also in a small amount, should exert a non-negligible effect. This means that manufacturers are called for a specific heat-treatment design, depending on the final properties required for different forged components.

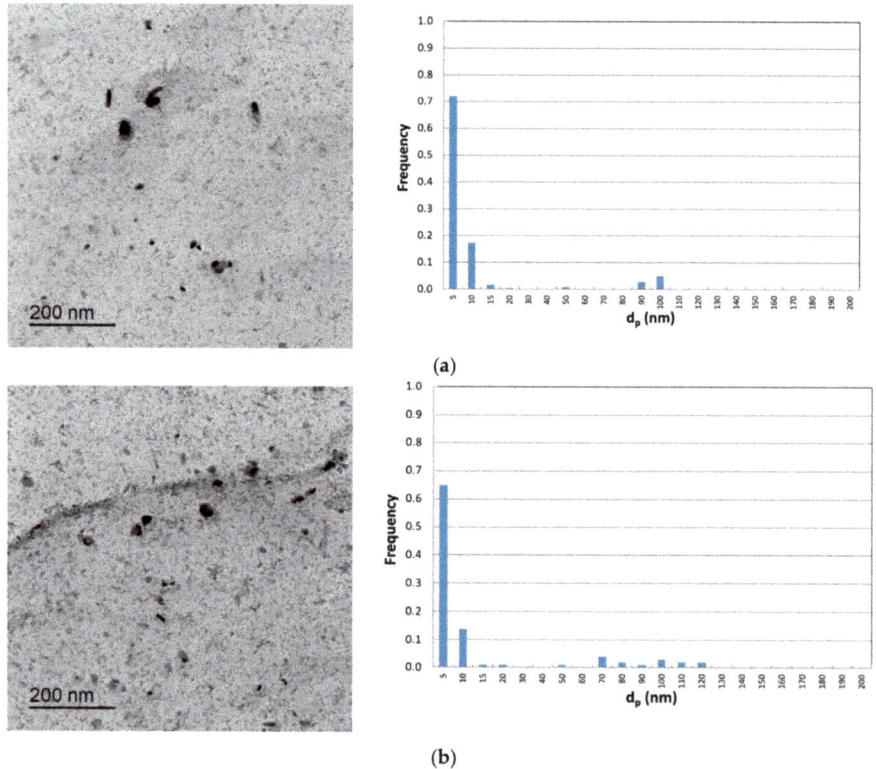

Figure 8. TEM images describing the precipitation state of (**a**) specimen PT-γ-3 and (**b**) specimen PT-γ-7.

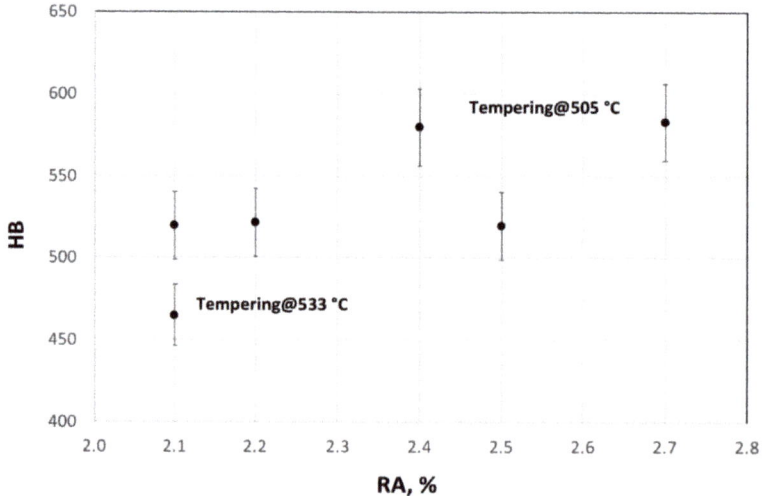

Figure 9. Hardness as a function of the retained austenite content.

4. Conclusions

In this paper, the heat treatment effect on a 7% Cr steel for forging was reported. Different heat treatments were considered, and microstructure evolution was analyzed. Heat treatments were designed with the aim of obtaining the following four microstructural families:

(1) Quenching and partitioning (specimen Q & P);
(2) Partially transformed austenite during a first quenching at 280 °C (specimens n. PT-γ-1 to PT-γ-4);
(3) Partially transformed austenite during a secondary cooling at 220 °C (specimens n. PT-γ-5 to PT-γ-8);
(4) 100% martensite (specimens Q just quenching, specimen Q & T quenching and tempering).

Results show the following:

(1) Retained austenite content is lower in double-tempered specimens at temperatures higher than 528 °C;
(2) The retained austenite disappeared after quenching down to room temperature and double tempering;
(3) The prior austenite grain size does not have an influence on the cooling behavior in this specific steel.

Author Contributions: Conceptualization, A.D.S.; methodology, A.D.S.; formal analysis, A.D.S. and C.T., formal analysis, A.D.S., C.T., and M.G. All authors have read and agreed to the published version of the manuscript.

Funding: This research received no external funding.

Conflicts of Interest: The authors declare no conflict of interest.

References

1. Changle, Z.; Hanguang, F.; Shengqiang, M.; Dawei, Y.; Jian, L.; Zhenguo, X.; Yongping, L. Effect of Mn content on microstructure and properties of wear-resistant bainitic steel. *Mater. Res. Express* **2019**, *6*, 1–22. [CrossRef]
2. Allain, S.; Bouaziz, O. Microstructure based modeling for the mechanical behavior of ferrite-pearlite steels suitable to capture isotropic and kinematic hardening. *Mater. Sci. Eng. A* **2008**, *496*, 329–336. [CrossRef]
3. De Andrés, C.G.; Capdevila, C.; Madariaga, I.; Gutierrez, I. Role of molybdenum in acicular ferrite formation under continuous cooling in a medium carbon microa-lloyed forging steel. *Scr. Mater.* **2001**, *45*, 709–716. [CrossRef]
4. Soffritti, C.; Merlin, M.; Vazsquez, R.; Fortini, A.; Garagnani, G.L. Failure analysis of worn valve train components of a four-cylinder diesel engine. *Eng. Fail. Anal.* **2018**, *92*, 528–538. [CrossRef]
5. Lee, K.S.; Yoon, D.H.; Kim, H.K.; Kwon, Y.-N.; Lee, Y.-S. Effect of annealing on the interface microstructure and mechanical properties of clad sheet. *Mater. Sci. Eng. A* **2012**, *556*, 319–330. [CrossRef]
6. Liu, C.Y.; Wang, Q.; Jia, Y.Z.; Jing, R.; Zhang, B.; Ma, M.Z.; Liu, R.P. Microstructures and mechanical properties of laminated composites prepared via warm roll bonding. *Mater. Sci. Eng. A* **2012**, *556*, 1–8. [CrossRef]
7. Di Schino, A.; Testani, C. Corrosion behavior and mechanical properties of AISI 316 stainless steel clad Q235 plate. *Metals* **2020**, *10*, 552. [CrossRef]
8. Morito, S.; Yoshida, H.; Maki, T.; Huang, X. Effect of block size on the strength of lath martensite in low carbon steels. *Mater. Sci. Eng. A* **2006**, *438*, 237–240. [CrossRef]
9. Fan, C.; Chen, M.M.; Chang, C.M.; Wu, W. Microstructure change caused by $(Cr,Fe)_{23}C_6$ carbides in high chromium Fe-Cr-C alloys. *Surf. Coat. Technol.* **2006**, *201*, 908–912. [CrossRef]
10. Akbarzadeh, A.; Naghdy, S. Hot workability of high carbon high chromium tool steel. *Mater. Des.* **2013**, *46*, 654–659. [CrossRef]
11. Di Schino, A.; Di Nunzio, P.E.; Turconi, G.L. Microstructure evolution during tempering of martensite in medium carbon steel. In *Materials Science Forum*; Trans Tech Publications Ltd.: Stafa-Zurich, Switzerland, 2007; Volume 558, pp. 1435–1441.
12. Di Schino, A. Manufacturing and application of stainless steels. *Metals* **2020**, *10*, 327. [CrossRef]

13. Di Schino, A. Analysis of heat treatment effect on microstructural features evolution in a micro-alloyed martensitic steel. *Acta Metall. Slovaca* **2016**, *22*, 266–270. [CrossRef]
14. Di Schino, A.; Kenny, J.M.; Abbruzzese, G. Analysis pf the recrystallization and grain growth processes in AISI 316 stainless steel. *J. Mater. Sci.* **2002**, *37*, 5291–5298. [CrossRef]
15. Di Schino, A.; Alleva, L.; Guagnelli, M. Microstructure evolution during quenching and tempering of martensite in a medium C steel. In *Materials Science Forum*; Trans Tech Publications Ltd.: Stafa-Zurich, Switzerland, 2007; Volume 715, pp. 860–865.
16. Mohammed, M.N.; Omar, M.Z.; Zubaidi, S. Microstructure and mechanical properties of tool steels. *Metals* **2018**, *8*, 316. [CrossRef]
17. Di Schino, A.; Di Nunzio, P.E. Metallurgical aspects related to contact fatigue phenomena in steels for back up rolling. *Acta Metall. Slovaca* **2017**, *23*, 62–71. [CrossRef]
18. Pezzato, L.; Gennari, C.; Chukin, D.; Toldo, M.; Sella, F.; Toniolo, M.; Zambon, A.; Brunelli, K.; Dabalà, M. Study of the effect of multiple tempering on the impact toughness of forged S690 structural steel. *Metals* **2020**, *10*, 507. [CrossRef]
19. Bhadeshia, H.K.D.H.; Honeycombe, R.W.K. *Steels: Microstructure and Properties*; Butterworths Heinemann (Elsevier): Aalborg, Denmark, 2006; ISBN 9780750680844.
20. Di Schino, A.; Valentini, L.; Kenny, J.M.; Gerbig, Y.; Ahmed, I.; Hefke, H. Wear resistance of high-nitrogen austenitic stainless steel coated with nitrogenated amorphous carbon films. *Surf. Coat. Technol.* **2002**, *161*, 224–231. [CrossRef]
21. Napoli, G.; Di Schino, A.; Paura, M.; Vela, T. Colouring titanium alloys by anodic oxidation. *Metalurgija* **2018**, *57*, 111–113.
22. Li, Z.; Jia, P.; Liu, Y.; Qi, H. Carbide Precipitation, Dissolution, and Coarsening in G18CrMo2–6 Steel. *Metals* **2019**, *9*, 916. [CrossRef]
23. Sharma, D.K.; Filipponi, M.; Di Schino, A.; Rossi, F.; Castaldi, J. Corrosion behaviour of high temeperature fuel cells: Issues for materials selection. *Metalurgija* **2019**, *58*, 347–351.
24. Qin, S.; Song, R.; Xiong, W.; Liu, Z.; Wang, Z.; Guo, K. Microstructure evolution and mechanical properties of grade E690 offshore platform steel. In *HSLA Steels 2015, Microalloying 2015 & Offshore Engineering Steels 2015*; John Wiley & Sons, Inc.: Hoboken, NJ, USA, 2015; Volume 2, pp. 1117–1123. ISBN 9781119223399.
25. Maropoulos, S.; Ridley, N.; Karagiannis, S. Structural variations in heat treated low alloy steel forgings. *Mater. Sci. Eng. A* **2004**, *380*, 79–92. [CrossRef]
26. Bregliozzi, G.; Ahmed, S.I.-U.; Di Schino, A.; Kenny, J.M.; Haefke, H. Friction and Wear Behavior of Austenitic Stainless Steel: Influence of Atmospheric Humidity, Load Range, and Grain Size. *Tribol. Lett.* **2004**, *17*, 697–704. [CrossRef]
27. Rufini, R.; Di Pietro, O.; Di Schino, A. Predictive simulation of plastic processing of welded stainless steel pipes. *Metals* **2018**, *8*, 519. [CrossRef]
28. Kobayashi, Y.; Tanaka, Y.; Matsuoka, K.; Kinoshita, K.; Miyamoto, Y.; Murata, H. Effect of forging ratio and grain size on tensile and fatigue strength of pure titanium forgings. *J. Soc. Mater. Sci.* **2005**, *54*, 66–72. [CrossRef]
29. Zhu, K.; Qu, S.; Feng, A.; Sun, J.; Shen, J. Microstructural evolution and refinement mechanism of alloy during multidirectional isothermal forging. *Materials* **2019**, *12*, 2496. [CrossRef] [PubMed]
30. Padap, A.K.; Chaudhari, G.P.; Nath, S.K.; Pancholi, V. Ultrafine-grained steel fabricated using warm multiaxial forging: Microstructure and mechanical properties. *Mater. Sci. Eng. A* **2009**, *527*, 110–117. [CrossRef]
31. Wang, B.; Liu, Z.; Li, J.; Chaudhari, G.P. Microstructure evolution in AISI201 austenitic stainless steel during the first compression cycle of multi-axial compression. *Mater. Sci. Eng. A* **2013**, *568*, 20–24. [CrossRef]
32. Soleymani, V.; Eghbali, B.; Chaudhari, G.P. Grain refinement in a low carbon steel through multidirectional forging. *J. Iron Steel Res. Int.* **2012**, *19*, 74–78. [CrossRef]
33. Padap, A.K.; Chaudhari, G.P.; Pancholi, V.; Nath, S.K. Warm multiaxial forging of AISI 1016 steel. *Mater. Des.* **2010**, *31*, 3816–3824. [CrossRef]
34. Ghosh, S.; Singh, A.K.; Mula, S. Effect of critical temperatures on microstructures and mechanical properties of Nb-Ti stabilized IF steel processed by multiaxial forging. *Mater. Des.* **2016**, *100*, 47–57. [CrossRef]
35. Zrnik, J.; Kvackaj, J.; Pongpaybul, A.; Sricharoenchai, P.; Vilk, J.; Vrchivinsky, V. Effect of thermomechanical processing on the microstructure and mechanical properties of Nb-Ti microalloyed steel. *Mater. Sci. Eng. A* **2001**, *319*, 321–325. [CrossRef]

36. De Ardo, A.I.; Garcia, C.I.; Hua, M. Micro-alloyed steels for high strength forgings. *Metall. Ital.* **2010**, *102*, 5–10.
37. Opiela, M. Effect of Thermomechanical Processing on the Microstructure and Mechanical Properties of Nb-Ti-V micro-alloyed Steel. *J. Mat. Eng. Perf.* **2014**, *23*, 3379–3388. [CrossRef]
38. Di Schino, A. Analysis of phase transformation in high strength low alloyed steels. *Metalurgija* **2017**, *56*, 349–352.
39. Zitelli, C.; Mengaroni, S.; Di Schino, A. Vanadium micro-alloyed high strength steels for forgings. *Metalurgija* **2017**, *56*, 326–328.
40. Mancini, S.; Langellotto, L.; Di Nunzio, P.E.; Zitelli, C.; Di Schino, A. Defect reduction and quality optimisation by modelling plastic deformation and metallurgical evolution in ferritic stainless steels. *Metals* **2020**, *10*, 186. [CrossRef]
41. Bendick, W.; Ring, M. Creep rupture strength of tungsten-alloyed 9–12% Cr steels for piping in power plant. *Steel Res.* **1996**, *67*, 382–397. [CrossRef]
42. Haarman, K.; Bendicl, W.; Arbaba, A. *The T91/P91 Book*, 2nd ed.; Vallourec and Mannesmann Tubes: Boulogne-Billancourt, France, 2002.
43. Garr, K.; Rhodes, C.; Kramer, D. Effects of microstructure on swelling and tensile properties of neutron irradiated Types 316 and 405 stainless steels. *ASTM Spec. Tech. Publ.* **1973**. [CrossRef]
44. Zitelli, C.; Folgarait, P.; Di Schino, A. Laser powder bed fusion of stainless-steel grades: A review. *Metals* **2019**, *9*, 731. [CrossRef]
45. Abe, F.; Horiuchi, T.; Taneike, M.; Sawada, K. Stabilization of martensitic microstructure in advanced 9Cr steel during creep at high temperature. *Mater. Sci. Eng. A* **2004**, *378*, 299–303. [CrossRef]
46. Caballero, F.; Alvarez, L.; Capdevila, C.; de Andrés, C.G. The origin of splitting phenomena in the martensitic transformation of stainless steels. *Scr. Mater.* **2003**, *49*, 315–320. [CrossRef]
47. Barlow, L.; Du Toit, M. Effect of austenitizing heat treatment on the microstructure and hardness of martensitic stainless steel AISI 420. *J. Mater. Eng. Perform.* **2012**, *1*, 1327–1336. [CrossRef]
48. Tao, X.G.; Gu, J.F.; Han, L.Z. Carbonitride dissolution and austenite grain growth in a high Cr ferritic heat-resistant steel. *ISIJ Int.* **2014**, *54*, 1704–1715. [CrossRef]
49. Chintan, S. Quenching and Partitioning Process of Grade P91 Steel. Master's Thesis, Texas University, Austin, TX, USA, 2018.
50. Pashangeh, S.; Zarchi, H.R.K.; Banadkouki, S.S.G.; Somani, M. Detection and Estimation of Retained Austenite in a High Strength Si-Bearing Bainite-Martensite-Retained Austenite Micro-Composite Steel after Quenching and Bainitic Holding (Q & B). *Metals* **2019**, *9*, 492.

© 2020 by the authors. Licensee MDPI, Basel, Switzerland. This article is an open access article distributed under the terms and conditions of the Creative Commons Attribution (CC BY) license (http://creativecommons.org/licenses/by/4.0/).

Article

Analytical Model to Compare and Select Creep Constitutive Equation for Stress Relief Investigation during Heat Treatment in Ferritic Welded Structure

Mengjia Hu [1], Kejian Li [1], Shanlin Li [1], Zhipeng Cai [1,2,3,*] and Jiluan Pan [1]

1. Department of Mechanical Engineering, Tsinghua University, Beijing 100084, China; hmj1078@163.com (M.H.); kejianli@mail.tsinghua.edu.cn (K.L.); shanlinli2015@163.com (S.L.); pjl-dme@mail.tsinghua.edu.cn (J.P.)
2. State Key Laboratory of Tribology, Tsinghua University, Beijing 100084, China
3. Collaborative Innovation Center of Advanced Nuclear Energy Technology, Beijing 100084, China
* Correspondence: caizhipeng92@outlook.com; Tel.: +86-010-6278-9568

Received: 9 April 2020; Accepted: 21 May 2020; Published: 23 May 2020

Abstract: The one-dimensional analytical model was promoted to help select the creep constitutive equation and predict heat treatment temperature in a ferritic welded structure, along with neglecting the impact of structural constraint and deformation compatibility. The analytical solutions were compared with simulation results, which were validated with experimental measurements in a ferritic welded rotor. The as-welded and post weld heat treatment (PWHT) residual stresses on the inner and outer cylindrical surfaces were measured with the hole-drilling method (HDM) for validation. Based on the one-dimensional analytical model, different effects of Norton and Norton-Bailey creep constitutive equation on stress relief during heat treatment in a ferritic welded rotor were investigated.

Keywords: residual stress; creep; stress relief; welded rotor

1. Introduction

Welding is widely used in the connection and manufacturing of thick-walled heavy mechanical components, considering the restraint of forging capacity and flexibility in design and fabricating [1]. However, due to the nonuniform heating and cooling processes encountered during welding, residual stress is bound to be generated in welded joints, which affects the yield strength [2,3], fatigue strength [4,5], and creep properties [6,7]. Heat treatment is an important approach to eliminate residual stress in large-scaled and heavy components. Nevertheless, it is expensive and inefficient to find the optimum heat treatment through an experimental measurement of residual stress. Thus, numerical simulation is widely employed to calculate the residual stress. The stress relief during heat treatment is considered as an effect of creep behavior [8], which can be estimated with creep constitutive equations [8–10]. The Norton equation or Norton-Bailey equation were widely employed in many researches [11–13]. It is found through an experiment that the Norton equation and Norton-Bailey equation have the same effect on eliminating residual stress at high temperatures [14]. However, there is a lack of detailed investigations on the different effects of Norton and Norton-Bailey equation on stress relief at different temperatures. In addition, the specific mechanism for the substitutability of creep equations at high temperatures is not clear and it is important to obtain the demarcation temperature, at which the two equations have substitutability in order to help select the appropriate creep equation in simulation depending on the employed PWHT temperature. Nevertheless, it is expensive and inefficient to obtain the demarcation temperature experimentally, which calls for another efficient method.

In this study, the one-dimensional analytical model was promoted to investigate the different effects of Norton and Norton-Bailey creep constitutive equation on stress relief during heat treatment in a ferritic welded rotor. The analytical solutions were compared with simulation results, which were validated with experimental measurements. It is found that the analytical model can be helpful to select the creep constitutive equation and predict the appropriate range of heat treatment temperature conveniently.

2. Material Properties

The base metal of welded ferritic rotor is 25Cr2Ni2MoV steel and its chemical compositions are listed in Table 1. The weld metal is similar to 25Cr2Ni2MoV steel in terms of chemical compositions and the same material property will be employed in simulation for the base metal and the weld metal. The material properties were measured experimentally, preparing for thermal analysis, mechanical analysis, and stress relief analysis corresponding to the welding process and heat treatment.

Table 1. Chemical compositions of 25Cr2Ni2MoV steel (wt. %).

C	Mn	Si	Ni	Mo	Cr	V	Co	Fe
0.25	0.65	0.07	0.73	1.10	2.40	0.30	—	Bal.

Thermal properties are shown in Figure 1, obtained with the JMatPro software (version 7.0, Sente Software Ltd., Guildford, UK). Figure 2 presents the shape and dimensions of the creep test specimens and tensile test specimens. Uniaxial tensile tests were conducted at temperatures from 25 to 850 °C. Phase transformation during the heating and cooling process was considered. A portion of tensile test specimens were heated up to 850 °C and cooled to the test temperature to obtain the microstructure corresponding to the cooling process before conducting a tensile test. Mechanical properties for the mechanical analysis are listed in Figure 3 and properties over 850 °C were inferred. The thermal strain in Figure 3a is composed of a thermal expansion and expansion from phase transformation. It is obvious that phase transformation occurred in the process of heating up and cooling down. The linear isotropic hardening rule was employed in the investigation.

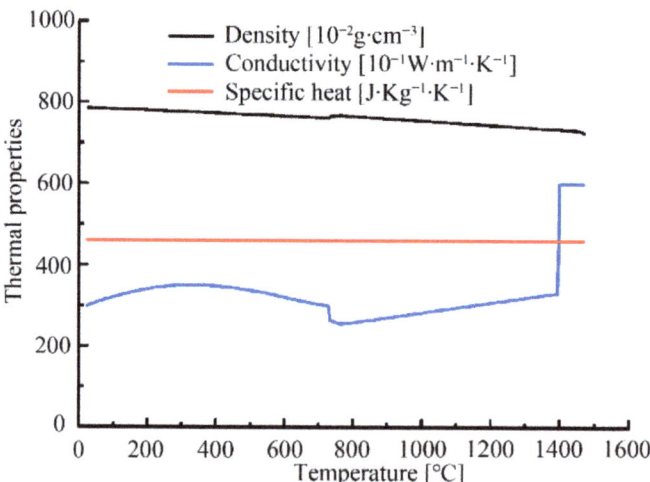

Figure 1. Thermal properties of 25Cr2Ni2MoV steel.

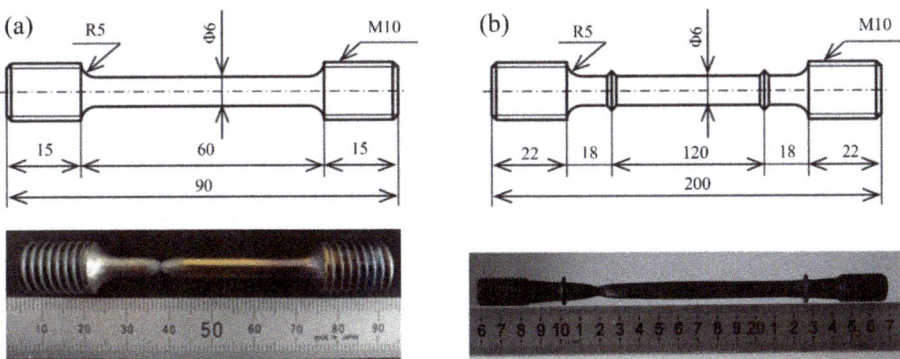

Figure 2. (a) Shape and dimensions of tensile test specimens and (b) creep test specimens of 25Cr2Ni2MoV steel (unit: mm).

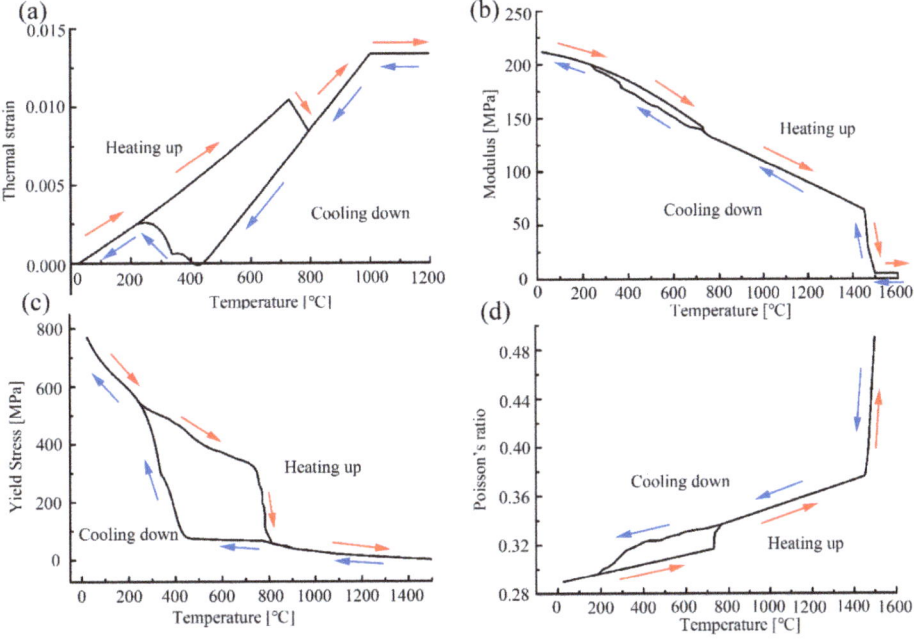

Figure 3. Mechanical properties of 25Cr2Ni2MoV steel versus temperature: (a) thermal strain; (b) Young's modulus; (c) yield stress and (d) Poisson's ratio.

Creep occurs at temperatures around and over $0.4T_m$ [15], where T_m corresponds to the melting temperature. The melting temperature of 25Cr2Ni2MoV is approximately 1360 °C. It is indicated that the stagnation temperature of 25Cr2Ni2MoV to creep is approximately 380 °C. It is also reported that parts of M/A constituents were deposited into an accumulated carbide at 650 °C [16]. Therefore, creep tests were carried out at 500, 560, 600, and 630 °C, preventing the M/A constituent deposition. Table 2 lists the conditions for creep tests. Material coefficients of the Norton equation and Norton-Bailey equation were extracted from experimental creep strain-time data, neglecting diffusional creep. The Norton equation is written as follows:

$$\dot{\varepsilon}_m = A\sigma^n \tag{1}$$

where $\dot{\varepsilon}_m$ refers to the minimum creep rate and σ is the effective stress in the specimen. The coefficients A and n are listed in Table 3. The Norton-Bailey equation is written as follows:

$$\dot{\varepsilon}_{cr} = B\sigma^u t^{-m} \tag{2}$$

where $\dot{\varepsilon}_{cr}$ refers to the creep rate and t is the total creep time. The coefficients B, u, and m are also listed in Table 3.

Table 2. Conditions for creep tests.

Temperature [°C]	Applied Stress [MPa]
500	200, 240, 280, 320
560	140, 180, 220, 260
600	90, 120, 150, 180
630	50, 80, 110, 140

Table 3. Material properties of 25Cr2Ni2MoV for the Norton and Norton-Bailey equation.

Temperature [°C]	A	n	B	u	m
500	2.31×10^{-27}	9.57	7.14×10^{-19}	6.61	0.67
560	4.03×10^{-21}	7.69	1.96×10^{-13}	4.56	0.52
600	1.22×10^{-17}	6.62	6.57×10^{-9}	2.32	0.48
630	3.86×10^{-15}	5.75	4.26×10^{-8}	1.98	0.42

3. One-Dimensional Analytical Model and Analytical Solutions

The dominant mechanism for stress relief in ferritic steel is creep strain relaxation and a little plastic strain occurs during the heat treatment [8]. Based on the creep strain relaxation mechanism, differences between Norton-Bailey and Norton equation in stress relaxation can be analyzed. The one-dimensional analytical model was promoted to neglect the impact of structural constraint and deformation compatibility, as shown in Figure 4. The simplified model was heated to a preset temperature T °C and a predetermined load σ was applied. Then, the end position was constrained and the stress inside was residual stress, which would be released with creeping at T °C. After sufficient heat treatment, the creep rate $\dot{\varepsilon}$ in any part was small enough that the creep strain under the action of finite time t was small and negligible, that is:

$$\dot{\varepsilon} t \propto 1/\infty \tag{3}$$

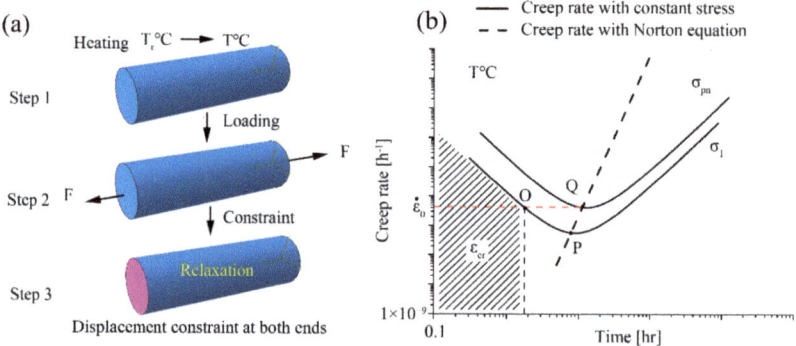

Figure 4. (a) The one-dimensional analytical model and (b) diagram of the small creep rate at the end of creep strain accumulation in heat treatment corresponding to Norton equation and Norton-Bailey equation using the creep rate data from a creep rupture test with constant stress.

The small creep rate was denoted as $\dot{\varepsilon}_0$. According to Equation (1), the residual stress after heat treatment under Norton equation in the one-dimensional analytical model (σ_{pn}) was solved as follows:

$$\sigma_{pn} = (\dot{\varepsilon}_0/A)^{1/n} \quad (\sigma \geq \sigma_{pn}) \tag{4}$$

Creep strain relaxation mechanism during heat treatment could be described as follows:

$$d\sigma = -Ed\varepsilon_{cr} \tag{5}$$

Combining Equations (1) and (5), the time required was formulated as follows:

$$t_n = \frac{\sigma^{1-n} - \sigma_{pn}^{1-n}}{(1-n)EA} \quad (\sigma \geq \sigma_{pn}) \tag{6}$$

Similarly, the residual stress after heat treatment under Norton-Bailey equation in the one-dimensional analytical model (σ_{pb}) could be solved according to Equations (2) and (5). The Norton-Bailey equation was rewritten in the form of strain-hardening, as illustrated in Figure 4b. The shaded region referred to the accumulated creep strain ε_{cr}. The equation in the form of strain hardening was as follows:

$$\dot{\varepsilon}_{cr} = (B\sigma^u)^{\frac{1}{1-m}} [(1-m)\varepsilon_{cr}]^{\frac{-m}{1-m}} \tag{7}$$

Then, the residual stress could be described as follows:

$$\sigma = \sigma_{pb} + \frac{E}{1-m} \left(B\sigma_{pb}^u\right)^{\frac{1}{m}} (\dot{\varepsilon}_0)^{\frac{1-m}{-m}} \tag{8}$$

The time required could also be formulated as follows:

$$t_b = \left[\frac{1-m}{(1-u)EB}\left(\sigma^{1-u} - \sigma_{pb}^{1-u}\right)\right]^{\frac{1}{1-m}} \tag{9}$$

Considering the engineering practice, the finite time for a negligible creep strain could be set as $t_{thr} = 20$ h. Taking the availability of stress relief into consideration, the deviation of stress relief at room temperature was set as $\sigma_{thr} \leq 10$ MPa. Therefore, $\dot{\varepsilon}_0$ could be obtained.

$$\dot{\varepsilon}_0 = \sigma_{thr}/(E \cdot t_{thr}) = 2.4 \times 10^{-6}/h \tag{10}$$

Based on Equations (4) and (8), the relationship between initial residual stress and PWHT residual stress under the control of Norton equation and Norton-Bailey equation can be obtained easily, which can help select the creep constitutive equation and predict the proper range of heat treatment temperature in a ferritic welded structure.

According to Equations (4) and (8), Figure 5 presents the relationship between initial stress σ and PWHT stress with Norton and Norton-Bailey equations in the one-dimensional analytical model. It indicates that stress relief will not occur when the initial stress is lower than σ_{pn} under Norton equation. When the initial stress is higher than σ_{pn}, the PWHT residual stress is the same, being σ_{pn} with Norton equation. Meanwhile, there is a positive correlation between PWHT residual stress and initial stress with Norton-Bailey equation being higher. It can be explained with the evolution of a dislocation structure and free dislocation density during heat treatment. In the process of stress relief by creep strain, a higher stress is required to obtain the same creep rate with dislocation entanglement and free dislocation density reduction, which means a greater creep resistance. On the other hand, creep strain releases stress, which weakens the driving force of creep. If the PWHT residual stress is the same, a greater creep strain corresponds to a higher initial stress, which means a greater creep resistance and smaller creep rate. Thus, only that a higher PWHT residual stress corresponds to a higher initial stress can ensure that the final creep rate is the same. It means that a higher initial stress leads to a higher PWHT residual stress with Norton-Bailey equation.

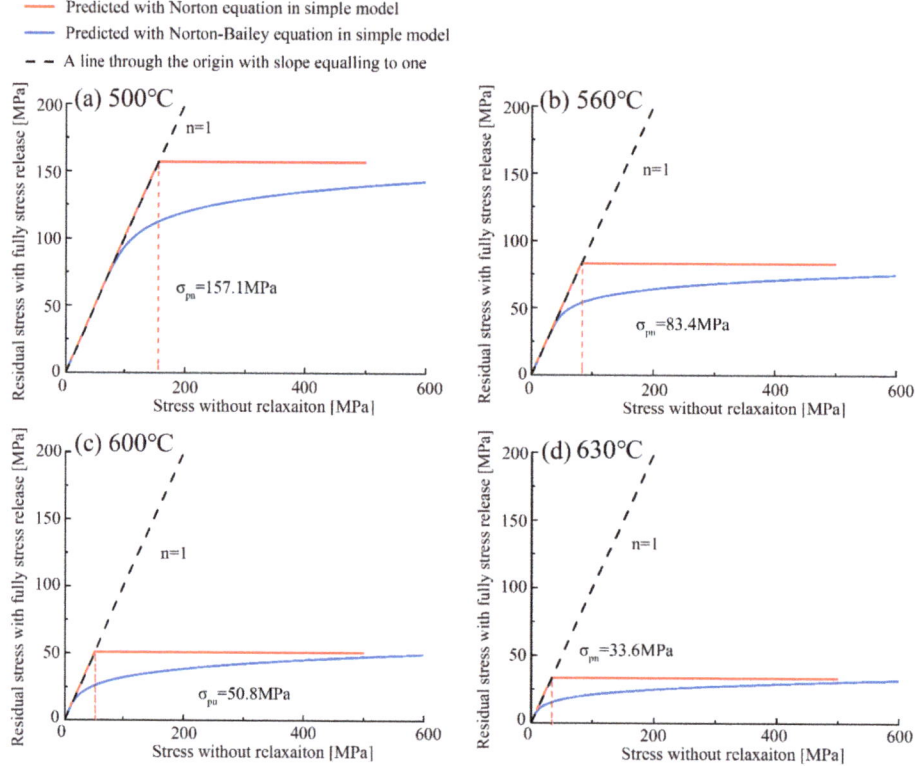

Figure 5. Relationship between initial stress and post heat-treatment stress with Norton and Norton-Bailey equation at different temperatures: (**a**) 500 °C; (**b**) 560 °C; (**c**) 600 °C and (**d**) 630 °C.

It is noted that the PWHT residual stress is higher with Norton equation than that with Norton-Bailey equation. It can be explained in Figure 4 that the PWHT residual stress is σ_1 with Norton-Bailey equation and σ_{pn} with Norton equation. It is obvious that σ_1 is lower than σ_{pn}.

With the PWHT temperature increasing, more stress will be released and the difference between Norton equation and Norton-Bailey equation decreases. When the PWHT temperature exceeds 600 °C, the difference is small enough to be neglected.

The analytical solutions above were compared with simulation results and experimental measurements for assessment and validation in Sections 6.2 and 6.3.

4. Simulation Procedure

4.1. Welding Simulation

A subsequent-coupled simulation was employed in the welding simulation, consisting of a thermal analysis and mechanical analysis. The temperature history from the thermal analysis was used as a predefined field for mechanical analysis. Then, the as-welded residual stress information from the mechanical analysis was used for stress relief analysis, corresponding to heat treatment. The axisymmetric model of the rotor was presented in Figure 6, composed of 6633 nodes and 6685 elements. The weld zone was meshed with 1342 elements and the size of elements was about

2 mm × 1 mm. The weld zone and heat affected zone (HAZ) were also meshed densely so as to calculate the distribution of stresses and strains in those zones. The weld was composed of 108 weld beads, corresponding to the actual situation. It was noted that the heat transfer element and axisymmetric stress element were employed in thermal analysis and mechanical analysis, respectively.

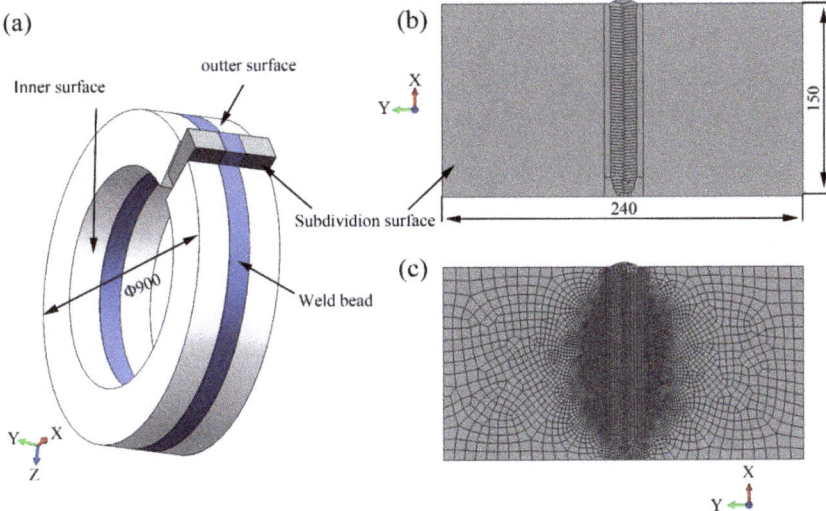

Figure 6. Simulation model of welded rotor (unit: mm): (**a**) Three-dimensional stereogram; (**b**) subdivision surface along the radial direction; (**c**) mapped meshing.

In thermal analysis, a double-ellipsoidal heat source [17] was employed to present the heat effect of welding arc. The front heat source and the rear heat source can be described with Equations (11) and (12), respectively.

$$q_1(x,y,z,t_w) = \frac{6\sqrt{3}f_1 Q}{a_f bc\pi \sqrt{\pi}} e^{\frac{-3(z-vt_w-z_0)^2}{a_f^2}} e^{\frac{-3x^2}{b^2}} e^{\frac{-3y^2}{c^2}} \quad (11)$$

$$q_2(x,y,z,t_w) = \frac{6\sqrt{3}f_2 Q}{a_r bc\pi \sqrt{\pi}} e^{\frac{-3(z-vt_w-z_0)^2}{a_r^2}} e^{\frac{-3x^2}{b^2}} e^{\frac{-3y^2}{c^2}} \quad (12)$$

$$Q = \eta UI \quad (13)$$

where f_1 and f_2 correspond to the fraction of heat, deposited in the front and rear zones. It is noted that $f_1 + f_2 = 2$. Parameters a_f, a_r, b, and c determine the shape of the heat source in a three-dimension space together. Q is the heat input power, v is the welding speed, t_w is the welding time, and z_0 is the offset distance in the welding direction. Heat efficiency η was assumed to be 88%, corresponding to submerged arc welding. Parameters, a_r, b, and c were obtained by an inverse method [18]. The values of a_f, a_r, b, and c were set as 5.8, 8.3, 5.7, and 11.4, respectively [19]. Heat convection and radiation were taken into consideration as the thermal boundary condition, which was set with the subroutine named FILM in ABAQUS (version6.14-1, Dassault Systemes Simulia Corp., Johnston, RI, USA) according to the sink temperature of cylindrical surface. In order to simplify the calculation, the heat exchange between the external cylindrical surface and the environment was described by an equivalent convection heat exchange [20]. The equivalent heat transfer coefficient h could be described as Equation (14).

$$h = \frac{h_c(T - T_0) + \varepsilon C_0\left[(T + 273.15)^4 - (T_0 + 273.15)^4\right]}{T - T_0} \tag{14}$$

where T is the temperature on the external cylindrical surface and T_0 is the temperature of the environment. Parameters h_c and ε describe the coefficient of heat convection and heat radiation, respectively. In this study, the values of h_c and ε were set as 80 W/m²·s and 0.8.

In mechanical analysis, mechanical properties of the material presented in Section 2 were employed in the way shown in [19]. The same simulation procedure was conducted to obtain the as-welded residual stress, which was used for stress relief analysis.

4.2. Stress Relief Analysis in Heat Treatment

A fully-coupled simulation was employed in stress relief analysis, which calculated the temperature field, elastic-plastic strain, and creep strain at the same time. The as-welded residual stress information from the mechanical analysis was preset initially by mapping residual stresses and plastic strains onto the same axisymmetric model in nodes with the Predefined-Field module in ABAQUS. The selected element type is CAX4RT (A 4-node thermally coupled axisymmetric quadrilateral element with bilinear displacement and temperature, reduced integration and hourglass control). Creep strain was considered during holding time when the temperature remained, reducing the calculating time and space. The sink temperature was set corresponding to the preset temperature curve in the furnace with subroutine FILM in ABAQUS. Heat convection and radiation were also set as the thermal boundary condition. Rigid body movement was restricted with an angular point constrained.

Creep equations estimating creep behavior were described and obtained in Section 2. The creep equation was employed in the form of strain-hardening with the Material-Property module in ABAQUS. The time index for Norton equation was zero, which neglected the strain hardening during creep strain accumulation. The simulation routine called variables including creep strains and stresses from each node to calculate the increment of creep strains with creep equations during every time increment in simulation, which contributed to strain coordination.

4.3. Numerical Experiments

Table 4 listed the experimental scheme for the study. Norton and Norton-Bailey equations were employed in simulation at 500, 560, 600, and 630 °C for comparison. A welded specimen was manufactured and heat-treated at 560 °C. As-welded and PWHT residual stresses were measured with the hole-drilling method in order to validate the simulation model.

Table 4. Experimental scheme for analysis.

Simulation and Experiment	500 °C	560 °C	600 °C	630 °C
Simulation with Norton	○	○	○	○
Simulation with Norton-Bailey	○	○	○	○
Experimental validation		√		

Note: ○—conducted simulation, √—conducted measurement

5. Experimental Validation

5.1. Welded Specimen and Heat Treatment

Two ring-shaped members were welded with submerged arc welding to manufacture the rotor, of which the outer diameter and inner diameter were approximately 900 and 600 mm respectively, as shown in Figure 7a. The axial length of the rotor was approximately 240 mm and the cap weld was approximately 30 mm in width. The base metal was 25Cr2Ni2MoV steel and the weld metal was similar to the base metal in terms of chemical compositions. Subsequently, the same thermal and elastic-plastic properties were employed in the numerical simulation, as shown in Section 2. The geometry of the welded rotor was determined according to one product from the Shanghai Turbine Plant. The voltage and electricity of submerged arc welding were 25–35 V and 450–500 A, respectively. The rotor was machined to eliminate the weld cap immediately after welding so as to facilitate stress measurement. The welded rotor was treated at 560 °C for 20 h in a furnace post welding. The temperature setting in the furnace is shown in Figure 8a. The temperature and operating time were determined based on the balance of stress relief and mechanical property improvement while considering time efficiency. The temperature should be 100–200 °C below the transformation temperature from ferrite to austenite [21], which is approximately 510–610 °C for 25Cr2Ni2MoV steel. The holding time is decided by the operability under the engineer condition, taking into account the evolution of the toughness and microstructure. Finally, a heat treatment at 560 °C for 20 h was employed in the investigation.

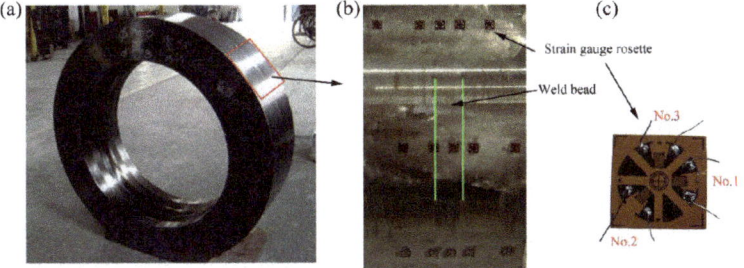

Figure 7. Welded rotor (**a**), deployment of strain gauge rosettes (**b**) and strain gauge rosette (**c**).

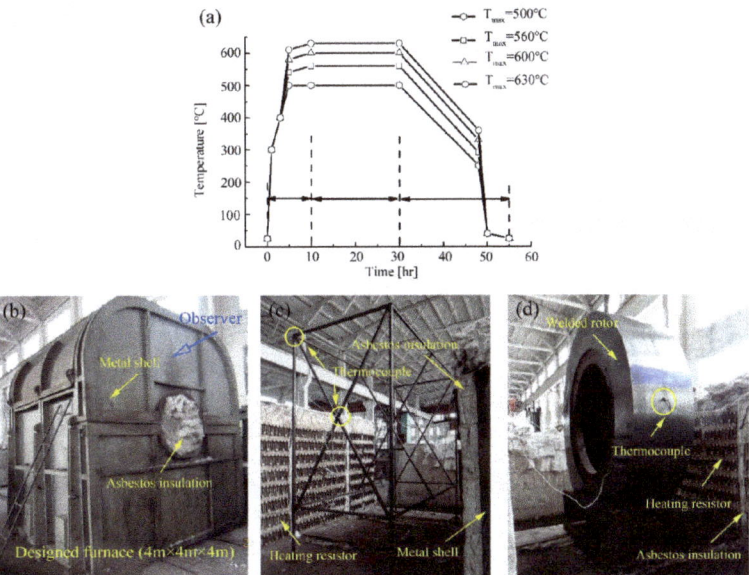

Figure 8. Temperature setting in furnace-based heat treatment for welded 25Cr2Ni2MoV steel rotor (**a**) and the furnace-based heat treatment: (**b**) The designed furnace (4 m × 4 m × 4 m); (**c**) interior of the designed furnace; (**d**) another welded rotor with a larger size in the furnace.

A furnace was employed, as shown in Figure 8b. It consists of seven assembled metal structures with asbestos insulations attached on the inner surface to block heat loss. Dimensions of the furnace are 4 m in length, 4 m in width, and 4 m in height. Heating resistors were attached on asbestos insulations on vertical walls to produce heat. The temperature was monitored with 12 thermocouples in the furnace and controlled with feedback regulation. As shown in Figure 8a, time for the heating up process was 10 h and holding time was 20 h. In the cooling process, the temperature evolved from 560 °C to room temperature in 25 h.

5.2. Residual Stress Measurement with HDM

The hole-drilling method (HDM) was employed to measure residual stresses on outer and inner cylindrical surfaces post weld and heat treatment. Stresses obtained with HDM are the average stress within the drilled hole [22], which is assumed to be in the plane stress condition. In this study, the depth and diameter of the hole were 2 and 1.5 mm, respectively. The diameter is the appropriate size, which was minimized such that the drill would not break and to reflect the residual stress accurately. The principal stresses and their direction can be expressed as follows:

$$\sigma_{max} = \frac{\varepsilon_1 + \varepsilon_3}{4A_r} + \frac{1}{4B_r}\sqrt{(\varepsilon_3 - \varepsilon_1)^2 + (\varepsilon_3 + \varepsilon_1 - 2\varepsilon_2)^2} \qquad (15)$$

$$\sigma_{min} = \frac{\varepsilon_1 + \varepsilon_3}{4A_r} - \frac{1}{4B_r}\sqrt{(\varepsilon_3 - \varepsilon_1)^2 + (\varepsilon_3 + \varepsilon_1 - 2\varepsilon_2)^2} \qquad (16)$$

$$\tan 2\gamma = \frac{\varepsilon_1 - 2\varepsilon_2 + \varepsilon_3}{\varepsilon_1 - \varepsilon_3} \qquad (17)$$

where σ_{max} and σ_{min} are principal stresses; ε_1, ε_2, and ε_3 present the relieved strain at each corresponding gauge; γ is the angle from σ_{max} to gauge no. 1. The stress relief coefficients A_r and B_r are 0.07255 and 0.1514, respectively. Meanwhile, the additional strain ε_m caused by machining is $-39~\mu\varepsilon$.

The strain gauge rosette used in the measurement was denoted TJ120-1.5-Φ1.5, a size of which was 10 mm × 10 mm as shown in Figure 7c. The strain gauge rosette was composed of three gauges in different directions on the circumference. The size of each gauge was 1.5 mm × 1.5 mm and the resistance was 120 Ω. The strain gauge rosette could be applied to HDM with a diameter of the hole being 1.5 mm. The sensitivity coefficient of each resistance strain gauge was 2.07 ± 0.01, which indicated that the maximum error was about 0.97%. The alcohol solution containing 4% nitric acid (volume fraction) was used to corrode the weld in order to locate the strain gauge rosette accurately. The weld seam and heat affected zone can be distinguished clearly on the corroded weld. The ruler and scriber were employed to help determine the location and direction of the strain gauge rosette. The strain gauge rosette was glued to the weld with gauge no. 1 along the axial direction and gauge no. 3 along the circumferential direction. The strain gauge rosettes were located at 0, 13, 15, and 65 mm away from the weld center line to visualize stress distribution, similar to those shown in Figure 6b. In order to reduce random error, the residual stress of each test point was measured three times and averaged. The measured results have also been presented together with the simulation results. The measured data indicated that the error was within 10 MPa, which resulted from the deviation of sensitivity coefficient, patch position, and direction of strain gauge rosette. Furthermore, a deviation of weld width could result in a deviation of patch position relative to the heated affected zone, although the weld width fluctuated to 30 ± 0.1 mm.

6. Results and Discussion

6.1. As-Welded Residual Stress

Contour plots of axial stress and hoop stress on the axisymmetric surface are presented in Figure 9. Axial stress and hoop stress in the cap weld are compressive, which may result from martensitic transformation at low temperatures [23]. The compressive stresses are balanced by tensile stress under the cap in the weld. It is also noted that the central region of the axisymmetric surface is characterized by compressive axial stresses and tensile hoop stresses. The compressive axial stresses are balanced by the stresses in the other part. Meanwhile, tensile hoop stresses in the weld bead are balanced by compressive stresses alongside the weld. It also indicates that axial stress and hoop stress on the inner surface in the weld are compressive and tensile, respectively. As shown in Figure 9, the hoop residual stress in the weld bead is extremely high, being approximately 891 MPa. However, the yield stress of 25Cr2Ni2MoV steel is approximately 780 MPa at room temperature, as shown in Figure 3c. It can be explained that the linear isotropic hardening rule was taken into consideration in the investigation.

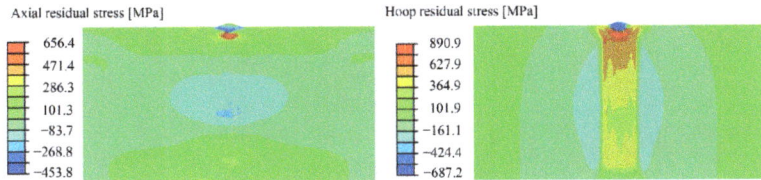

Figure 9. Contour plots of as-welded axial stress and hoop stress.

Comparing simulation results with experimental data, residual stress distribution from the simulation is also plotted along experimental scan lines in Figure 10. The stress distribution confirms general trends presented in Figure 9. It is also noted that simulation results are validated with experimental data in Figure 10. Some discrepancies come out between simulation results and experimental data, which may result from experimental measurement errors and the axisymmetric model [24,25]. The axisymmetric model neglects restriction from a colder region in front of the arc and after the arc.

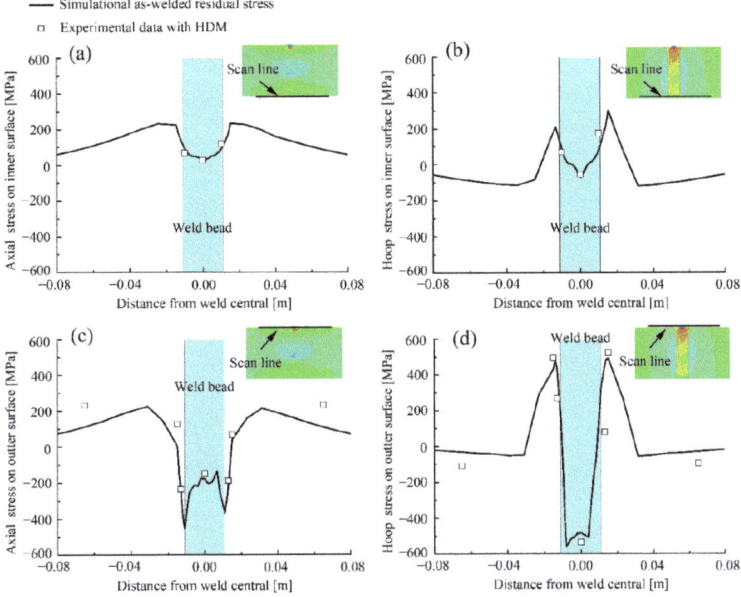

Figure 10. Comparison between simulation results and experimental data along the inner surface: (**a**) As-welded axial stress; (**b**) as-welded hoop stress; and along the outer surface: (**c**) As-welded axial stress; (**d**) as-welded hoop stress.

6.2. Stress Relaxation during PWHT

Figure 11 presents contour plots of the PWHT axial and hoop residual stress on axisymmetric faces, including the PWHT residual stress with Norton equation (Figure 11a) and Norton-Bailey equation (Figure 11b). The PWHT residual stress distribution is similar to the as-welded residual stress (Figure 9) and is numerically close to a stress-free state. As shown in Figure 9, the as-welded residual stress is concentrated in the weld and heat affected zone, where stress relaxation is induced during heat treatment. Compared with the as-welded residual stress, stress relaxation reduced the tensile hoop residual stress in the weld and tensile hoop residual stress appeared in the vicinity of the outer surface of the base metal owing to a mechanical equilibrium, as shown in Figure 11. The axial residual stress distribution trend remained almost the same as the as-welded residual stress. The axial stress in the weld on the inner surface evolved from tensile stress to compressive stress after PWHT, implying the strain coordination in PWHT.

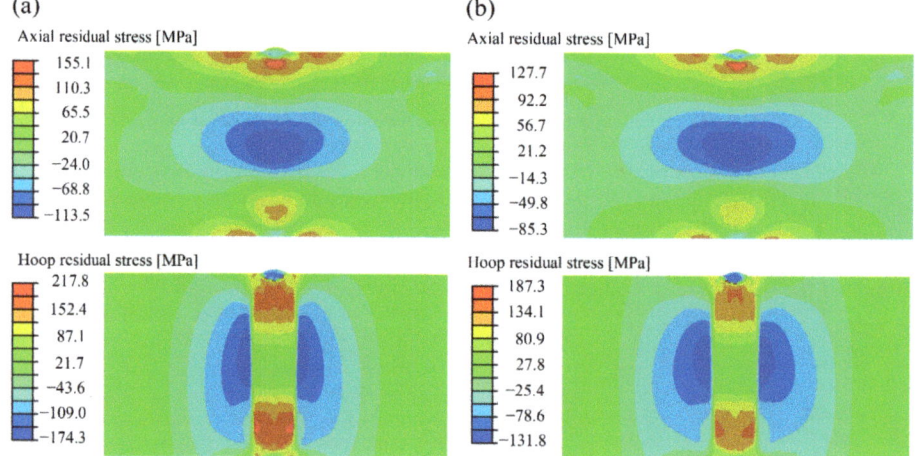

Figure 11. Contour plots of the post weld heat treatment (PWHT) axial stress and hoop stress: (**a**) With Norton equation and (**b**) with Norton-Bailey equation at 560 °C.

Comparing simulation results and experimental data, residual stress distributions from the simulation with Norton equation and Norton-Bailey equation are plotted along lines on the outer and inner surfaces in Figure 12. Simulation results with Norton-Bailey equation are more accurate, which is validated by experimental data.

Based on the numerical simulation, an effect of temperature on eliminating residual stress during heat treatment can be analyzed. Considering that the von-Mises stress is a yield criteria and the main driving force for creep, as shown in Equation (18) corresponding to Norton-Bailey creep model.

$$\dot{\varepsilon}_{ij} = \frac{3}{2} A \bar{\sigma}^{n-1} s_{ij} t^m \tag{18}$$

where $\dot{\varepsilon}_{ij}$ is the deviator strain rate, $\bar{\sigma}$ is the von-Mises stress (MISES), and s_{ij} is the deviator stress. Figure 13 presents the PWHT residual stress at different temperatures in the form of MISES along the outer line. It indicates that residual stresses simulated with Norton equation is higher in number than that simulated with Norton-Bailey equation. The difference between residual stresses simulated with Norton equation and Norton-Bailey equation decreases as the heat treatment temperature increases. The residual stresses in weld bead are almost the same at 600 and 630 °C. Meanwhile, it is obvious that more stress is eliminated when the heat treatment temperature is higher. All of the abovementioned phenomena are also presented with analytical solutions, as shown in Figure 5.

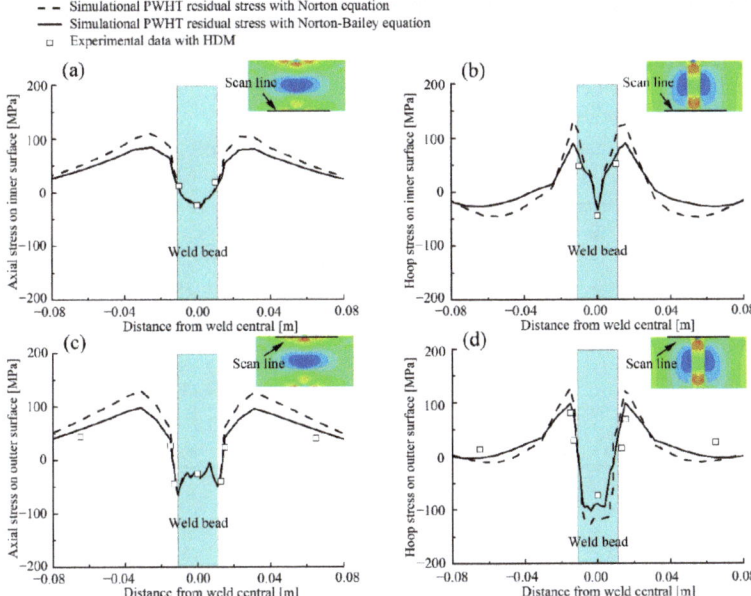

Figure 12. Comparison between PWHT simulation results and experimental data along the inner cylindrical surface: (**a**) PWHT axial stress; (**b**) PWHT hoop stress; and along the outer cylindrical surface: (**c**) PWHT axial stress; (**d**) PWHT hoop stress.

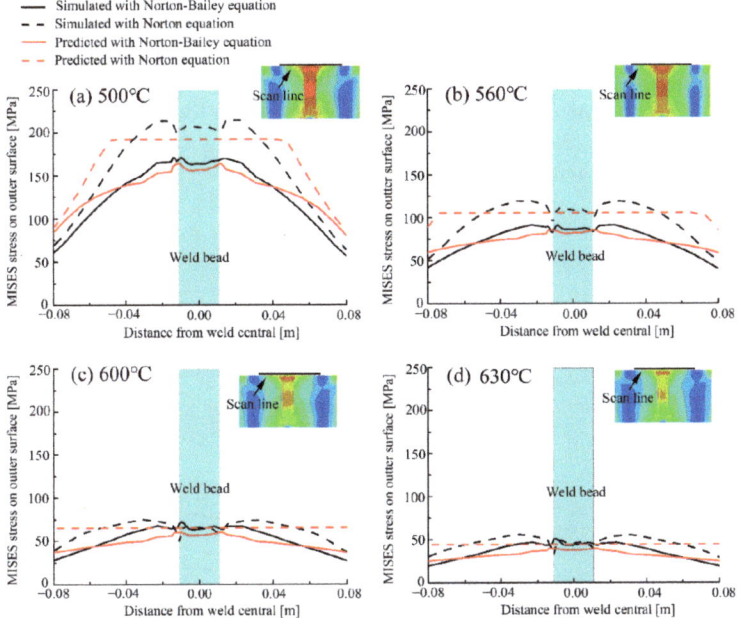

Figure 13. PWHT residual stress in the form of MISES along the outer line post heat treatment at different temperatures: (**a**) 500 °C; (**b**) 560 °C; (**c**) 600 °C and (**d**) 630 °C.

Table 5 lists the calculating time and space employed in simulations with Norton equation and Norton-Bailey equation. It indicates that simulations with Norton equation have a great superiority in both computing time and space. It is predicted that the superiority will be more evident in large-scaled and complex structures. Considering the discussion above, Norton-Bailey equation should be employed at low temperatures to ensure the calculation accuracy. Meanwhile, Norton equation should be used at high temperatures to reduce the calculation complexity. It seems that Norton equation should be employed at temperatures higher than 600 °C in this study. For other cases, it is important to find the demarcation temperature conveniently, which can be obtained easily with the one-dimensional analytical model, as shown in Section 3.

Table 5. Calculating time and space employed in simulations.

Property	Creep Equation	500 °C	560 °C	600 °C	630 °C
Calculating time [h]	Norton equation	2.32	2.35	2.41	2.37
	Norton-Bailey equation	4.21	4.14	4.29	4.25
Calculating space [Gb]	Norton equation	1.65	1.71	1.86	1.82
	Norton-Bailey equation	3.51	3.26	3.72	3.63

6.3. Comparison between Analytical Solutions and Simulation Results

The analytical model was assessed with numerical simulation and experimental results. The distribution on the outer cylindrical surface was predicted with the analytical model, as shown in Figure 13 with an imaginary line. Figure 13 indicates that stresses predicted with the analytical model in the weld bead and adjacent areas are lower than those simulated results with Norton equation or Norton-Bailey equation, separately. While stresses predicted with the analytical model in the base metal are higher than those simulated results. It can be explained that the stresses are released out of proportion in a three-dimensional simulation, which leads to compatible deformation. The deformation coordination contributes to that a higher initial stress corresponds to a higher residual stress. Figure 14 presents the relationship between initial stresses and residual stresses at different temperatures, which are all divided by Young's modulus to obtain the elastic strain. As shown in Figure 14a, the elastic strains for creep initiation are smaller at a higher temperature, which will also lead to deformation coordination during the process of heating up. Considering the discussion above, the results predicted with the analytical model will introduce deviation owing to neglecting deformation coordination. It is necessary to obtain accurate results with numerical simulation to take more relevant factors into account. However, the analytical model can help investigate the approximate temperature range for heat treatment, taking σ_{pn} as the largest PWHT residual stress, as shown in Figures 5 and 13.

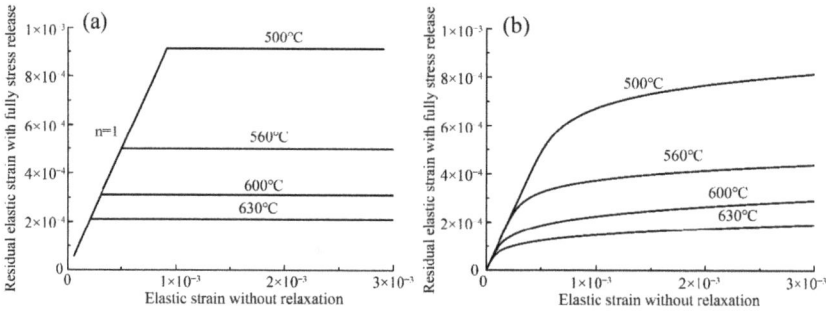

Figure 14. Relationship between initial stresses and residual stresses at different temperatures, which are all divided by Young's modulus to obtain the elastic strain: (**a**) With Norton equation and (**b**) with Norton-Bailey equation.

6.4. Effect of PWHT Temperature on Stress Relaxation and Creep Equation Selection

As shown in Figure 13, deviation between residual stresses predicted with Norton equation and Norton-Bailey equation decreases as temperature increases. With the temperature increasing, atomic diffusion plays a greater role in creeping rather than stress, which contributes to a lower PWHT residual stress and decreases the deviation. Figure 13 indicates that the deviation is very small at 600 and 630 °C, which is also present in Figure 5. In order to help select creep equation and investigate the approximate temperature range for heat treatment, Figure 14 can be plotted in a graph as shown in Figure 15. The lines corresponding to Norton and Norton-Bailey equation at 600 and 630 °C are close to each other. Therefore, the lines corresponding to Norton-Bailey equation have been neglected. Given that the as-welded residual stress and the expected PWHT residual stress are known as the Node M, as shown in Figure 15, it is obvious that the heat treatment temperature should be between 500 and 560 °C. Meanwhile, Norton-Bailey equation should be employed in the simulation to obtain PWHT residual stresses in a three-dimensional model.

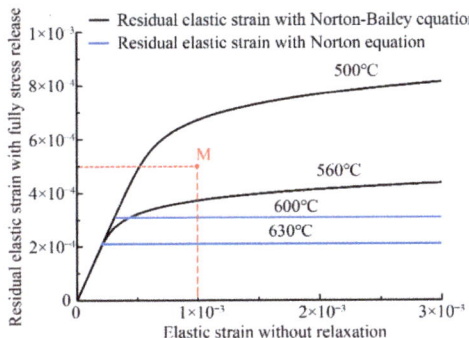

Figure 15. Relationship between initial elastic strains and residual elastic strains at different temperatures, combining Norton equation and Norton-Bailey equation.

Figure 16 obtains the evolution of MISES stress and creep strain at Node A during the process of heat treatment at 560 °C. It indicates that creep strain was accumulated and stress was released in a short time. Taking Figures 13 and 16 into consideration, it indicates that the heat treatment temperature is more effective than the holding time. Therefore, the holding time for heat treatment should be determined based on the requirement of mechanical properties and evolution of microstructures during heat treatment.

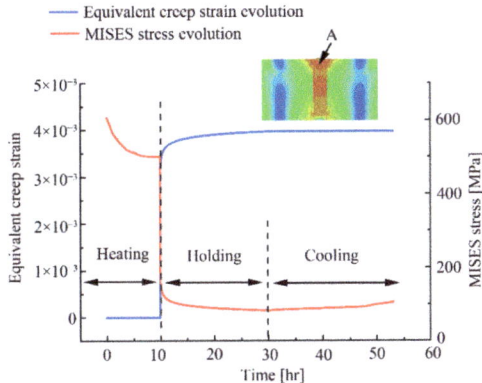

Figure 16. Evolution of MISES stress and creep strain at Node A during heat treatment at 560 °C.

7. Conclusions

The one-dimensional analytical model was promoted to investigate different effects of Norton and Norton-Bailey creep constitutive equation on stress relief during heat treatment in a ferritic welded rotor. The analytical solutions were compared with simulation results, which were validated with experimental measurements. The following conclusions can be drawn:

(1) Norton equation should be employed in the stress relief simulation instead of Norton-Bailey equation to reduce the calculation complexity exceeding 600 °C.
(2) The PWHT residual stress calculated with Norton equation is higher than that with Norton-Bailey equation.
(3) With the PWHT temperature increasing, the deviation between Norton equation and Norton-Bailey equation decreases.
(4) The one-dimensional analytical model was promoted to help select a creep constitutive equation and predict an appropriate range of heat treatment temperature, along with neglecting the impact of structural constraint and deformation compatibility.
(5) Deformation coordination plays an important role in stress relaxation in heat treatment.

Author Contributions: Conceptualization, M.H., K.L., and Z.C.; methodology, M.H. and S.L.; resources, Z.C. and J.P.; writing—original draft preparation, M.H.; writing—review and editing, K.L. and Z.C.; project administration, K.L., Z.C., and J.P. All authors have read and agreed to the published version of the manuscript.

Funding: The research work presented in this paper has been supported by the Shanghai Turbine Plant and National Natural Science Foundation of China (No. 51775300 and No. 51901113). They are profoundly appreciated. The authors also express their gratitude to Wangteng Cai (Fox Valley Lutheran High School) for help in the writing of the article.

Conflicts of Interest: The authors declare no conflict of interest.

References

1. Li, Y.; Li, K.; Cai, Z.; Pan, J.; Liu, X.; Wang, P. Alloy design of welding filler metal for 9Cr/2.25Cr dissimilar welded joint and mechanical properties investigation. *Weld. World* **2018**, *62*, 1137–1151. [CrossRef]
2. Webster, G.; Ezeilo, A. Residual stress distributions and their influence on fatigue lifetimes. *Int. J. Fatigue* **2001**, *23*, 375–383. [CrossRef]
3. Marshall, D.B.; Lawn, B.R. Residual stress effects in sharp contact cracking. *J. Mater. Sci.* **1979**, *14*, 2001–2012. [CrossRef]
4. Cheng, X. Residual stress modification by post-weld treatment and its beneficial effect on fatigue strength of welded structures. *Int. J. Fatigue* **2003**, *25*, 1259–1269. [CrossRef]
5. Bussu, G.; Irving, P.E. The role of residual stress and heat affected zone properties on fatigue crack propagation in friction stir welded 2024-T351 aluminum joints. *Int. J. Fatigue* **2003**, *25*, 77–88. [CrossRef]
6. Turski, M.; Bouchard, P.J.; Steuwer, A.; Withers, P.J. Residual stress driven creep cracking in AISI Type 316 stainless steel. *Acta Mater.* **2008**, *56*, 3598–3612. [CrossRef]
7. Francis, J.; Mazur, W.; Bhadeshia, H.K.D.H. Review Type IV cracking in ferritic power plant steels. *Mater. Sci. Technol.* **2006**, *22*, 1387–1395. [CrossRef]
8. Dong, P.; Song, S.; Zhang, J. Analysis of residual stress relief mechanisms in post-weld heat treatment. *Int. J. Press. Vessel. Pip.* **2014**, *122*, 6–14. [CrossRef]
9. Wang, J.; Lu, H.; Murakawa, H. Mechanical behavior in local post weld heat treatment (report i): Visco-elastic-plastic fem analysis of local pwht (mechanics, strength & structure design). *Trans. JWRI* **1998**, *27*, 83–88.
10. Ueda, Y.; Fukuda, K. Analysis of welding stress relieving by annealing based on finite element method. *Trans. JWRI* **1975**, *4*, 39–45.
11. Yanagida, N.; Ogawa, K.; Saito, K.; Kingston, E. Study on Residual-Stress Redistributions During the Process of Manufacture of a Vessel Penetration Set-On Joint. In Proceedings of the ASME 2009 Pressure Vessels and Piping Conference, Prague, Czech Republic, 26–30 July 2009.

12. Ogawa, K.; Okuda, Y.; Saito, T.; Hayashi, T.; Sumiya, R. Welding residual stress analysis using axisymmetric modeling for shroud support structure. In Proceedings of the ASME 2008 Pressure Vessels and Piping Conference, Chicago, IL, USA, 27–31 July 2008.
13. Udagawa, M.; Katsuyama, J.; Onizawa, K. Effects of residual stress by weld overlay cladding and PWHT on the structural integrity of RPV during PTS. In Proceedings of the ASME 2007 Pressure Vessels and Piping Conference, San Antonio, TX, USA, 22–26 July 2007.
14. Takazawa, H.; Yanagida, N. Effect of creep constitutive equation on simulated stress mitigation behavior of alloy steel pipe during post-weld heat treatment. *Int. J. Press. Vessel. Pip.* **2014**, *117*, 42–48. [CrossRef]
15. *Creep-Resistant Steel*; Abe, F.; Torsten-Ulf, K.; Ramaswamy, V. (Eds.) Elsevier: Amsterdam, The Netherlands, 2008; pp. 10–11.
16. Li, C.; Han, L.; Yan, G.; Liu, Q.; Luo, X.; Gu, J. Time-dependent temper embrittlement of reactor pressure vessel steel: Correlation between microstructural evolution and mechanical properties during tempering at 650 °C. *J. Nucl. Mater.* **2016**, *480*, 344–354. [CrossRef]
17. Goldak, J.; Chakravarti, A.; Bibby, M. A new finite element model for welding heat sources. *Met. Mater. Trans. A* **1984**, *15*, 299–305. [CrossRef]
18. Bai, X.; Zhang, H.; Wang, G. Improving prediction accuracy of thermal analysis for weld-based additive manufacturing by calibrating input parameters using IR imaging. *Int. J. Adv. Manuf. Technol.* **2013**, *69*, 1087–1095. [CrossRef]
19. Hu, M.; Li, K.; Cai, Z.; Pan, J. A new weld material model used in welding analysis of narrow gap thick-walled welded rotor. *J. Manuf. Process.* **2018**, *34*, 614–624. [CrossRef]
20. Yaghi, A.; Hyde, T.; Becker, A.; Sun, W.; Williams, J. Residual stress simulation in thin and thick-walled stainless steel pipe welds including pipe diameter effects. *Int. J. Press. Vessel. Pip.* **2006**, *83*, 864–874. [CrossRef]
21. *Metal Materials and Heat Treatment*; Cui, Z.; Liu, H. (Eds.) Central South University Press: Changsha, China, 2010.
22. Vangi, D. Data Management for the Evaluation of Residual Stresses by the Incremental Hole-Drilling Method. *J. Eng. Mater. Technol.* **1994**, *116*, 561–566. [CrossRef]
23. Neubert, S.; Pittner, A.; Rethmeier, M. Influence of non-uniform martensitic transformation on residual stresses and distortion of GMA-welding. *J. Constr. Steel Res.* **2017**, *128*, 193–200. [CrossRef]
24. Murthy, Y.; Rao, G.; Iyer, P. Numerical simulation of welding and quenching processes using transient thermal and thermo-elasto-plastic formulations. *Comput. Struct.* **1996**, *60*, 131–154. [CrossRef]
25. Dike, J.; Cadden, C.; Corderman, R.; Schultz, C.; McAninch, M. Finite element modeling of multipass GMA welds in steel plates. In Proceedings of the 4th International Conference on Trends in Welding Research, Gatlinburg, TN, USA, 5–8 June 1995; pp. 57–65.

© 2020 by the authors. Licensee MDPI, Basel, Switzerland. This article is an open access article distributed under the terms and conditions of the Creative Commons Attribution (CC BY) license (http://creativecommons.org/licenses/by/4.0/).

Article

Influence of Vacuum Heat Treatments on Microstructure and Mechanical Properties of M35 High Speed Steel

Chiara Soffritti [1], Annalisa Fortini [1,*], Ramona Sola [2], Elettra Fabbri [1], Mattia Merlin [1] and Gian Luca Garagnani [1]

1. Department of Engineering, University of Ferrara, via Saragat 1, 44122 Ferrara, Italy; chiara.soffritti@unife.it (C.S.); elettra.fabbri@unife.it (E.F.); mattia.merlin@unife.it (M.M.); gian.luca.garagnani@unife.it (G.L.G.)
2. Department of Industrial Engineering, University of Bologna, viale del Risorgimento 4, 40139 Bologna, Italy; ramona.sola@unibo.it
* Correspondence: annalisa.fortini@unife.it; Tel.: +39-0532-974914

Received: 6 April 2020; Accepted: 11 May 2020; Published: 15 May 2020

Abstract: Towards the end of the last century, vacuum heat treatment of high speed steels was increasingly used in the fabrication of precision cutting tools. This study investigates the influence of vacuum heat treatments at different pressures of quenching gas on the microstructure and mechanical properties of taps made of M35 high speed steel. Taps were characterized by optical microscopy, scanning electron microscopy with energy dispersive spectroscopy, X-ray diffraction, apparent grain size and Vickers hardness measurements, and scratch tests. Failure analysis after tapping tests was also performed to determine the main fracture mechanisms. For all taps, the results showed that microstructures and the values of characteristics of secondary carbides, retained austenite, apparent grain size and Vickers hardness were comparable to previously reported ones for vacuum heat treated high speed steels. For taps vacuum heat treated at six bar, the highest plane strain fracture toughness was due to a higher content of finer small secondary carbides. In contrast, the lowest plane strain fracture toughness of taps vacuum heat treated at eight bar may be due to an excessive amount of finer small secondary carbides, which may provide a preferential path for crack propagation. Finally, the predominant fracture mechanism of taps was quasi-cleavage.

Keywords: high speed steel; vacuum heat treatment; microstructure; plane strain fracture toughness

1. Introduction

High speed steels (HSS) are a complex class of tool steels continuously improved for applications in the global cutting tool market as wear-resistant materials for drills, taps, milling cutters, broaches, slotting tool and hobs. These steels must withstand abrasive or adhesive wear and have sufficient hardness, toughness and ductility to prevent chipping, galling, cracking, etc. [1].

In the fabrication of precision cutting tools, heat treatment of HSS in a salt bath has been replaced by vacuum heat treatment with uniform high-pressure gas quenching, since this procedure avoids waste and the need for washing the heat treated tools. A typical vacuum heat treatment includes several pre-heat, high heat, quenching and tempering stages. The final microstructure consists of large primary carbides in a tempered martensite matrix, hardened by the precipitation of uniformly distributed secondary carbides [2]. Nevertheless, the vacuum heat treatment has its limitations, especially concerning the number of pre-heats and holding time at high heat. Moreover, steel suppliers recommend only general guidelines about the quench rate of many grades of HSS in vacuum heat treatment [3].

The role of microstructural features is well-known in enhancing the mechanical properties of steels [2,4,5]. Over the past few decades, many studies have been carried out on wear resistance and performance of high speed steels after vacuum heat treatment. The effects of different austenitizing and tempering temperatures were evaluated, and the microstructure was optimized for increasing fracture toughness, hardness, fatigue and wear resistance of the tool [6–11]. With respect to fracture toughness, several authors used circumferentially notched and fatigue pre-cracked tensile test specimens, according to previously published methods [12,13]. The advantage of such specimens is their radial symmetry, which makes them specifically suitable for studying the influence of the microstructure on fracture toughness of metals. Due to the radial symmetry of heat transfer during heat treatment, the microstructure forming along a circumferential area is completely uniform. Moreover, the fatigue pre-crack can be created before the heat treatment without detrimental effects on the crack tip blunting. However, this method is not only difficult to apply, but also time consuming and expensive [14]. The most common alternative method is the Vickers indentation fracture test, where the fracture toughness is determined through a Vickers probe and by expressions accounting for indentation load, Young's modulus, hardness, residual stresses, plastic dissipation inside the material and the nature of cracks produced during indentation [15–17]. Recently, Sola et al. proposed a novel technique to measure the fracture toughness of tool steels by scratch test [18]. Generally, the scratch test consists of pulling a probe across the surface of the material under a controllable applied load. According to the authors, the failure mode (fracture or plastic yielding) depends on material properties and geometry of the scratching tool. It is therefore possible to link the forces acting on the scratch tip and on the tool geometry to the plane strain fracture toughness (K_C) by an equation accounting for the tangential force necessary to move the indenter, the indenter tip width and the measured penetration depth [16].

The present study investigates microstructure and mechanical properties of taps made of M35 high speed steel and vacuum heat treated at increasing pressures of quenching gas for improving their fracture toughness and durability. For this purpose, the microstructure and plane strain fracture toughness of taps undergoing conventional vacuum heat treatment were characterized by optical microscopy (OM), scanning electron microscopy with energy dispersive spectroscopy (SEM/EDS), X-ray diffraction (XRD), apparent grain size and Vickers hardness measurements, and scratch tests. Tapping tests and failure analyses were also performed on the same taps to determine the main fracture mechanisms involved during normal operating conditions. Finally, a new set of taps was vacuum heat treated at increasing pressures of quenching gas, and the resulting microstructure and plane strain fracture toughness were evaluated by the above mentioned techniques.

2. Materials and Methods

2.1. Chemical Composition and Conventional Heat Treatment of Taps

The material used in this investigation was commercially available M35 high speed steel with the following chemical composition (wt.%): 0.935% C, 0.362% Si, 0.233% Mn, 0.012% P, 0.001% S, 5.018% Co, 3.803% Cr, 4.643% Mo, 1.830% V, 6.613% W, balance Fe. The material was shaped in the form of cylindrical taps 10 mm in diameter and 100 mm in length.

The conventional heat treatment (indicated as CHT_5) of taps was carried out in a horizontal vacuum tube furnace with uniform high-pressure gas quenching, using nitrogen gas at a pressure of 5 bar and quenching speed of 10 °C/s. Quenching gas flew through nozzles located circumferentially on the walls of the furnace. After three preheat stages at 550, 850 and 1000 °C, the taps were heated at 2 °C/min to the austenitizing temperature of 1195 °C and maintained at this temperature for 10 min. Taps were then gas quenched from the austenitizing temperature to a temperature of 30 °C. Finally, triple tempering at 560 °C for 2 h were performed in the same furnace. All heat treatment parameters were selected on the basis of the recommendations of the steel supplier. The scheme in Figure 1 shows the CHT_5 treatment applied to the taps.

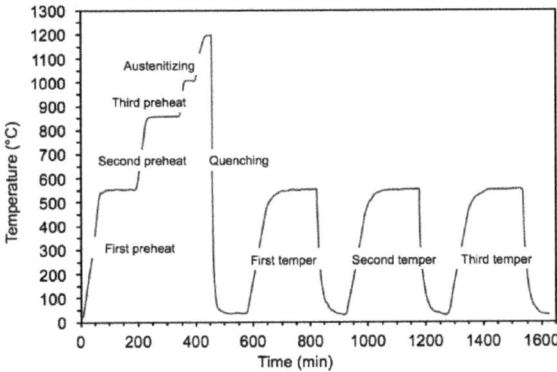

Figure 1. Conventional heat treatment (CHT_5) applied to the taps.

2.2. Material Characterization

Material characterization was performed on taps pulled out of the horizontal vacuum furnace after quenching and tempering. To determine the microstructure of alloy, longitudinal sections (parallel to the metal surface) and cross-sections (perpendicular to the metal surface) of the taps were prepared, mounted in resin, polished and examined after chemical etching by Picral (4 g picric acid in 100 mL ethanol) through a Leica MEF4M optical microscope (Leica, Wetzlar, Germany) and a Zeiss EVO MA 15 (Zeiss, Oberkochen, Germany) scanning electron microscope (SEM), equipped with an Oxford X-Max 50 (Oxford Instruments, Abingdon-on-Thames, UK) energy dispersive microprobe for semi-quantitative analyses (EDS). The SEM micrographs were recorded in secondary electron imaging (SEI-SEM) and back-scattered electron (BSE-SEM) mode. The carbide particles visible in the matrix were classified as primary carbides (PCs: size > 5.0 µm) and secondary carbides (SCs: size ≤ 5.0 µm). The secondary carbides were further sub-classified as large secondary carbides (LSCs: 1.0 µm < size ≤ 5.0 µm) and small secondary carbides (SSCs: 0.1 µm ≤ size ≤ 1.0 µm) [19]. The micrographs of the cross-sections after tempering were processed by Leica Application Suite (LAS V4.12.0, Leica, Wetzlar, Germany) image analysis software to evaluate some characteristics of large and small secondary carbides, such as area fraction, population density and size, according to a previously described procedure [20]. A mean of 1000 secondary carbides was considered for each tap to achieve appropriate statistical reliability associated with these measurements.

On the cross-sections after tempering, the apparent grain size was determined in agreement with the EN ISO 643:2020 standard [21], whereas the type of carbide particles and retained austenite content (RA) were evaluated by X-ray diffraction (XRD) analyses of the bulk taps with a Philips X'PERT PW3050 diffractometer (Philips, Amsterdam, The Netherlands), using Cu-Kα radiations and diffracted beam monochromator, with an intensity scanner versus diffraction angle between 38° and 105° (0.01° step size, 2 s/step scanner velocity and 1.5 grid), a 40 kV voltage and a 40 mA filament current. The calculation of RA was performed according to the ASTM E975-13 standard [22].

The Vickers hardness measurements under 1000 g_f load and 15 s loading time (HV1) were carried out in triplicate on the cross-sections by a Future-Tech FM-110 Vickers microindenter (Future-Tech Corp., Kawasaki, Japan). The Vickers hardness values were then converted in Rockwell C hardness values (HRC) in agreement with the ASTM E140-12B standard [23].

Finally, the fracture toughness was determined by scratch tests on the cross-sections after tempering and in agreement with a previously published method [18,24,25]. Prior to the tests, the surfaces of the samples were prepared according to a previously published procedure [26]. The scratch test produced a scratch in the material using a probe drawn across the sample under a constant vertical load and with constant speed. Compared to the standard fracture toughness tests, this technique provided the following advantages: (i) It was non-destructive; (ii) it did not require specific dimensions of the samples, nor an initial notch, pre-crack or fatigue crack; (iii) it allowed the sampling of the fracture properties at different locations in the material; (iv) it had a minimal cost and (v) could be performed on a limited supply of the material. The scratch tests were carried out by a Revetest Scratch Tester (CSM Instrument, Needham, MA, USA), using a 200 µm Rockwell C diamond probe, a scratching speed of 6 mm/min, an applied load of 100 N and a scratch length of 6 mm. In agreement with a previously published method [25], the plane strain fracture toughness K_C was determined by the following equation,

$$K_C = \frac{F_t}{\sqrt{2wd(w+2d)}} \tag{1}$$

where F_t is the tangential force necessary to move the indenter (measured by the equipment), w is the indenter tip width and d is the penetration depth. The penetration depth was directly calculated by the scratch tester. The equation was derived from previous studies on fracture mechanics framework [25,26].

2.3. Tapping Tests

Tapping tests were performed by a TH6563x63 (Shenyang Machine Tool Co., Ltd., Liaoning, China) horizontal tester machine. A grounded workpiece in 30CrMnTi steel, 700 × 500 × 48 mm³ in volume and with hardness equal to 210 HBW 2.5/187.5/30 was inserted and fixed inside the machine. Holes arranged in rows were made inside the workpiece. Taps were then rotated and moved into holes with a cutting speed of 7.5 m/min. A cutting fluid based on an emulsion of mineral oil at 5% was applied at a rate of 5 L/min. The criterion for end of tap life was its catastrophic failure. A scheme of the tapping test is shown in Figure 2.

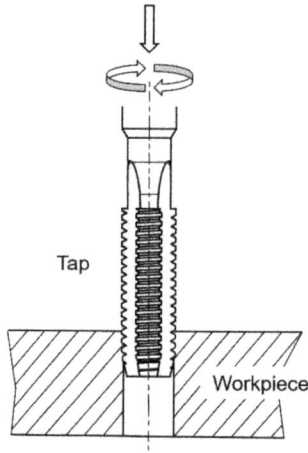

Figure 2. Scheme of the tapping test.

2.4. Failure Analysis

After tapping tests, fracture surfaces of taps were first observed by a Leica MZ6 (Leica) stereomicroscope. To identify the main fracture mechanisms, morphological observations were performed by the Zeiss EVO MA 15 (Zeiss) scanning electron microscope with the Oxford X-Max 50 (Oxford Instruments) energy dispersive microprobe. Metallographic analyses of properly polished cross-sections of the fracture surfaces were conducted after chemical etching by Picral under the Leica MEF4M optical microscope (Leica).

2.5. Vacuum Heat Treatments

A new set of taps with the same chemical composition of those in Section 2.1 was heat treated in the horizontal vacuum furnace according to the treatment schematized in Figure 1, but using nitrogen gas at pressures of 6 and 8 bar (indicated as CHT_6 and CHT_8, respectively). During each heat treatment, the material was characterized following the procedure described in Section 2.2.

3. Results

3.1. Microstructural Characterization and Hardness Variation

Optical and BSE-SEM micrographs of the microstructure observed on the cross-section of taps, after CHT_5 heat treatment, are shown in Figure 3. The microstructure exhibited a non-uniform distribution of large and dendritic-type primary carbides (marked by PCs), and a fairly uniform distribution of secondary carbides in a tempered martensite matrix, with retained austenite and resolved prior austenite grain boundaries. The two types of secondary carbides, classified according to their size, appeared either as well-defined white regions (marked by LSCs) or tiny white patches (marked by SSCs) (Figure 3a). In Figure 3b, a semi-continuous distribution of proeutectoid carbides was also visible along prior austenitic grain boundaries. The mean value of area fraction of large secondary carbides measured on taps cross-sections after CHT_5 heat treatment was $2.76 \pm 0.23\%$, whereas those of population density and size were $4.0 \pm 0.5 \times 10^3/mm^2$ and 2.78 ± 0.12 μm, respectively. The mean value of area fraction of small secondary carbides measured on the same regions was $2.53 \pm 0.15\%$, while that of population density was $134 \pm 4 \times 10^3/mm^2$. The mean value of size of small secondary carbides was 0.74 ± 0.08 μm.

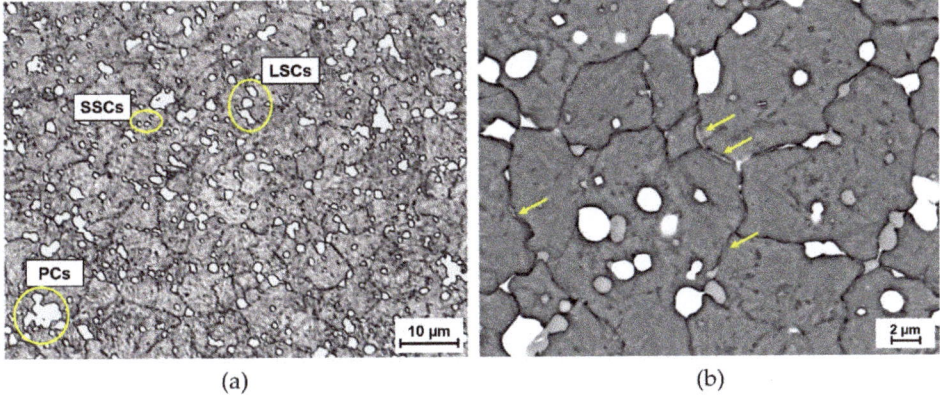

Figure 3. (a) Optical and (b) BSE-SEM micrographs of the microstructure observed on the cross-section of taps after CHT_5. Yellow circles in (a) indicate primary carbides (PCs), large secondary carbides (LSCs) and small secondary carbides (SSCs). Yellow arrows in (b) show the semi-continuous distribution of proeutectoid carbides visible along prior austenitic grain boundaries.

Comparing the EDS semi-quantitative analyses of carbide particles (Figure 4), after quenching the primary carbides were rich in V or in W and Mo (spectra 1 and 2 in Figure 4), and changed only marginally their chemical composition after triple tempering at 560 °C. After quenching, secondary carbides and proeutectoid carbides were rich in Fe and poor in Cr, and maintained their chemical composition until the end of the heat treatment (spectrum 3 in Figure 4). Nevertheless, the EDS results proved that during tempering there was a precipitation of new Cr-rich secondary carbides. The XRD diffractograms recorded on the bulk taps after CHT_5 heat treatment revealed the presence of martensite (indexed as ferrite, since the resolution of the diffractometer was not sufficient to resolve the tetragonality of martensite), alloyed cementite (indexed as Fe$_3$C), MC, M$_6$C, M$_7$C$_3$ and RA (Figure 5). The mean value of RA in the matrix measured by XRD was 12.4 ± 0.5%.

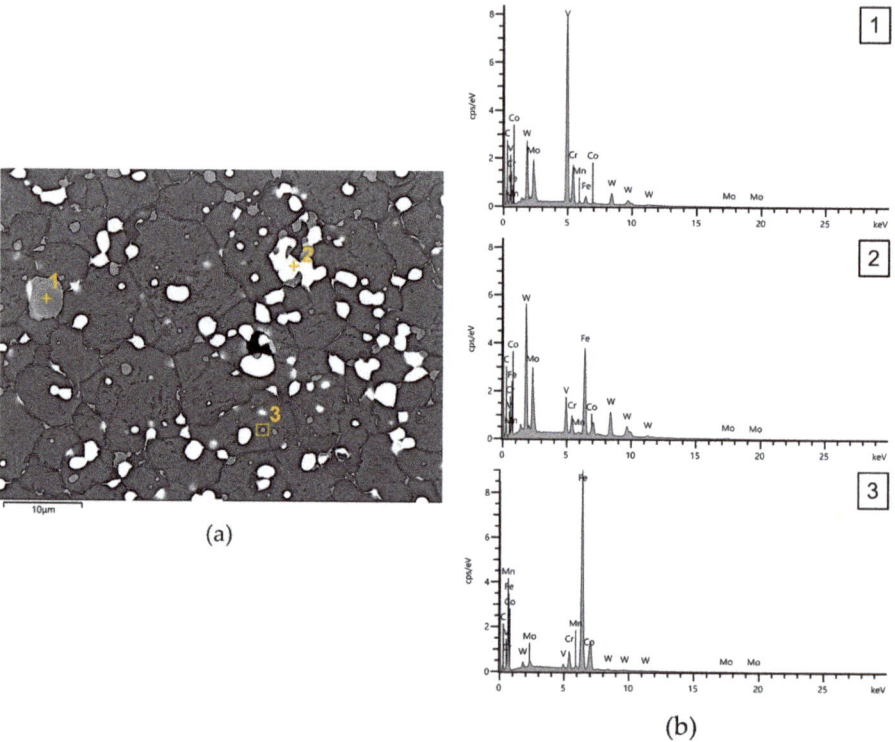

Figure 4. (a) BSE-SEM micrograph of the microstructure observed on the cross-section of taps after quenching and (b) semi-quantitative EDS analyses (wt.%) of different carbide particles in the same microstructure.

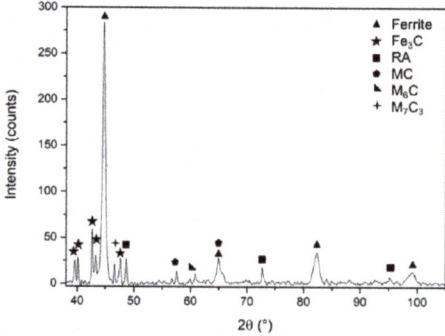

Figure 5. XRD diffractogram recorded on the bulk taps after CHT_5.

The mean value of apparent grain size measured on tap cross-sections after CHT_5 heat treatment was 2.6 ± 0.1 µm (G14, according to the EN ISO 643:2020 standard), whereas that of plane strain fracture toughness, calculated by Equation (1), was 31 ± 5 MPa·m$^{1/2}$. After quenching, the mean value of Vickers hardness measured on the same regions was 914 HV1 (corresponding to 67 HRC). Triple tempering at 560 °C resulted in Vickers hardness of 892 HV1, corresponding to 66 HRC.

3.2. Failure Analysis after Tapping Tests

After tapping tests, the stereomicroscopy images of the fracture surfaces of taps had a smooth and flat appearance, typical of brittle materials. At higher magnification, the SEI-SEM analyses revealed that the morphology of the fracture surfaces (Figure 6a) was characterized by two features: classical cleavage fracture of primary carbides (marked as C in Figure 6b) and quasi-cleavage fracture of the matrix (marked as the yellow rectangle QC in Figure 6b). In the first case, cracking of primary carbides was visible, and microcracks occurred at the interface between the primary carbide and the matrix. The EDS semi-quantitative analyses on the cleavage regions of the fracture surfaces confirmed that these regions corresponded to primary carbides. In the second case, a large number of microvoids were formed by decohesion of secondary carbides (yellow arrows in Figure 6c). No cracking of secondary carbides was detected. The optical micrograph of the fracture surfaces observed on the cross-sections of taps is shown in Figure 6d. These surfaces were mostly irregular. Fractures were deflected and partially surrounded the numerous secondary carbides detected close to the fracture surfaces, also suggesting that these carbide particles provided a preferential path for crack propagation.

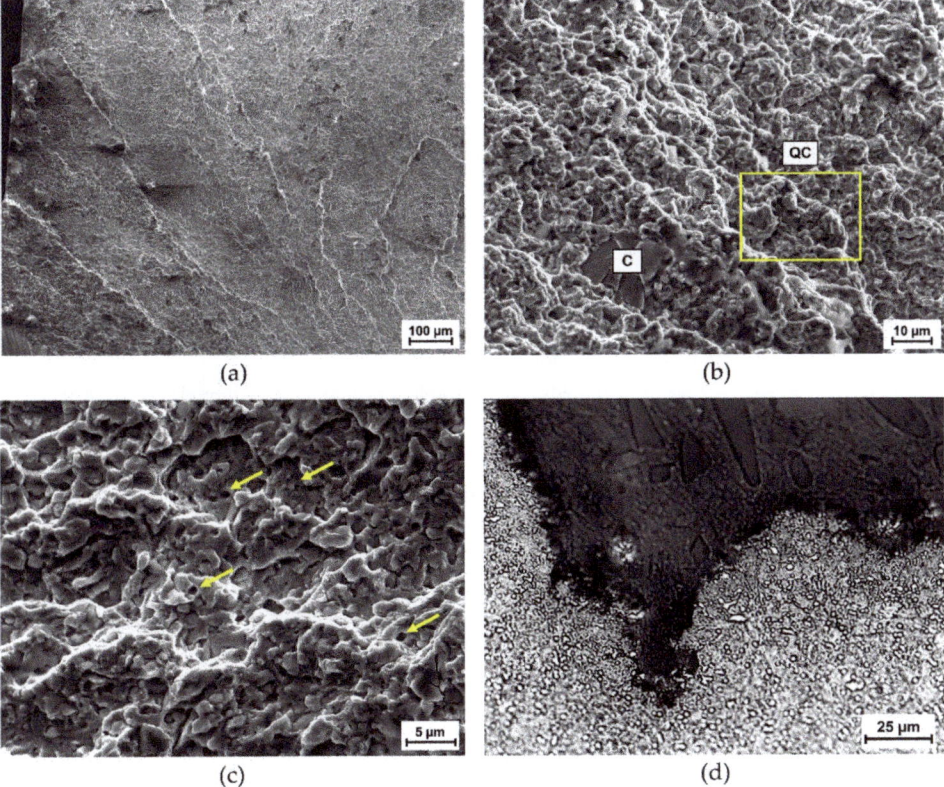

Figure 6. (**a**–**c**) SEI-SEM micrographs of the fracture surfaces of taps after tapping test showing, in the order, an overview of fracture surface, details of cleavage (C) and a quasi-cleavage (QC, yellow rectangle) region, and microvoids formed by decohesion of secondary carbides (yellow arrows); (**d**) optical micrograph of the fracture surfaces observed on the cross-sections of taps.

3.3. Influence of Vacuum Heat Treatments on Microstructure and Fracture Toughness

Optical and scanning electron micrographs of the microstructure observed on the cross-sections of taps after CHT_6 and CHT_8 heat treatments (Figure 7a,c, respectively) were similar to those after CHT_5 (Figure 3a), but the extent of the semi-continuous distribution of proeutectoid carbides along prior austenitic grain boundaries decreased with increasing pressure of quenching gas (Figure 7b,d). During CHT_6 and CHT_8 heat treatments, primary and secondary carbides, detected by EDS and XRD, were unchanged with respect to those observed after CHT_5.

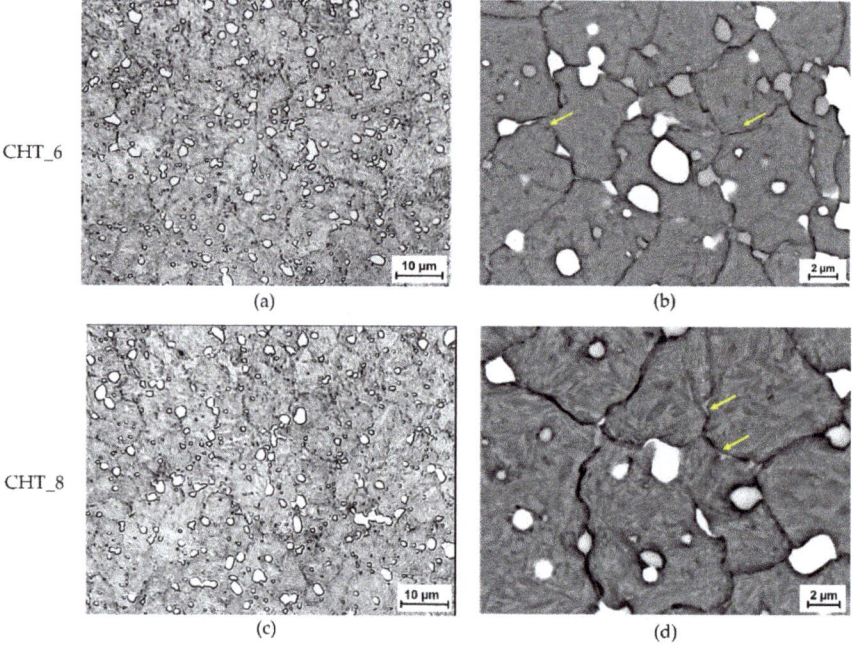

Figure 7. (a,c) Optical and (b,d) BSE-SEM micrographs of the microstructure observed on the cross-section of taps after CHT_6 and CHT_8 heat treatments. Yellow arrows in (b,d) show the semi-continuous distribution of proeutectoid carbides visible along prior austenitic grain boundaries.

The characteristics of large and small secondary carbides at increasing pressure of quenching gas are shown in Figure 8a–c. The mean values of area fraction (Figure 8a) and population density (Figure 8b) of large and small secondary carbides progressively increased with increasing pressure of quenching gas. The mean value of area fraction of large secondary carbides increased up to the final value of 3.21 ± 0.19% (corresponding to an overall increase of about 16%), while that of small secondary carbides reached the final value of 3.49 ± 0.17% (corresponding to an overall increase of about 38%). The mean value of population density of large secondary carbides raised up to the final value of $9.0 ± 0.3 \times 10^3/mm^2$ (corresponding to an overall increase of about 125%), whereas that of small secondary carbides reached the final value of $167 ± 3 \times 10^3/mm^2$ (corresponding to an overall increase of about 25%). Conversely, the mean value of size of the same carbide particles was reduced with an increase in pressure of quenching gas (Figure 8c). The mean value of size of large secondary carbides decreased up to the final value of 2.17 ± 0.18 µm (corresponding to an overall decrease of about 22%), while that of small secondary carbides reached the final value of 0.58 ± 0.07% (corresponding to an overall decrease of about 21%).

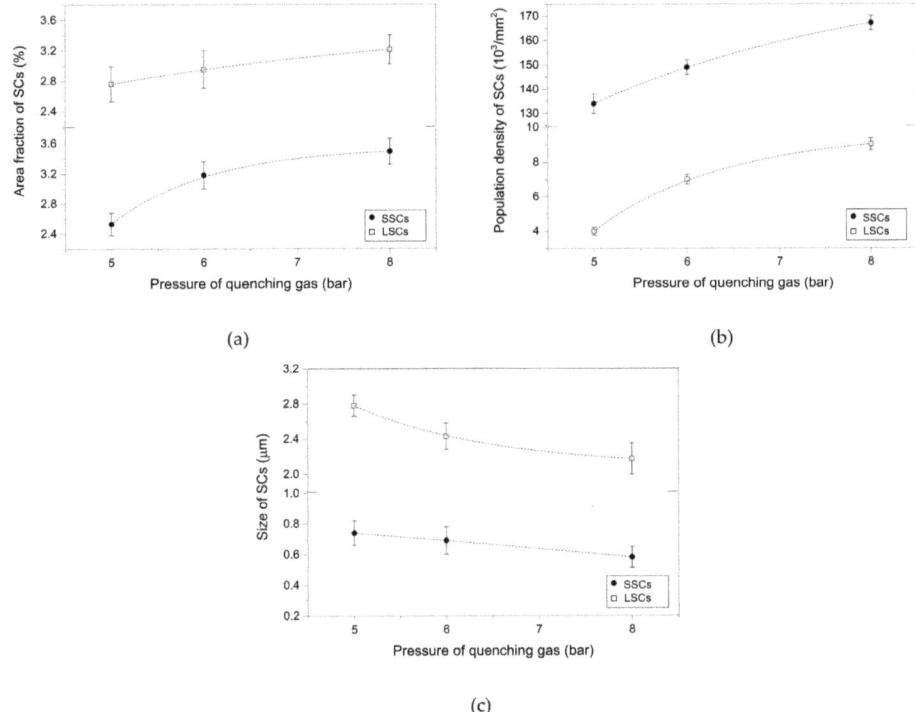

Figure 8. Characteristics of large and small secondary carbides at increasing pressure of quenching gas: mean values of (**a**) area fraction, (**b**) population density and (**c**) size.

The mean values of retained austenite, apparent grain size, plane strain fracture toughness and Vickers hardness of taps after the different heat treatments are reported in Table 1. The mean value of RA decreased with increasing pressure of quenching gas, reaching the final value of 10.8 ± 0.4% in taps undergoing the CHT_8 heat treatment. Irrespective of the type of heat treatment, there were almost no changes in mean apparent grain size, which ranged from 2.6 to 2.7 µm (G14, according to the EN ISO 643:2020 standard). The mean plane strain fracture toughness of taps initially increased up to 32 ± 6 MPa·m$^{1/2}$ with increasing pressure of quenching gas, but then decreased with a further increase in the same parameter, reaching a minimum value of 26 ± 4 MPa·m$^{1/2}$. Concerning Vickers hardness, the mean values ranged from 892 ± 11 (corresponding to 66 HRC) to 906 ± 4 (corresponding to 67 HRC) with varying pressure of quenching gas.

Table 1. Mean values of retained austenite (RA), apparent grain size, plane strain fracture toughness (K_C) and Vickers hardness (HV1) of taps after the different heat treatments.

Parameter	Heat Treatment		
	CHT_5	CHT_6	CHT_8
RA (%)	12.4 ± 0.5	11.2 ± 0.2	10.8 ± 0.4
Apparent grain size (µm)	2.6 ± 0.1	2.7 ± 0.2	2.6 ± 0.2
Kc (MPa·m$^{1/2}$)	31 ± 5	32 ± 6	26 ± 4
HV1	892 ± 11	906 ± 4	898 ± 7

The relationships between plane strain fracture toughness and the characteristics of small secondary carbides are shown in Figure 9. Considering the CHT_5 and CHT_6 heat treatments, plane strain fracture toughness increased with increasing area fraction (Figure 9a) and population density

(Figure 9b), but increased with decreasing of the size of small secondary carbides (Figure 9c). On the contrary, for taps undergoing the CHT_8 heat treatment, the higher area fraction and population density and the smaller size of small secondary carbides resulted in the lowest plane strain fracture toughness.

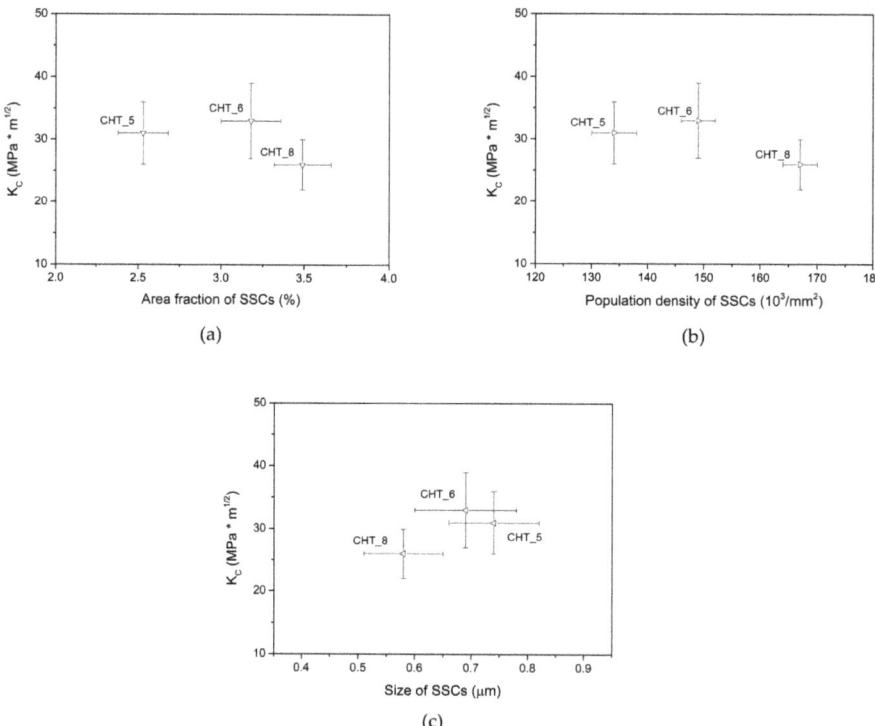

Figure 9. Relationships between plane strain fracture toughness and (**a**) area fraction, (**b**) population density and (**c**) size of small secondary carbides.

4. Discussion

The microstructure and mechanical properties of taps made of M35 high speed steel and vacuum heat treated at increasing pressures of quenching gas were studied for improving their fracture toughness and durability. The extensive metallographic investigation of taps, vacuum heat treated with pressure of quenching gas of five bar, shows a microstructure similar to that of high speed steels [2]. After tempering, the microstructure consists of a non-uniform distribution of large and dendritic-type MC and M_6C primary carbides and a fairly uniform distribution of alloyed cementite and M_7C_3 secondary carbides in a tempered martensite matrix. Previous research showed that the V-rich MC primary carbides are known to be the hardest phase, main responsible for the wear resistance of HSS, whereas the W- and Mo-rich M_6C primary carbides derive from the conversion of metastable M_2C in more stable M_6C during reheating for hot-forging or hot-rolling [27–29]. Concerning the Cr-rich M_7C_3 secondary carbides, studies on the effects of tempering conditions on secondary hardening of high speed steels proved that at tempering temperatures above 550 °C, these carbides gradually precipitate from the edge of the M_6C primary carbides towards the interior [30]. After this process, M_7C_3 and M_6C can coexist in the matrix, as supported by our SEM/EDS and XRD examinations. Based on the results of the image analysis, the area fraction, population density and size of large secondary carbides are between 2.53 and 2.99%, 3.5 and $4.5 \times 10^3/mm^2$, 2.66 and 2.90 µm, respectively. Considering the characteristics of small secondary carbides, area fraction is between 2.38 and 2.68%, while population density ranges from 130 to 138 × 10³/mm². The size is between 0.66 and 0.82 μm. All values of both large and small secondary carbides are comparable to those previously reported for vacuum heat treated high speed steels [18,31,32].

In the tempered martensite matrix, an average of 12.4% of retained austenite is detected, which represents good average value for commercial practice [33]. A certain amount of proeutectoid carbides is also visible along prior austenitic grain boundaries due to quench embrittlement [2]. This phenomenon involves the precipitation of carbide particles during cooling from austenitizing temperature, but when the material is still in the austenite field. The SEM/EDS analyses confirm that the precipitation occurs at the austenite grain boundaries since those regions are preferential sites for nucleation in terms of space and thermodynamic energy [2]. Overall, the presence of proeutectoid carbides does not affect the hardness of steels, but may favor crack propagation, and therefore, reduce the final fracture toughness [2]. In this study, not only the mean values of apparent grain size and Vickers hardness after tempering but also those of plane strain fracture toughness are comparable to those of high speed steels undergoing conventional vacuum heat treatments, despite the occurrence of the quench embrittlement phenomenon [9,33].

After tapping tests, failure analysis of vacuum heat treated taps by optical and scanning electron microscopy shows that, in normal operating conditions, the predominant fracture mechanism is quasi-cleavage [31,32]. Fracture initiates by cracking of primary carbides and cracking at the primary carbides/matrix interfaces, because these carbides are large and prone to fracture. Moreover, primary carbides are found in the interdendritic regions and such non-uniform distribution facilitates their cracking from the matrix at low-stress levels [20]. The nucleation of microvoids by decohesion of secondary carbides then occurs, followed by cleavage fracture. Unlike primary carbides, no cracking of secondary carbides is detected. Formation of microvoids by decohesion of secondary carbides is quite common in the fracture processes of tool steels [18,34–36]. The criteria for void nucleation is based on the dislocation pile-up model [31]. In detail, void nucleates when the local interfacial stresses, occurring on a carbide particle, reach a critical interfacial strength, which is a characteristic of the material.

Vacuum heat treatment of taps at pressures of quenching gas of six and eight bar to improve their fracture toughness and durability produces microstructures similar to those of taps vacuum heat treated at five bar, but with lower amounts of retained austenite and proeutectoid carbides along prior austenitic grain boundaries. An increase in the pressure of quenching gas also modifies the precipitation behavior of secondary carbides, as proved by the rise in their area fraction and population density, and by the decrease in size of large and small secondary carbides [31,32]. Slight differences are found in the mean apparent grain size and Vickers hardness of all taps.

With respect to plane strain fracture toughness, it is known that the microstructure has a great influence on this parameter [33]. In our taps, among all microstructural constituents primary carbides appear identical regardless of the type of heat treatment, since their characteristic are controlled only by the temperature and time of austenitization [32]. Accordingly, any variation of the measured plane strain fracture toughness is supposed to be due to alterations of the secondary carbides, and more specifically, the small secondary carbides. By comparing the plane strain fracture toughness of conventionally heat treated taps with that of taps vacuum heat treated at six bar, it was observed that the plane strain fracture toughness slightly increased with increasing area fraction and population density, but also with decreasing size of small secondary carbides. Based on the metallographic analyses by optical microscopy of our fracture surfaces, this may be due to the fact that small secondary carbides, more uniformly distributed, are able to deflect more effectively cracks propagation, leading to more energy absorption associated with cracking. Moreover, some authors also demonstrated that the precipitation of a higher amount of small secondary carbides reduced the contents of dissolved carbon and alloying elements in the matrix, thus raising its ductility and fracture toughness [31]. Nevertheless, taps vacuum heat treated at a pressure of quenching gas of eight bar showed the lowest plane strain fracture toughness, despite the highest area fraction and population density, and the smallest size of small secondary carbides. This is probably due to an excessive amount of finer small secondary carbides, which may reduce the plane strain fracture toughness of our high speed steel by providing a preferential path for crack propagation.

5. Conclusions

This study investigates the microstructure and mechanical properties of taps made of M35 high speed steel and vacuum heat treated at increasing pressures of quenching gas, in order to improve their fracture toughness and durability. Based on the results, the following conclusions can be drawn:

- The microstructure of taps undergoing vacuum heat treatment at a pressure of quenching gas of five bar is similar to that of high speed steels. The microstructure mostly consists of primary carbides (MC and M_6C), alloyed cementite and secondary carbides (M_7C_3) in a tempered martensite matrix, with retained austenite and proeutectoid carbides along prior austenite grain boundaries. All values of the characteristics of secondary carbides, as well as those of retained austenite, apparent grain size, Vickers hardness and plane strain fracture toughness are comparable to previously reported data for vacuum heat treated high speed steels;
- in normal operating conditions the predominant fracture mechanism of taps is quasi-cleavage. Fracture initiates by cracking of primary carbides at the primary carbides/matrix interfaces and by nucleation of microvoids by decohesion of secondary carbides, followed by cleavage fracture;
- vacuum heat treatment of taps at pressures of quenching gas of six and eight bar produces microstructures similar to those of taps vacuum heat treated at a pressure of quenching gas of five bar, but with lower amounts of retained austenite and proeutectoid carbides along prior austenitic grain boundaries. An increase in pressure of quenching gas also modifies the precipitation behavior of secondary carbides, whereas slight differences are found in the mean apparent grain size and Vickers hardness of all taps. In relation to plane strain fracture toughness, taps vacuum heat treated at six bar shows the highest values of this parameter thanks to a higher content of finer small secondary carbides, which deflect crack propagation. Conversely, the lowest plane strain fracture toughness of taps vacuum heat treated at eight bar may be due to an excessive amount of finer small secondary carbides, which may provide a preferential path for crack propagation.

Author Contributions: In the frame of the present investigation, C.S. conceived the experiments and wrote the manuscript. A.F. contributed to data analyses and to the manuscript. R.S. and E.F. supported the experimental investigations. M.M. and G.L.G. supervised the experiments and aided the results discussion. All authors have read and agreed to the published version of the manuscript.

Funding: This research was funded by the Department of Engineering, University of Ferrara (Ferrara, Italy), grant number 2238/2010.

Acknowledgments: The authors would like to dedicate this study to the memory of Vincenzo Gabrielli, whose invaluable contribution to the experimental activity and whose guidance have been essential to achieve the present results. The authors also thank Milvia Chicca for help in revising the manuscript.

Conflicts of Interest: The authors declare no conflict of interest.

References

1. Statharas, D.; Papageorgiou, D.; Sideris, J.; Medrea, C. Preliminary examination of the fracture surfaces of a cold working die. *Int. J. Mater. Form.* **2008**, *1*, 431–434. [CrossRef]
2. Mesquita, R.A. High-speed steels. In *Tool Steels: Properties and Performance*, 1st ed.; CRC Press: Boca Raton, FL, USA, 2017; pp. 217–239.
3. Rousseau, A.; Doyle, E.; McCulloch, D. Vacuum heat treatment of high speed steel cutting tools. *Int. Heat Treat. Surf. Eng.* **2013**, *7*, 110–114. [CrossRef]
4. Di Schino, A.; Di Nunzio, P.E.; Lopez Turconi, G. Microstructure Evolution during Tempering of Martensite in a Medium-C Steel. *Mater. Sci. Forum* **2007**, *558–559*, 1435–1441. [CrossRef]
5. Di Schino, A. Analysis of heat treatment effect on microstructural features evolution in a micro-alloyed martensitic steel. *Acta Metall. Slovaca* **2016**, *22*, 266–270. [CrossRef]

6. Leskovšek, V.; Šuštaršič, B.; Jutriša, G. The influence of austenitizing and tempering temperature on the hardness and fracture toughness of hot-worked H11 tool steel. *J. Mater. Process. Technol.* **2006**, *178*, 328–334. [CrossRef]
7. Podgornik, B.; Leskovšek, V. Microstructure and Origin of Hot-Work Tool Steel Fracture Toughness Deviation. *Metall. Mater. Trans. A* **2013**, *44*, 5694–5702. [CrossRef]
8. Leskovšek, V.; Podgornik, B. Vacuum heat treatment, deep cryogenic treatment and simultaneous pulse plasma nitriding and tempering of P/M S390MC steel. *Mater. Sci. Eng. A* **2012**, *531*, 119–129. [CrossRef]
9. Leskovšek, V.; Ule, B. Improved vacuum heat-treatment for fine-blanking tools from high-speed steel M2. *J. Mater. Process. Technol.* **1998**, *82*, 89–94. [CrossRef]
10. Suchánek, J.; Kuklík, V. Influence of heat and thermochemical treatment on abrasion resistance of structural and tool steels. *Wear* **2009**, *267*, 2100–2108. [CrossRef]
11. Podgornik, B.; Leskovšek, V.; Tehovnik, F.; Burja, J. Vacuum heat treatment optimization for improved load carrying capacity and wear properties of surface engineered hot work tool steel. *Surf. Coat. Technol.* **2015**, *261*, 253–261. [CrossRef]
12. Ule, B.; Leskovšek, V.; Tuma, B. Estimation of plain strain fracture toughness of AISI M2 steel from precracked round-bar specimens. *Eng. Fract. Mech.* **2000**, *65*, 559–572. [CrossRef]
13. Podgornik, B.; Leskovšek, V. Experimental Evaluation of Tool and High-Speed Steel Properties Using Multi-Functional KIc -Test Specimen. *Steel Res. Int.* **2013**, *84*, 1294–1301. [CrossRef]
14. Leskovšek, V.; Ule, B.; Liščić, B. Relations between fracture toughness, hardness and microstructure of vacuum heat-treated high-speed steel. *J. Mater. Process. Technol.* **2002**, *127*, 298–308. [CrossRef]
15. Quinn, G.D.; Bradt, R.C. On the Vickers Indentation Fracture Toughness Test. *J. Am. Ceram. Soc.* **2007**, *90*, 673–680. [CrossRef]
16. Harding, D.S.; Oliver, W.C.; Pharr, G.M. Cracking During Nanoindentation and its Use in the Measurement of Fracture Toughness. *MRS Online Proc. Libr.* **1994**, *356*, 663. [CrossRef]
17. Widjaja, S.; Yip, T.H.; Limarga, A.M. Measurement of creep-induced localized residual stress in soda-lime glass using nano-indentation technique. *Mater. Sci. Eng. A* **2001**, *318*, 211–215. [CrossRef]
18. Sola, R.; Giovanardi, R.; Parigi, G.; Veronesi, P. A Novel Method for Fracture Toughness Evaluation of Tool Steels with Post-Tempering Cryogenic Treatment. *Metals* **2017**, *7*, 75. [CrossRef]
19. Das, D.; Dutta, A.K.; Ray, K.K. On the refinement of carbide precipitates by cryotreatment in AISI D2 steel. *Philos. Mag.* **2009**, *89*, 55–76. [CrossRef]
20. Fukaura, K.; Yokoyama, Y.; Yokoi, D.; Tsujii, N.; Ono, K. Fatigue of cold-work tool steels: Effect of heat treatment and carbide morphology on fatigue crack formation, life, and fracture surface observations. *Metall. Mater. Trans. A Phys. Metall. Mater. Sci.* **2004**, *35*, 1289–1300. [CrossRef]
21. ISO. *EN ISO 643: Steels—Micrographic Determination of the Apparent Grain Size*; ISO International: Geneva, Switzerland, 2020.
22. ASTM International. *ASTM Standard E975-13: Standard Practice for X-Ray Determination of Retained Austenite in Steel with Near Random Crystallographic Orientation*; ASTM International: West Conshohocken, PA, USA, 2013. [CrossRef]
23. ASTM International. *ASTM Standard E140-12B(2019)e1: Standard Hardness Conversion Tables for Metals Relationship Among Brinell Hardness, Vickers Hardness, Rockwell Hardness, Superficial Hardness, Knoop Hardness, Scleroscope Hardness, and Leeb Hardness*; ASTM International: West Conshohocken, PA, USA, 2019. [CrossRef]
24. Akono, A.-T.; Ulm, F.-J. Scratch test model for the determination of fracture toughness. *Eng. Fract. Mech.* **2011**, *78*, 334–342. [CrossRef]
25. Akono, A.-T.; Ulm, F.-J. An improved technique for characterizing the fracture toughness via scratch test experiments. *Wear* **2014**, *313*, 117–124. [CrossRef]
26. Akono, A.-T.; Randall, N.X.; Ulm, F.-J. Experimental determination of the fracture toughness via microscratch tests: Application to polymers, ceramics, and metals. *J. Mater. Res.* **2012**, *27*, 485–493. [CrossRef]
27. Sackl, S.; Leitner, H.; Clemens, H.; Primig, S. On the evolution of secondary hardening carbides during continuous versus isothermal heat treatment of high speed steel HS 6-5-2. *Mater. Charact.* **2016**, *120*, 323–330. [CrossRef]
28. Pan, F.; Wang, W.; Tang, A.; Wu, L.; Liu, T.; Cheng, R. Phase transformation refinement of coarse primary carbides in M2 high speed steel. *Prog. Nat. Sci. Mater. Int.* **2011**, *21*, 180–186. [CrossRef]

29. Wießner, M.; Leisch, M.; Emminger, H.; Kulmburg, A. Phase transformation study of a high speed steel powder by high temperature X-ray diffraction. *Mater. Charact.* **2008**, *59*, 937–943. [CrossRef]
30. Liu, B.; Qin, T.; Xu, W.; Jia, C.; Wu, Q.; Chen, M.; Liu, Z. Effect of Tempering Conditions on Secondary Hardening of Carbides and Retained Austenite in Spray-Formed M42 High-Speed Steel. *Materials* **2019**, *12*, 3714. [CrossRef]
31. Das, D.; Sarkar, R.; Dutta, A.K.; Ray, K.K. Influence of sub-zero treatments on fracture toughness of AISI D2 steel. *Mater. Sci. Eng. A* **2010**, *528*, 589–603. [CrossRef]
32. Yan, X.G.; Li, D.Y. Effects of the sub-zero treatment condition on microstructure, mechanical behavior and wear resistance of W9Mo3Cr4V high speed steel. *Wear* **2013**, *302*, 854–862. [CrossRef]
33. Roberts, G.; Krauss, G.; Kennedy, R. High speed steels. In *Tool Steels*, 5th ed.; ASM International: Materials Park, OH, USA, 1998; pp. 251–290.
34. Imbert, C.A.C.; McQueen, H.J. Flow curves up to peak strength of hot deformed D2 and W1 tool steels. *Mater. Sci. Technol.* **2000**, *16*, 524–531. [CrossRef]
35. Johnson, A.R. Fracture toughness of AISI M2 and AISI M7 high-speed steels. *Metall. Trans. A* **1977**, *8*, 891–897. [CrossRef]
36. Sola, R.; Veronesi, P.; Giovanardi, R.; Merlin, M.; Garagnani, G.L.; Soffritti, C.; Morri, A.; Parigi, G. Influence of a post-tempering cryogenic treatment on the toughness of the AISI M2 steel. In Proceedings of the 7th International Congress on Science and Technology of Steelmaking, Venice, Italy, 13–15 June 2018.

 © 2020 by the authors. Licensee MDPI, Basel, Switzerland. This article is an open access article distributed under the terms and conditions of the Creative Commons Attribution (CC BY) license (http://creativecommons.org/licenses/by/4.0/).

Article

Corrosion Behavior and Mechanical Properties of AISI 316 Stainless Steel Clad Q235 Plate

Andrea Di Schino [1],* and Claudio Testani [2]

1. Engineering Department, University of Perugia, Via G. Duranti 93, 06125 Perugia, Italy
2. CALEF- ENEA CR Casaccia, Via Anguillarese 301, Santa Maria di Galeria, 00123 Rome, Italy; claudio.testani@consorziocalef.it
* Correspondence: andrea.dischino@unipg.it

Received: 23 March 2020; Accepted: 21 April 2020; Published: 24 April 2020

Abstract: This paper deals with carbon steel and stainless steel clad-plate properties. Cladding is performed by the submerged-arc welding (SAW) overlay process. Due to element diffusion (Fe, Cr, Ni, and Mn), a 1.5 mm wide diffusion layer is formed between the stainless steel and carbon steel interface of the cladded plate affecting corrosion resistance. Pitting resistance is evaluated by measuring the critical-pitting temperature (CPT), as described in the American Society for Testing and Materials (ASTM) G-48 standard test. Additionally, Huey immersion tests, in accordance with ASTM A262, Type C, are carried out to evaluate the intergranular corrosion resistance. Some hardness peaks are detected in microalloyed steel close to the molten interface line in the coarse-grained heat-affected zone (CGHAZ). Results show that stress-relieving treatments are not sufficient to avoid hardness peaks. The hardness peaks in the CGHAZ of the microalloyed steel disappear after quenching and tempering (Q and T).

Keywords: steel-clad plate; element diffusion; microstructure; mechanical properties

1. Introduction

Metal-clad plates for different applications have been developed in many industrial sectors [1–3]. Among them, carbon-steel plates cladded by stainless steel are one of the most widely used products since they allow one to combine the mechanical properties of carbon steel with the excellent corrosion resistance of the adopted clad.

In fact, following their peculiar corrosion-resistance properties, stainless steels are nowadays used in many applications that are high targets of corrosion resistance [4,5]. In particular, they are employed in automotive [6], construction and building [7,8], energy [9–11], aeronautics [12], food [13–15], and three-dimensional (3D) printing [16] applications. Their poor mechanical resistance does not allow them to be directly applied in structural applications. The combination of carbon steel adopted as a substrate coated by stainless steel is, therefore, well-suited for several applications calling for high-mechanical targets coupled with good-corrosion resistance (especially in petroleum, petrochemical, shipbuilding, and pressure-vessel industries). In particular, in the case of stainless steel clad plates, the cladding material satisfies the corrosion resistance requirement. In this case, the cladding material thickness just corresponds to about 10–20% of the clad plate. Therefore, stainless steel clad plates show a chromium and nickel content reduction if compared to plates fully manufactured by stainless steel. This reduces about 30–50% of the costs following the carbon steel low cost [17,18]. Many technologies are utilized to produce metal-clad plates, for instance, explosion welding [19], roll bonding, and welding overlay [20–22]. Among the welding overlay methods, currently, several welding technologies have been adopted in the welding of stainless steel clad plates. Shielded-metal-arc welding (SMAW) [23], submerged-arc welding (SAW) [24], tungsten-inert-gas welding (TIG) [25],

CO$_2$ arc welding [26,27], and laser welding [28] were successfully applied in the case of AISI 304 stainless steel adopted as the clad for Q235 steel, thus improving the corrosion resistance of the original carbon steel plate. However, under some conditions, stainless steel could also undergo corrosion [29]. It is, for example, known that carbide precipitates on the grain boundaries of AISI 304 steel, forming a chromium-depleted region if it is exposed to the temperature range of 550–850 °C. Following this formation, the steel becomes susceptible to attack in a corrosive medium. Similarly, stainless and carbon steel bimetal plates are no exception [30]. Moreover, for the stainless and carbon steel bimetal plate, the effect of element diffusion should also be taken into consideration in the study of its corrosion-resistance evolution. Many previous works have shown that there is a diffusion area between stainless steel and carbon steel that makes the stainless steel and carbon steel bond well together [31]. In particular, the composite mechanism and interface microstructure of metal-clad plates have been largely analyzed [32]. It was found that element diffusion occurring near the interface of the clad plate appears to be the key to the bonding of metal-clad plates [33,34]. This strongly depends on the cladding process that also affects the interface microstructure and hardness. Compared to carbon and stainless steel, the interface layer retains different microstructures and mechanical properties, which can affect the mechanical performance of the clad plate. In particular, it was found that due to element diffusion (Fe, Cr, Ni, and Mn), a 20 µm thick diffusion layer forms between AISI 304 stainless steel and carbon steel clad plates. The diffusion layer is characterized by stable mechanical performance, and that microstructure does not show any grain growth [35], with internal mechanical properties showing a gradual change in the thickness direction [36]. Therefore, such a transition layer appears to be beneficial to a strong bond between stainless steel and carbon steel, also guaranteeing a stable transition of mechanical performance in the thickness direction. Carburization of the stainless steel with a thickness of 150 µm is found, and decarburization carbon steel with a thickness of 80 µm is formed on the carbon steel side [37]. Many efforts have been made in order to assess the corrosion rate at the inner surface of ultrasupercritical boilers [38]. Moreover, often a hard peak is detected at stainless and carbon steel interfaces following austenite grain growth in the carbon steel side during welding [39]. Such peaks should be carefully taken and kept as low as possible. Since it is often difficult to avoid such peaks during the welding process, proper heat treatments are necessary in order to optimize interface microstructure and hardness values [40,41].

In this paper, the microstructure characterization of stainless steel clad as obtained by the welding overlay is shown. The postwelding overlay heat treatment effect is analyzed on the interface of in terms of microstructure and hardness.

2. Materials and Methods

The stainless steel clad plate was manufactured by weld overlaying with the submerged-arc strip cladding method, which utilizes an arc that runs back and forth at high speeds along the strip. Two feed hoppers were necessary to guarantee the complete protection of the electrical arc. A simplified scheme of the adopted overlay welding process is shown in Figure 1. The process was conducted by means of a protective slag. Q235 steel was chosen as a substrate, and the AISI 316 strip was chosen as cladding material (strip of 30 mm wide and thickness of 0.5 mm). Q235 steel was chosen as a substrate. Q235 steel samples were received in the form of 16 mm thick plates after hot-rolling and quenching and tempering (Q and T) at an industrial scale. Such steel was chosen since it is characterized by good plasticity, toughness, weldability, and good cold-bending performance, making it widely used in construction and engineering welding structures. AISI 316L was chosen as the clad for its excellent corrosion performances. The two actual steel chemical composition are shown in Table 1.

Table 1. Main chemical composition of the considered materials (mass, %).

Alloy	Cr	Ni	Mo	Mn	C	Si	P	S	Fe
AISI 316 L	17.9	8.0	1.1	1.5	0.02	0.12	0.020	0.10	Balance
Q235	0	0	0	1.0	0.20	0.35	0.040	0.40	Balance

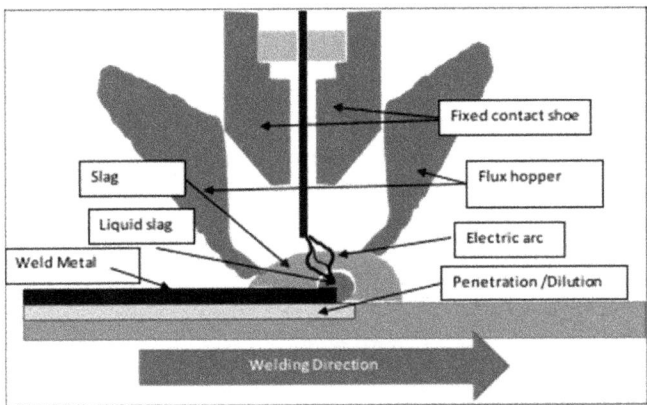

Figure 1. Schematic view of the adopted overlay process.

Plates sizing 500 mm × 500 mm × 16 mm were clad by an external supplier. The welding supplier declared the plate manufacturing process parameters as weld metal dilution of 40%, 7 kg/h deposition, and a deposit efficiency of 95%. Two layers were deposited, and the main welding parameters are shown in Table 2.

Table 2. Adopted main welding parameters.

Welding Parameters	Values
Welding Current	450–550 A
Arc Voltage	28–33 V
Travel Speed	2–3 mm/s
Stick-out	35 mm
Overlap	8–11 mm
Heat Input Max	5.3 (kJ/mm)
Preheat temperature	About 150 °C

2.1. Microstructure and Hardness

Longitudinal and transverse metallographic specimens taken from plates were prepared by a standard procedure. In particular, the steel-corrosion resistant alloy interface was examined by light microscopy (LM, Nikon) and scanning electron microscopy (SEM). The low-alloy steel was etched by Nital 2%, while the microstructure of the AISI 316L weld overlay revealed etching with 50% HNO_3 + 50% H_2O. In order to assess that the distribution was of various chemical elements (in particular Ni, Cr, Fe, and Mo) in the through-thickness direction, quantitative energy dispersive spectrometry (EDS) was performed by the scanning electron microscope field emission gun (SEM-FEG), LEO 1550 (Zeiss, Oberkochen, Germany). Hardness through-thickness profiles were measured by means of 10 kg Vickers hardness indenter (HV_{10}) using steps of 0.3 and 0.5 mm at various locations on longitudinal sections.

2.2. Corrosion Testing

Pitting resistance was evaluated by measuring the critical-pitting temperature (CPT) by the American Society for Testing and Materials (ASTM) G-48 test. Additionally, Huey immersion tests in accordance with ASTM A262, Type C, were carried out to evaluate the intergranular corrosion resistance.

- ASTM G-48 test (method C). Coupons of size 20 mm × 30 mm × 1.5–2.5 mm thickness were taken from the AISI 316L clad layer. The carbon steel was machined and removed in order to obtain a flat surface. The other surface of the stainless steel coupons was practically left in the as-received condition, except for slight grinding and polishing. Duplicate specimens were considered. The ASTM G-48 test (method C) consisted of the determination of the CPT in a highly acidic media, e.g., ferric chloride. CPT temperature is defined as the lowest temperature at which pitting corrosion occurs. Test samples were exposed for 72 h to 6% ferric chloride solution. If no pitting corrosion was detected, the testing temperature was increased. The test performed was duplicated.

- ASTM A262 test (method C, Huey). This test consisted of a cyclic exposure to nitric acid of coupons (30 mm × 20 mm), taken from the AISI 316L clad layer, for 5 periods of 48 h each. The test performed was duplicated.

2.3. Heat Treatments

Specimens (100 mm × 20 mm) were taken from the clad plate and heat-treated at laboratory scale by means of an FM77H (chamber size 250 mm × 190 mm × 500 mm) muffle able to maintain 20 °C difference for 2 h. The effect of final heat-treatment conditions on the properties of both the welded layer and substrate were investigated by performing stress-relieving treatments on specimens taken from the clad plate. The original specimens were also fully retreated to simulate a Q and T treatment after cladding of the as-rolled plates. The heat-treated samples were examined in terms of microstructure and hardness indentation profiles. The following cases were considered (temperatures and holding times are indicated) (Table 3).

Table 3. Heat treatment conditions on the performed tests.

Material State	Performed Tests	
	Hardness	Microstructure
Q&T+ cladding	√	√
Stress relieved at 640 °C × 2 h	√	
Stress relieved at 660 °C × 2 h	√	√
Q&T (920 °C × 1 h + 670 °C × 2 h)	√	√
Q&T (980 °C × 1 h + 670 °C × 2 h)	√	√
Q&T (1000 °C × 1 h + 670 °C × 2 h)	√	√

The 640–660 °C stress-relieving temperature range was chosen high enough (higher than the standard one for such class of material) to investigate its effect on the eventual hardness of peaks at the interface. Three different fully Q and T cycles (including austenitization step) were chosen to simulate the possibility of cladding the as-rolled plate before submitting it to a Q and T treatment.

2.4. Chemical Composition by X-ray Fluorescence (XRF)

In order to have a more quantitative assessment of the iron content in the cladding material, a through-thickness profile was measured by X-ray fluorescence (XRF), starting from the second pass layer and moving towards the base steel on 30 mm × 20 mm specimens. Because a flat surface was required for the analysis, a section perpendicular to the plate was prepared by grinding the first 0.5 mm layer at the external clad surface. After performing the chemical analysis, the other 0.5 mm was ground and a chemical analysis carried out again. This procedure was repeated for a total depth of 3.5 mm.

3. Results

3.1. As-Received Material

3.1.1. Light Microscopy Investigation

No lack of fusion or bond, cracks, or other defects are observed. The individual runs are recognized. The first weld overlay pass has a thickness of 1.2 to 2 mm, while the second pass has a thickness from 2 to 2.5 mm (Figure 2). The dendritic structure of the weld overlay (second pass) is clearly shown in Figure 3. A detail of the coarse-grained heat-affected zone (CGHAZ) is shown in Figure 4. The heat-affected zone (HAZ), 1.0 to 1.5 mm thick, can be clearly distinguished in the low-alloy steel after etching (Figure 5). Additionally, the coarse-grained heat-affected zone is revealed between adjacent runs of the first weld overlay, close to the fusion line.

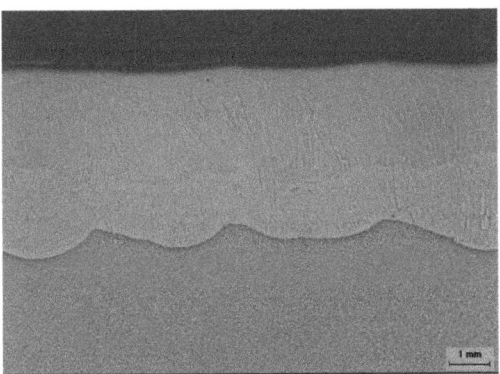

Figure 2. As-clad material (polished section).

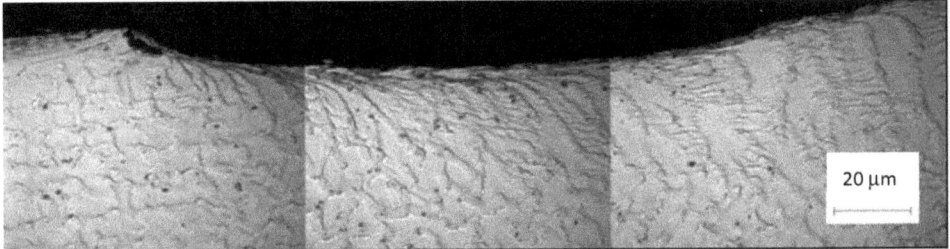

Figure 3. AISI 316L weld overlay (etching: 50% HNO_3, and 50% H_2O).

Figure 4. Q235 substrate (Q and T material).

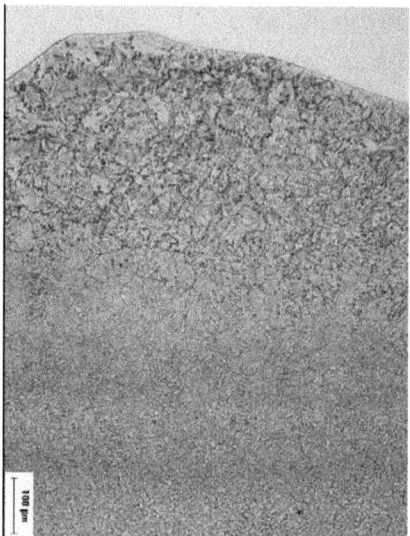

Figure 5. Detail of in the coarse-grained heat-affected zone (CGHAZ) (2% Nital etching).

3.1.2. Hardness Profiles

Examples of indentation array used to measure HV_{10} hardness are shown in Figure 6. Three indentation profiles acquired in three different positions are shown in Figure 7 (profiles 1–3). Figure 7 shows that the hardness peaks (e.g., 250 to 270 HV_{10}) are detected in the Q235 steel close to the fusion line in the CGHAZ.

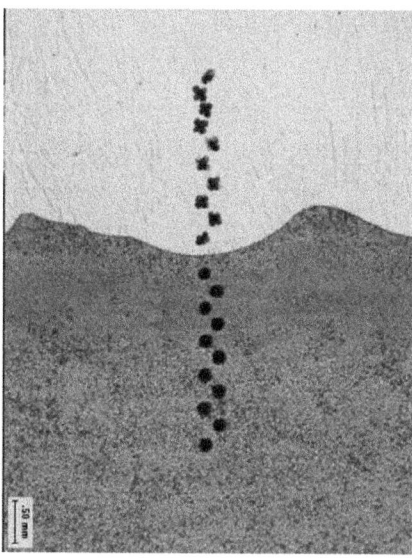

Figure 6. Examples of indentation across the Q235 and AISI 316L interface.

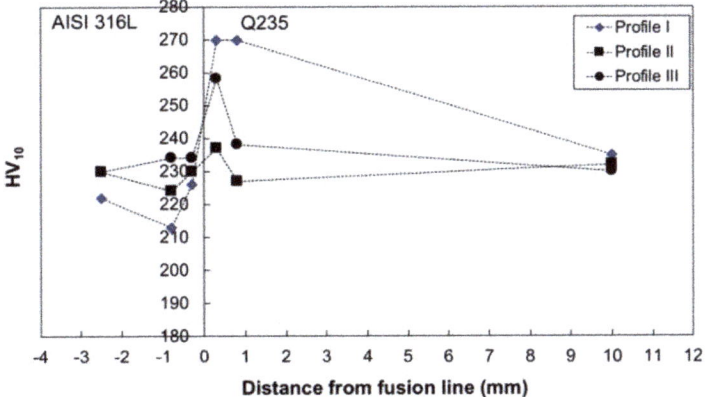

Figure 7. Hardness profiles across the Q235-AISI 316L interface.

3.1.3. SEM-EDS Investigation

On the basis of the light microscopy results, four zones are selected (Figure 8) and examined by SEM-EDS. The average values of the EDS area analysis are shown in Table 4 for the various zones.

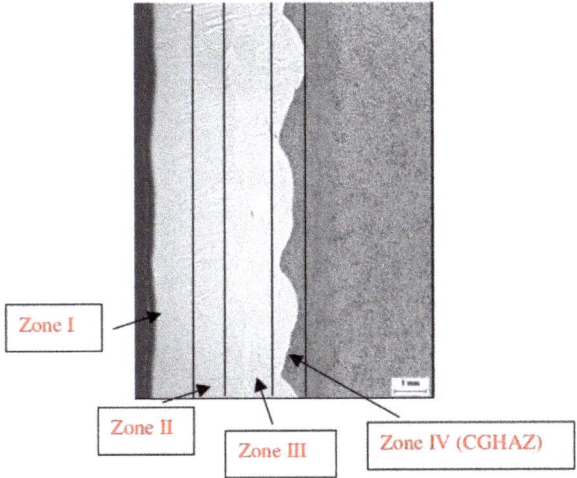

Figure 8. Identification at the light microscope of the zones examined by SEM-EDS.

Table 4. Quantitative SEM-EDS microanalysis (mass, %).

	Cr %	Ni %	Mn %	Mo %	Fe, %
Zone IV (CGHAZ)	0.05	0.1	0.9	0.05	98.9
Zone III	13.2	5.0	1.0	0.50	80.3
Zone II	18.8	7.7	0.9	1.1	71.5
Zone I	17.9	7.8	1.0	1.2	72.1

Zone I is chosen as the external layer zone, zone II as the interface between the two-layer passes, zone III as second layer pass, and zone IV as the carbon steel-stainless steel interface.

Fe increases and is detected in the weld overlay. Due to the dilution phenomena, iron is detected about 80% closer to the microalloyed steel (first pass, zone III) and about 72% closer to the second overlay pass (surface to be in contact with sour fluid, zone I), in the AISI 316L weld. In zone IV, the CGHAZ is observed (Figure 9). Austenite grains reached a size greater than 50 µm. The hardness of the peaks is attributed to increased local hardenability caused by grain-coarsening.

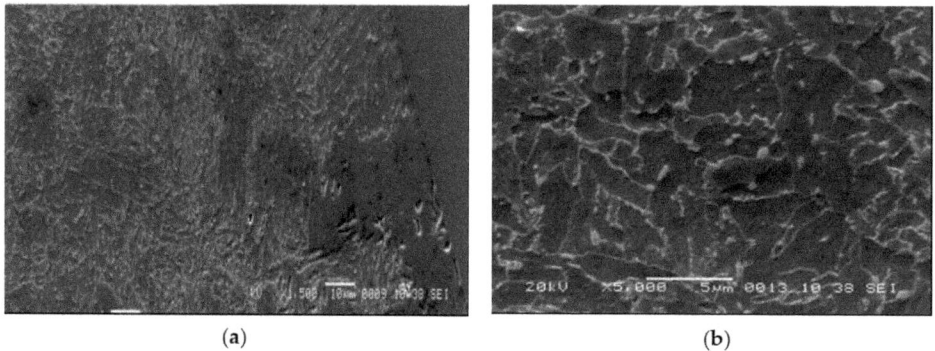

(a) (b)

Figure 9. Detail of the CGHAZ at the Q235-AISI 316L interface.

3.1.4. Corrosion Resistance of Clad Layer

Corrosion tests on the as-received cladding, i.e., determination of CPT by the ASTM G-48 test, were not promising (Table 5). This is because at 10 °C, severe pitting corrosion was exhibited on one face when the weld overlay specimen was machined considering all its thickness (both first and second welding pass). This behavior was likely due to excessive Fe content (>15%) in the corrosion-resistant alloy layer. Also, the Huey (ASTM A262 Type C) immersion test to evaluate the intergranular corrosion resistance gave unsatisfactory results with corrosion rates greater than 60 mm/yr in the first immersion. Later, after the third immersion, when the cladding with the lowest iron content remained, the corrosion rate decreased to 2.6 mm/year. When cladding coupons were predominantly sampled from the second overlay pass, the corrosion resistance significantly improved (Table 4) with CPT > 10 °C, and the corrosion rate in the Huey solution was about 2.5 mm/yr, although slightly below that expected for standard AISI 316L.

Table 5. The corrosion resistance of the clad layer.

Cladding Specimens	CPT (°C) ASTM G-48	Corrosion Rate (mm/year) ASTM A262-C
2.5 mm thickness (almost total weld overlay)	Failed at 20 °C (heavy corrosion)	61 mm/year (first cycle) 2.6 mm/year (third cycle)
1.5 mm thickness (second pass of the weld overlay)	Failed at 20 °C (5 pits)	2.4 mm/year (first cycle) 2.4 mm/year (third cycle)
1.5 mm thickness (second pass of the weld overlay)	Passed at 10 °C	-

3.1.5. Chemical Composition Profiles

In order to have a more quantitative assessment of the iron content in the cladding, a through-thickness profile is measured by XRF on the clad material, following the procedure described in Section 2.4, for a total depth of 3.5 mm (Figure 10). The first 2 mm thick layer of the cladding shows a uniform composition of about 70%. However, depths greater than 2 mm give an iron content >70%, with values that increase almost linearly, reaching 80% at a 3.0 mm depth (Figure 9). At this position, Cr and Mo contents are about 13.1% and 0.9%, respectively. Of course, given this through-thickness profile of iron, it is very difficult to take corrosion coupons from cladding having Fe content not affected by diffusion from Q235 steel.

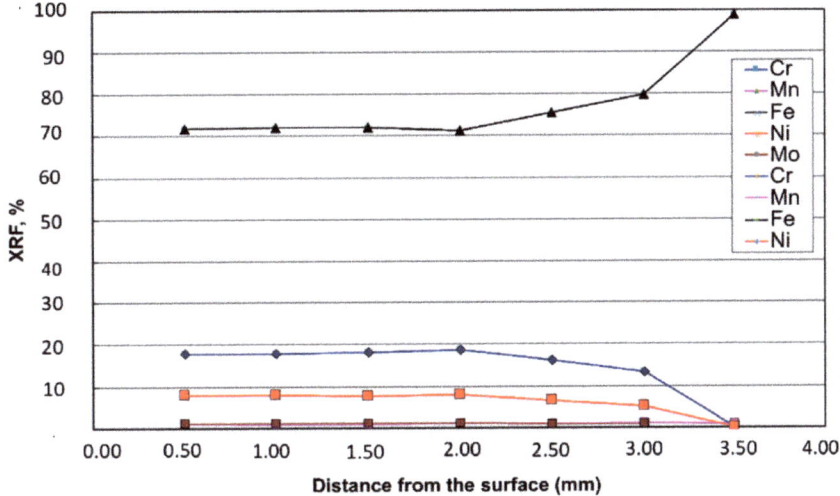

Figure 10. Chemical composition profiles (mass pct) through the thickness of the AISI 316L overlay by X-ray fluorescence (XRF) measurements.

3.2. Stress-Relieving Effect

Because stress-relieving treatments do not give significant microstructural changes that can be revealed by light microscopy and SEM, only results in terms of hardness will be shown. The hardness profiles performed on the clad material after stress relieving at 640 and 660 °C for a holding time of 2 h are shown in Figures 11 and 12, respectively. The hardness peaks (e.g., 255 to 260 HV_{10}) still remain in the CGHAZ of the microalloyed steel, close to the fusion line (0.3 mm distance), although slightly reduced compared to the as-clad material.

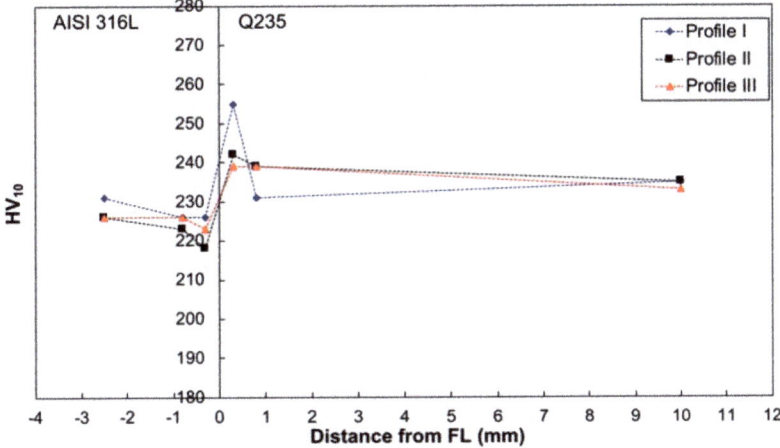

Figure 11. Hardness profiles across the Q235-AISI 316L interface after a stress-relieving heat treatment at 640 °C for a 2 h holding time.

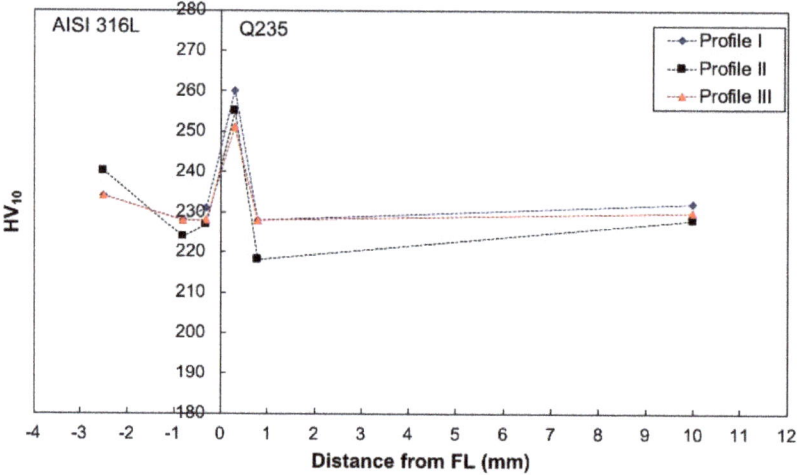

Figure 12. Hardness profiles across the Q235-AISI 316L interface after a stress-relieving heat treatment at 660 °C for a 2 h holding time.

This means that stress-relieving treatments are not sufficient to avoid hardness on the higher peaks at the interface. A possible alternative in the production route is to use the as-rolled (green) plates, which are clad and submitted to Q and T treatment later. In order to investigate the possibility to follow such a route, specimens were treated considering three austenitizing temperatures, 920 °C, 980 °C, and 1000 °C, respectively, and one tempering condition (670 °C × 2 h). The lower austenitizing temperature is typical of standard (unclad) Q235 plates, while the other temperatures were selected because they were recommended for the heat treatment of AISI 316L, which is still practiced in the present industrial furnaces. The hardness profiles performed on the clad material after the Q and T laboratory treatments are shown in Figure 13.

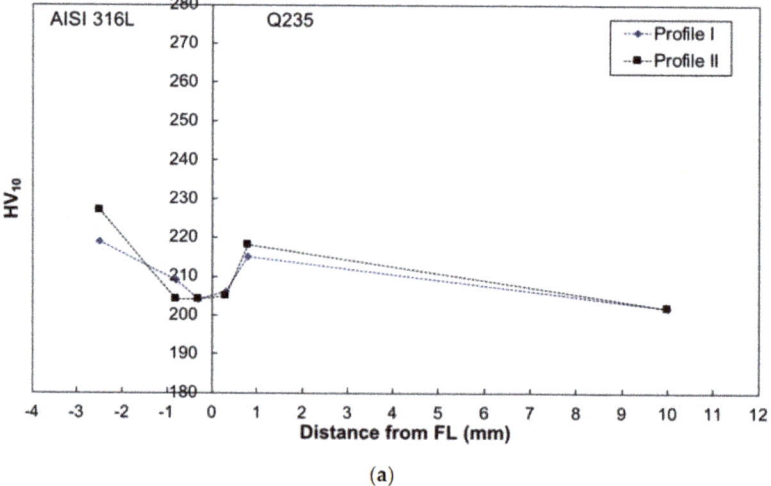

(a)

Figure 13. *Cont.*

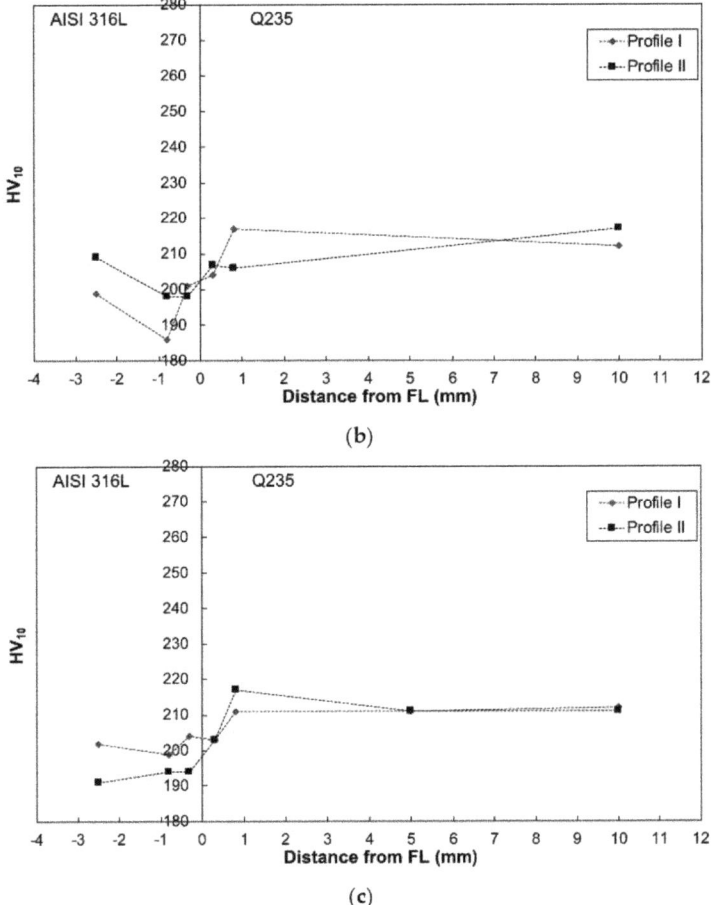

Figure 13. Hardness profiles across the Q235-AISI 316L interface after stress–relieving heat treatment at 920 °C × 1 h + 670 °C × 2 h (**a**), 980 °C × 1 h + 670 °C × 2 h and (**b**) 1000 °C × 1 h + 670 °C × 2 h (**c**).

The hardness peaks in the CGHAZ of the microalloyed steel, close to the fusion line (0.3 mm distance), disappeared; they composed all values lower than 220 HV_{10}. Reaustenitizing acts at the interface microstructure by refining austenite grain size with respect to the as-clad material (Figure 14), thus lowering local hardenability, with consequent lower hardness values at the interface. No significant effects are found depending on austenitization temperature variation in the range of 920–1000 °C. The hardness of the base material after Q and T appears slightly decreased compared to the as-clad and Q and T materials. However, this is not a critical aspect and can be balanced using a suitable tempering temperature.

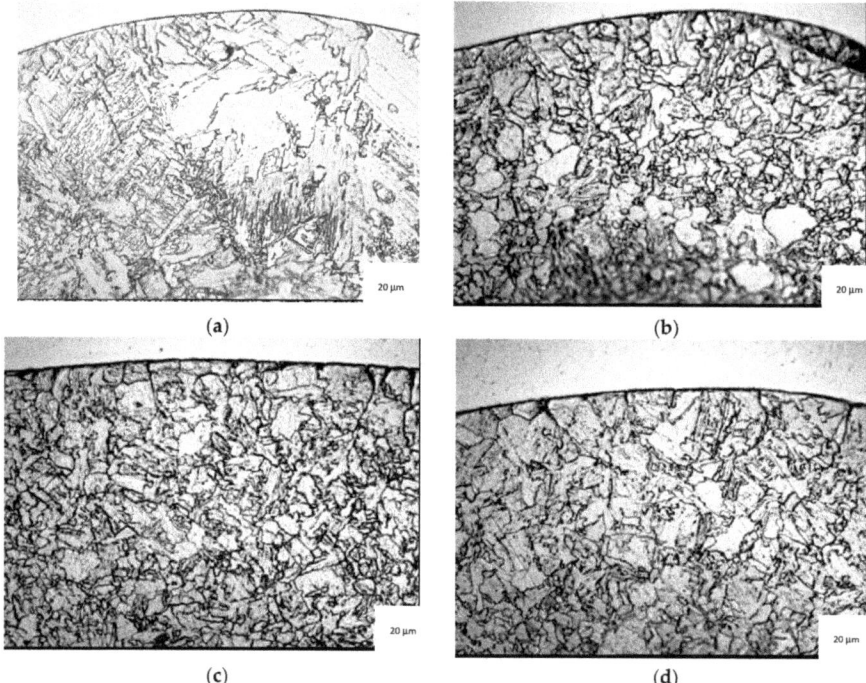

Figure 14. Q and T effect on the microstructure at the interface. (**a**) As-received material, (**b**) 920 °C × 1 h + 670 °C × 2 h, (**c**) 980 °C × 1 h + 670 °C × 2 h, and (**d**) 1000 °C × 1 h + 670 °C × 2 h.

4. Conclusions

The following conclusions can be drawn from the above results:

(1) Corrosion tests on the clad and Q and T material, i.e., the determination of CPT by the ASTM G-48 test and Huey (ASTM A262 Type C) immersion tests to evaluate the intergranular corrosion resistance, were not promising when almost all weld overlays were sampled, likely due to excessive Fe content in the CRA layer. When cladding coupons were predominantly sampled from the second overlay pass, the corrosion resistance was significantly improved, although it remained slightly below what was expected for the standard AISI 316L steel.

(2) The above results are related to the interface chemical composition. In particular, the results showed that the first 2 mm thick layer of the cladding showed a uniform composition at about 70% Fe content. However, depths greater than 2 mm gave an iron content >70%, with values that increased almost linearly, reaching 80% at 3.0 mm depth due to dilution phenomena.

(3) Some hardness peaks (e.g., 250 to 270 HV_{10}) were detected in the microalloyed steel close to the fusion line in the CGHAZ.

(4) The hardness peaks in the CGHAZ of the microalloyed steel, close to the fusion line (0.3 mm distance), disappeared after Q and T; they composed all values lower than 220 HV_{10}. This means that the reaustenitizing treatment, at temperatures below those experienced at 0.3 mm from the fusion line, produces a refinement of austenite grains, sufficient to decrease hardenability and hardness after tempering.

Author Contributions: Conceptualization, A.D.S. and C.T.; methodology, A.D.S. and C.T.; formal analysis, A.D.S. and C.T. All authors have read and agreed to the published version of the manuscript.

Funding: This research received no external funding.

Conflicts of Interest: The authors declare no conflict of interest.

References

1. Lee, K.S.; Yoon, D.H.; Kim, H.K.; Kwon, Y.-N.; Lee, Y.-S. Effect of annealing on the interface microstructure and mechanical properties of a STS–Al–Mg 3-ply clad sheet. *Mater. Sci. Eng. A* **2012**, *556*, 319–330. [CrossRef]
2. Liu, C.Y.; Wang, Q.; Jia, Y.Z.; Jing, R.; Zhang, B.; Ma, M.Z.; Liu, R.P. Microstructures and mechanical properties of Mg/Mg and Mg/Al/Mg laminated composites prepared via warm roll bonding. *Mater. Sci. Eng. A* **2012**, *556*, 1–8. [CrossRef]
3. Hang, S.U.; Luo, X.-B.; Chai, F.; Shen, J.-C.; Sun, X.-J.; Lu, F. Manufacturing Technology and Application Trends of Titanium Clad Steel Plates. *J. Iron Steel Res. Int.* **2015**, *22*, 977–982.
4. Marshall, P. *Austenitic Stainless Steels: Microstructure and Mechanical Properties*; Elsevier Applied Science Publisher: Amsterdam, The Netherlands, 1984.
5. Di Schino, A. Manufacturing and application of stainless steels. *Metals* **2020**, *10*, 327. [CrossRef]
6. Rufini, R.; Di Pietro, O.; Di Schino, A. Predictive simulation of plastic processing of welded stainless steel pipes. *Metals* **2018**, *8*, 519. [CrossRef]
7. Di Schino, A.; Di Nunzio, P.E.; Turconi, G.L. Microstrucure evolution during tempering of martensite in medium carbon steel. *Mat. Sci. For.* **2007**, *558–559*, 1435–1441.
8. Di Schino, A.; Kenny, J.M.; Abbruzzese, G. Analysis pf the recrystallization and grain growth processes in AISI 316 stainless steel. *J. Mat. Sci.* **2002**, *37*, 5291–5298. [CrossRef]
9. Di Schino, A.; Porcu, G.; Longobardo, M.; Lopez Turconi, G.; Scoppio, L. Metallurgical design and development of C125 grade for mild sour service application. In *NACE-International Corrosion Conference Series, San Diego, CA, USA, 10–14 September 2006*; NACE International: Houston, TX, USA; pp. 061251–0612514.
10. Di Schino, A. Analysis of heat treatment effect on microstructural features evolution in a micro-alloyed martensitic steel. *Acta Met. Slovaca* **2016**, *22*, 266–270. [CrossRef]
11. Sharma, D.K.; Filipponi, M.; Di Schino, A.; Rossi, F.; Castaldi, J. Corrosion behaviour of high temeperature fuel cells: Issues for materials selection. *Metalurgija* **2019**, *58*, 347–351.
12. Cianetti, F.; Ciotti, M.; Palmieri, M.; Zucca, G. On the evaluation of surface fatigue strength of stainless steel aeronautical component. *Metals* **2019**, *9*, 455. [CrossRef]
13. Di Schino, A.; Valentini, L.; Kenny, J.M.; Gerbig, Y.; Ahmed, I.; Hefke, H. Wear resistance of high-nitrogen austenitic stainless steel coated with nitrogenated amorphous carbon films. *Surf. Coat. Technol.* **2002**, *161*, 224–231. [CrossRef]
14. Bregliozzi, G.; Ahmed, S.I.-U.; Di Schino, A.; Kenny, J.M.; Haefke, H. Friction and wear behavior of austenitic stainless steel: Influence of atmospheric humidity, load range, and grain size. *Tribol. Lett.* **2004**, *17*, 697–704. [CrossRef]
15. Valentini, L.; Di Schino, A.; Kenny, J.M.; Gerbig, Y.; Haefke, H. Influence of grain size and film composition on wear resistance of ultrafine grained AISI 304 stainless steel coated with amorphous carbon films. *Wear* **2002**, *253*, 458–464. [CrossRef]
16. Zitelli, C.; Folgarait, P.; Di Schino, A. Laser powder bed fusion of stainless-steel grades: A review. *Metals* **2019**, *9*, 731. [CrossRef]
17. Di Schino, A.; Di Nunzio, P.E. Metallurgical aspects related to contact fatigue phenomena in steels for back up rolling. *Acta Met. Slovaca* **2017**, *23*, 62–71. [CrossRef]
18. Li, C.; Quin, G.; Tang, Y.; Zhang, B.; Lin, S.; Geng, P. Microstructures and mechanical properties of stainless steel clad plate joint with diverse filler metals. *J. Mat. Res. Technol.* **2020**, *9*, 2522–2534. [CrossRef]
19. Wang, Y.; Li, X.; Wang, X.; Yan, H. Fabrication of a thick copper-stainless steel clad plate for nuclear fusion equipment by explosive welding. *Fusion Eng. Des.* **2018**, *137*, 91–96. [CrossRef]
20. Qin, Q.; Zhang, D.-T.; Zang, Y.; Guan, B. A simulation study on the multi-pass rolling bond of 316L/Q345R stainless clad plate. *Adv. Mech. Eng.* **2015**, *7*, 1–13. [CrossRef]
21. Li, Y.-W.; Liu, H.-T.; Wang, Z.-J.; Zhang, X.-M.; Wang, G.-D. Suppression of edge cracking and improvement of ductility in high borated stainless steel composite plate fabricated by hot-roll-bonding. *Mater. Sci. Eng. A* **2018**, *731*, 377–384. [CrossRef]
22. Mousawi, A.; Barrett, S.A.A.; Al-Hassani, S.T.S. Explosive welding of metal plates. *J. Mater. Process. Technol.* **2008**, *202*, 224–239.

23. Tian, M.J.; Chen, H.; Zhang, Y.K.; Wang, T.; Zhu, Z.Y. Welding process of composite plate for construction and its welded microstructure and properties. *Heat Treat. Met.* **2015**, *40*, 110–115.
24. Qiu, T.; Wu, B.X.; Chen, Q.Y.; Chen, W.J. Analysis on welded joint properties of stainless clad steel plates. *Electric Weld. Mach.* **2013**, *43*, 83–87.
25. Liao, H.M.; Song, K.Q.; Cao, Y.P.; Zeng, M. Welding process and welded joint microstructure of austenitic stainless clad steel plate. *Hot Work. Technol.* **2012**, *41*, 148–150.
26. Wang, W.X.; Wang, Y.F.; Liu, M.C.; Cheng, F.C.; Wu, W. Microstructure and corrosion resistance of butt joint of 1Cr18Ni9Ti+Q235 composite plate. *Trans. China Weld. Inst.* **2010**, *31*, 89–92.
27. Zhang, J.J.; Ju, Z.P. Investigation on welding performance of 321–Q345q–D composite steel plate used in railway bridge. *Steel Constr.* **2012**, *27*, 48–53.
28. Napoli, G.; Di Schino, A.; Paura, M.; Vela, T. Colouring titanium alloys by anodic oxidation. *Metalurgija* **2018**, *57*, 111–113.
29. Xin, J.; Song, Y.; Fang, J.; Wei, J.; Huang, C.; Wang, S. Evaluation of inter-granular corrosion susceptibility in 304LN austenitic stainless steel weldments. *Fusion Eng. Des.* **2018**, *133*, 70–76. [CrossRef]
30. Aydogdu, G.H.; Aydinol, M.K. Determination of susceptibility to intergranular corrosion and electrochemical reactivation behaviour of AISI 304L type stainless steel. *Corros. Sci.* **2006**, *8*, 3565–3583. [CrossRef]
31. Li, H.; Zhang, L.; Zhang, B.; Zhang, Q. Microstructure characterization and mechanical properties of stainless steel clad plate. *Materials* **2019**, *12*, 509. [CrossRef]
32. Li, Z.; Wang, X.; Luo, Y. Equivalent properties of transition layer based on element distribution in laser bending of 304 stainless steel/Q235 carbon steel laminated plate. *Materials* **2018**, *11*, 2326. [CrossRef]
33. Jing, Y.-A.; Qin, Y.; Zang, X.; Shang, Q.; Song, H. A novel reduction-bonding process to fabricate stainless steel clad plate. *J. Alloys Compd.* **2014**, *617*, 688–698. [CrossRef]
34. Liu, B.X.; Wei, J.Y.; Yang, M.X.; Yin, F.X.; Xu, K.C. Effect of heat treatment on the mechanical properties of copper clad steel plates. *Vacuum* **2018**, *154*, 250–258. [CrossRef]
35. Dhib, Z.; Guermazi, N.; Ktari, A.; Gasperini, M.; Haddar, N. Mechanical bonding properties and interfacial morphologies of austenitic stainless steel clad plates. *Mater. Sci. Eng. A* **2017**, *696*, 374–386. [CrossRef]
36. Yu, T.; Jing, Y.-A.; Yan, X.; Li, W.; Pang, Q.; Jing, G. Microstructures and properties of roll-bonded stainless/medium carbon steel clad plates. *J. Mater. Process. Technol.* **2019**, *266*, 264–273.
37. Chen, C.X.; Liu, M.Y.; Liu, B.X.; Yin, F.X.; Dong, Y.C.; Zhang, X.; Zhang, F.Y.; Zhang, Y.G. Tensile shear sample design and interfacial shear strength of stainless steel clad plate. *Fusion Eng. Des.* **2017**, *125*, 431–441. [CrossRef]
38. Gabrel, J.; Coussement, C.; Verelest, L.; Blum, R.; Chen, Q.; Testani, C. Superheaters materials testing for USC Boilers: Steam side oxidation rate of 9 advanced materials in industrial conditions. *Mat. Sci. For.* **2001**, *369*, 931–938. [CrossRef]
39. Zhang, S.; Xiao, H.; Xie, H.; Gu, L. The preparation and property research of the stainless steel/iron scrap clad plate. *J. Mater. Process. Technol.* **2014**, *214*, 1205–1210. [CrossRef]
40. Di Schino, A.; Alleva, L.; Guagnelli, M. Microstructure evolution during quenching and tempering of martensite in a medium C steel. *Mat. Sci. For.* **2012**, *715–716*, 860–865. [CrossRef]
41. Song, H.; Shin, H.; Shin, Y. Heat-treatment of clad steel plate for application of hull structure. *Ocean Eng.* **2016**, *122*, 278–287. [CrossRef]

 © 2020 by the authors. Licensee MDPI, Basel, Switzerland. This article is an open access article distributed under the terms and conditions of the Creative Commons Attribution (CC BY) license (http://creativecommons.org/licenses/by/4.0/).

Article
Compound Layer Design for Deep Nitrided Gearings

Stefanie Hoja *, Matthias Steinbacher and Hans-Werner Zoch

Leibniz-Institut für Werkstofforientierte Technologien-IWT, Badgasteiner Straße 3, 28359 Bremen, Germany and MAPEX, Center for Materials and Processes, Universität Bremen, Bibliothekstraße 1, 28359 Bremen, Germany; steinbacher@iwt.uni-bremen.de (M.S.); zoch@iwt-bremen.de (H.-W.Z.)
* Correspondence: shoja@iwt-bremen.de; Tel.: +49-421-218-51395

Received: 2 March 2020; Accepted: 27 March 2020; Published: 31 March 2020

Abstract: Deep nitriding is used to obtain a nitriding hardness depth beyond 0.6 mm. The long nitriding processes, which are necessary to reach the high nitriding hardness depths, mostly have a negative influence on the hardness and strength of the nitrided layer as well as on the bulk material. The compound layer often is considered less, because in most practical cases, it is removed mechanically after nitriding, to avoid spalling in service. However, in former investigations, it was shown, that thick and compact compound layers have the potential for high flank load capacity of gears. The investigations focus on the simultaneous formation of a high nitriding depth and a thick and compact compound layer. Beside the preservation of the strength, a challenge is to control the porosity of the compound layer, which should be as low as possible. The investigations were carried out using the common nitriding and heat treatable mild steel 31CrMoV9, which is often used for gear applications. The article gives an insight on the development of multistage nitriding processes studied by short- and long-term experiments aiming for a specific compound layer build-up with low porosity and high strength of the nitride layer and core material.

Keywords: nitriding; nitrocarburizing; two-stage treatment; process design; compound layer; white layer; nitriding hardness depth

1. Introduction

High performance gears are usually subjected to heat treatment in order to achieve optimum performance for the respective application. Thermochemical surface layer heat treatment, such as case hardening, carbonitriding, nitriding or nitrocarburizing are common treatments, since the highest stress is in the surface layer of the teeth and a tough core is positive for the service behavior [1]. Studies have shown that nitriding treatment can improve the load-bearing properties of gears with regard to different stresses. König et al. provide an overview of nitriding applications for gearings [2].

The necessary hardening depth of the surface layer depends on the normal modulus of the gearing [3]. For nitriding, a nitriding hardening depth of approximately 0.6 mm is regarded as the economic limit value. Therefore case hardening is frequently used for larger gears, since higher case depths can be achieved in a shorter time. Nevertheless, nitriding offers some process-related advantages compared to case hardening, such as better protection against scuffing, wear and corrosion, higher surface hardness, a temperature-stable surface layer for use at higher temperatures, as well as less dimensional and shape changes and thus generally no need for reworking [4].

Deep nitriding differs from conventional nitriding in aiming higher nitriding hardness depths of approximately 0.8–1.0 mm. In order to achieve those nitriding hardness depths, long nitriding periods and/or high nitriding temperatures are necessary, which leads to losses of strength in the nitriding layer and in the core. In order to minimize these losses in strength caused by tempering effects during nitriding, the maximum possible nitriding temperature is limited depending on the material [5].

Investigations on deep nitriding were already carried out in the 1980s [6], when the nitriding plant technology was not yet sufficiently developed to enable controlled and reproducible nitriding treatments to be carried out in gas or plasma. For this reason, more recent investigations concern also nitriding processes for achieving high nitriding hardness depths [4,7–9]. In [4] material-oriented two-stage processes for deep nitriding were developed with the aim of achieving high nitriding hardness depths with the highest possible strength. These processes were carried out according to the two-stage principle developed by Floe [10,11], where a closed compound layer was formed during the first stage. The temperature was kept low at 520 °C in order to obtain finely distributed precipitates in the diffusion layer. In the second stage at elevated temperature, diffusion was carried out into the depth and the growth of the compound layer was restricted by a low nitriding index. By this process control, negative effects on the strength of the material could be kept within limits and stress-compatible nitriding layers with high nitriding hardness depths with sufficient strength of the base material could be reliably and reproducibly produced by modern process technology.

Nitriding increases the rolling contact fatigue life of gears, whereas the generated compressive residual stresses within the diffusion layer are responsible for the enhancement of the fatigue strength [12,13]. Until now, the primary objective of nitriding process development was to achieve high nitriding hardness depths with economical process duration and little attention has been paid to the compound layer during. In random tests on the tooth flank load-bearing capacity, one of the investigated variants, which had a comparatively thick, compact compound layer due to a higher nitriding temperature, showed particularly good load-bearing capacity behavior. It was remarkable that the compound layer remained almost undamaged after the load-tests at high load levels, whereas in the other tested variants with a thinner compound layer, defects in the compound layer occurred after only a few load cycles [14]. Additionally the authors of [15] showed in their work, that a control of the nitriding parameters is necessary because the crystalline structure profile of the compound layer is important for the gear performance.

The composition of the compound layer (γ'-nitride/ε-nitride/alloying element nitrides) and the shape of the porous zone also play a significant role in the tribological load-bearing capacity [16]. The ε-nitride has a higher hardness than the γ'-nitride and due to its hexagonal structure, the ε-nitride has a low number of sliding systems, which prevents the wear partners from approaching each other over an area down to atomic distances. For this reason, the adhesion tendency of ε-nitride layers is low. The pore content and the pore distribution and size also have an influence on the wear behavior. Depending on the wear load, sometimes γ'-layers show better wear properties and sometimes thin, slightly porous ε-compound layers [17–19].

The current investigations deal with the simultaneous achievement of high nitriding hardness depths and a thick, compact compound layer. In addition to maintaining the strength properties, another challenge is to ensure that the compound layer does not become too porous.

2. Materials and Methods

2.1. Reference State

The deep nitrided material 32CDV13 (0.32% C, 0.33% Si, 0.55% Mn, 3.00% Cr, 0.90% Mo and 0.29% V) showed a remarkable load-bearing potential in [14]. The gears were deep nitrided in a material-oriented two-stage treatment in a controlled gas nitriding process with steel retort. In the first stage at 520 °C in 15 h with a nitriding potential of $K_N = 1$, a compound layer and fine nitride precipitates were formed in the first approximately 0.25 mm of the diffusion zone. In the second stage, the temperature was increased to 570 °C for 130 h and at the same time the nitriding potential was reduced to $K_N = 0.5$. The second stage was mainly used for rapid diffusion of the nitrogen into the material in order to achieve the desired nitriding hardness depth.

The compound layer of the reference material 32CDV13 formed on the tooth flank is shown in Figure 1. Table 1 shows the compound layer thickness (CLT), the thickness of the porous zone (CLT_P)

and nitriding hardness depth (NHD) averaged over the tooth area. Despite the low nitriding potential of $K_N = 0.5$ in the diffusion phase, a relatively thick compound layer has been formed due to the comparatively high temperature and long duration required for nitriding.

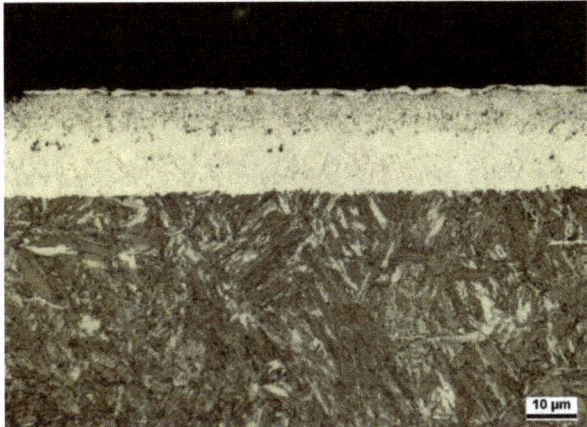

Figure 1. Compound layer of the reference state 32CDV13 deep nitrided in a two-stage treatment.

Table 1. Average compound layer thickness (CLT), thickness of the porous zone (CLT_P) and nitriding hardness depth (NHD) of the toothing of the deep nitrided steel 32CDV13 in [12].

Reference 32CDV13	CLT	CLTP	NHD
	15–17 µm	5–7 µm	ca. 1 mm

The nitrogen and carbon profiles of the reference compound layer determined by Glow Discharge Optical Emission Spectroscopy (GD-OES) are shown in Figure 2. In the area of the porous zone, the nitrogen concentration of approximately 9–9.5 mass-% was slightly higher than in the compact area of the compound layer below. Here the nitrogen concentration was about 6.5 mass-%. Taking into account the alloying elements, which also bind nitrogen as nitride, it can be assumed that the compound layer consists mainly of γ'-nitride and that in the area of the porous zone ε-nitrides have also been formed. A X-ray phase analysis with Cr K_α radiation (surface measurement) showed a compound layer composition of approximately 14% ε-nitride and approximately 86% γ'-nitride, which confirms the considerations made on the basis of the nitrogen profile.

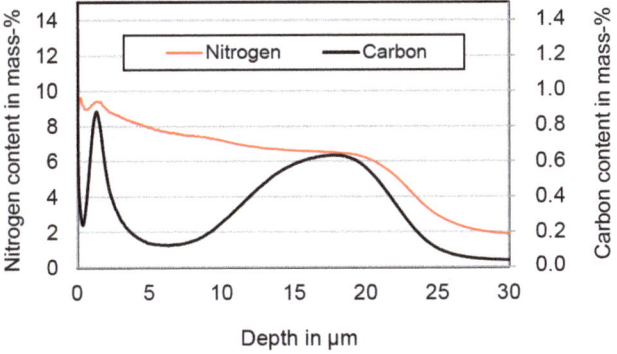

Figure 2. GD-OES nitrogen and carbon depth profiles of the deep nitrided reference state Compound layer of the deep nitrided reference state.

2.2. Investigated Material

The frequently used nitriding and tempering steel 31CrMoV9 (0.33% C, 0.29% Si, 0.57% Mn, 2.41% Cr, 0.16% Mo and 0.12% V) served as the investigation material. The material was quenched and tempered prior to nitriding. For this purpose, the material was austenitized at 950 °C, quenched in oil and tempered for 2 h at 630 °C. The hardening temperature, which is untypical for the material 31CrMoV9, was selected in accordance with the practice for the reference material 32CDV13 in order to produce a similar microstructural state to that of the reference material. Compared to a classical hardening at 870 °C and tempering at 630 °C, the higher austenitizing temperature resulted in a hardness that was about 30 HV1 higher. The reason for the higher hardness was assumed to be that more carbon was dissolved at the higher austenitizing temperature due to the dissolution of low chromium-containing precipitates, and that the carbides were finely precipitated again during subsequent tempering.

2.3. Nitriding Treatments

The nitriding and nitrocarburizing treatments were carried out in a gas nitriding plant with controlled nitriding and carburizing potential. The heating was carried out under an ammonia atmosphere to activate the sample surface. The cooling after nitriding was also carried out with the addition of ammonia (phase-controlled cooling) in order to avoid nitrogen effusion and subsequent denitration at the end of the process. Table 2 gives an overview of the parameter variations carried out in the treatments.

Table 2. Nitriding and nitrocarburizing treatments.

Short-Term Treatments		Long-Term Treatments	
Nitriding	Nitrocarburizing	One-Stage	Two-Stage
520 °C 15 h $K_N = 1$	520 °C 15 h $K_N = 1$ $K_C^B = 0.1$	530 °C 170 h $K_N = 0.5$	520 °C 15 h $K_N = 5$ 530 °C 170 h $K_N = 0.5$
520 °C 15 h $K_N = 3$	520 °C 15 h $K_N = 5$ $K_C^B = 0.1$	530 °C 170 h $K_N = 1$	530 °C 170 h $K_N = 0.5$ 520 °C 15 h $K_N = 5$
520 °C 15 h $K_N = 5$	550 °C 15 h $K_N = 5$ $K_C^B = 0.1$	530 °C 170 h $K_N = 3$	530 °C 170 h $K_N = 0.5$ 550 °C 15 h $K_N = 5$
		550 °C 120 h $K_N = 1$	530 °C 170 h $K_N = 0.5$ 550 °C 15 h $K_N = 5$ $K_C^B = 0.1$

With the aim of gaining an understanding of the relationship between nitriding and carburizing potential and compound layer structure and phase composition, short-term treatments were initially carried out. By a specific variation of the nitriding and carburizing potential different compound layers could be produced by nitriding and nitrocarburizing. The temperature of 520 °C and the duration of 15 h were chosen for the short-term tests analogous to the first stage of the two-stage treatment of the reference material. In one treatment, the temperature of 520 °C was increased to 550 °C for nitrocarburizing, as nitrocarburizing is more likely to be carried out at temperatures of 550–590 °C. The short-term treatments were intended to provide assistance in selecting the process parameters for the long-term treatments.

The long-term treatments were carried out with two different strategies: In one-stage treatments, a continuous built-up compound layer structure was aimed, whereas in the two-stage treatments the compound layer was designed in an extra process stage prior to or after the diffusion layer was built up. The nitriding hardness depth was aimed to be approximately 0.8 mm after all long-term treatments.

The one-stage treatment with the nitriding potential $K_N = 0.5$ corresponds to the diffusion phase of the process developed for the material 31CrMoV9 in [4,14] and was also used as a diffusion phase in the two-stage treatments. In two further one-stage treatments, a thicker compound layer was aimed at by increasing the nitriding potential. The tempering parameter according to Hollomon-Jaffe [20] was

H = 17.9 for all 170-h treatments at 530 °C and was thus still below the critical value of about H = 18, which, according to Hoja et al. would lead to a decrease of strength [4,5]. Since in the tests carried out in the past with the material 32CDV13 the thick compound layer was formed at a higher temperature (570 °C), a one-stage treatment at 550 °C was also carried out. With the temperature increase, the nitriding time was shortened accordingly to achieve a similar nitriding hardness depth. With H = 18.2, the tempering parameter of the 120-hour single-stage test at 550 °C was slightly above H = 18, which means that after this process a decrease in strength in the nitriding layer and in the base material is to be expected.

In the previous investigations [4], the two-stage treatments used the principle developed by Floe, according to which a compound layer is first built up and its growth is suppressed or restricted in the further process [10,11]. A focus of the present investigations was therefore on the sequence of the process stages. The compound layer build-up after the diffusion layer was built up had the goal of forming a more compact, less porous compound layer.

2.4. Characterization of the Compound Layer

The thickness of the compound layer and the porous zone was evaluated according to the test specification of the Arbeitsgemeinschaft Wärmebehandlung + Werkstofftechnik e. V. (AWT) [21] on the metallographic cross-section taken from coupon samples separated by spark erosion. All sections were etched with Nital and images were taken at 1000× magnification via light microscope Leica DM6000 (Leica, Wetzlar, Germany). The average compound layer and thickness of the porous zone were determined by measuring at ten points on three micrographs. In addition to documenting the compound layer, the nitriding hardness depth was also determined on the cross section in accordance with DIN 50190-3. The composition of the compound layers was determined using glow discharge spectroscopy (GD-OES, LECO GDS 750A, SN 3030, St. Joseph, MI, USA) and X-ray methods. For this purpose, nitrogen and carbon depth profiles were recorded to a depth of about 50–70 μm. Based on the nitrogen and carbon content, the thickness and composition of the compound layer can be estimated, since the (carbo-)nitride phases have different nitrogen and carbon contents. In addition, X-ray diffraction spectra with Cr-K_α radiation were recorded and the proportions of ε- and γ'-nitride were determined using the Rietveld method. The Rietveld method involves fitting the entire spectrum. The diffraction spectrum of a polycrystalline substance is regarded as a mathematical function of the angle of diffraction, which also depends on the crystal structure. Starting from an initial model of the atomic arrangement, structural and instrumental parameters are continually refined. Since the penetration depth of the X-rays, defined as the depth to which the radiation intensity drops to 1/e, is limited to about 10 μm and the information about the phase fractions according to an exponential function weighted according to depth was included in the measured value, the information obtained in this way could be seen as an average phase composition for thicker compound layers in addition to the GD-OES measurements and the thermodynamic calculations.

3. Results and Discussion

3.1. Short-Term Nitriding Treatments

The short-term treatments initially carried out for nitriding and nitrocarburizing were intended to give a clue for the design of the process stages of the two-stage treatments in order to influence the compound layer structure. Table 3 shows the compound layer and thicknesses of the porous zone achieved in the short-term treatments. During nitriding, only a nitriding potential of K_N = 5 was sufficient to achieve a compound layer thickness that approximates the thickness of the compound layer of the reference state in 15 h at 520 °C.

Table 3. Compound layer thickness (CLT) and thickness of the porous zone (CLT_P) after the short time nitriding and nitrocarburizing treatments.

Nitriding			Nitrocarburizing		
520 °C 15 h $K_N = 1$	520 °C 15 h $K_N = 3$	520 °C 15 h $K_N = 5$	520 °C 15 h $K_N = 1$ $K_C^B = 0.1$	520 °C 15 h $K_N = 5$ $K_C^B = 0.1$	550 °C 15 h $K_N = 5$ $K_C^B = 0.1$
CLT = 2 µm	CLT = 7 µm	CLT = 12 µm	CLT = 3 µm	CLT = 6 µm	CLT = 19 µm
CLT_P = 1 µm	CLT_P = 3 µm	CLT_P = 5 µm	CLT_P = 2 µm	CLT_P = 5 µm	CLT_P = 7 µm

During nitrocarburizing, where the aim was to achieve a faster built-up of the compound layer due to the simultaneous diffusion of nitrogen and carbon, similar compound layers were formed at 520 °C as in the corresponding nitriding treatments with the same nitriding potential. An increase in the nitrocarburizing temperature to 550 °C resulted in a thick, relatively compact compound layer of 19 µm, which is comparable in dimensions to that of the reference.

The phase composition of the compound layers was determined using GD-OES element depth profiles, Thermo-Calc calculations and X-ray diffraction spectra. Figure 3 shows the phase diagram calculated for the material 31CrMoV9 as a function of nitrogen content. It can be seen that in the temperature range of nitriding from a nitrogen content of about 7 mass-% the α-phase (ferrite) disappears and a closed nitride layer forms. This compound layer mainly consists of γ'-nitride with iron chromium nitrides and silicon nitride as well as increasing amounts of ε-nitride with increasing nitrogen concentration. From a concentration of around 11 mass-% nitrogen, the compound layer no longer contains γ'-nitrides.

The nitrided samples all had a nitrogen content of about 7 mass-% in the area of the compact compound layer despite different nitriding potentials. Slightly higher nitrogen contents were present towards the edge. On the basis of the calculated phase diagram it can therefore be assumed that the compound layers after nitriding consist mainly of γ'-nitride, with small amounts of ε-nitride in the near-surface area of the compound layer. This is also confirmed by the results of the X-ray phase analysis in Table 4.

After nitrocarburizing, the sum of nitrogen and carbon was considered to evaluate the phase composition, since both γ'-nitride and ε-nitride can dissolve certain amounts of carbon. In the face-centered cubic γ'-nitride up to 0.8-at% nitrogen can be replaced by carbon, in the hexagonal ε-nitride 25–33 at% nitrogen [22].

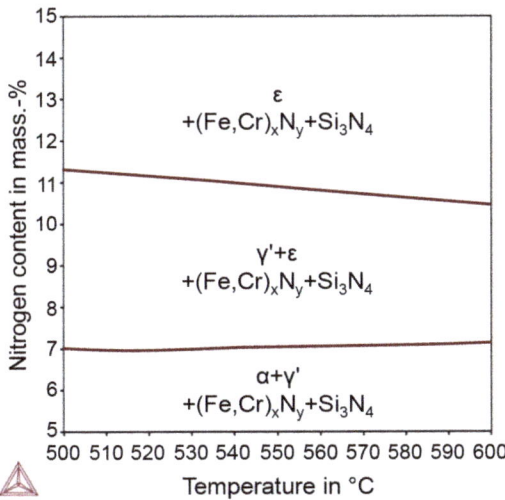

Figure 3. 31CrMoV9-nitrogen phase diagram calculated with Thermo-Calc.

Table 4. Average phase composition of the nitride layers after the short time treatments measured by X-ray diffraction.

Nitriding	γ′-Nitride in %	ε-Nitride in %	Nitro-Carburizing	γ′-Nitride in %	ε-(Carbo-)Nitride in %
520 °C 15 h $K_N = 1$	ca. 74	ca. 26	520 °C 15 h $K_N = 1$ $K_C^B = 0.1$	ca. 54	ca. 46
520 °C 15 h $K_N = 3$	ca. 79	ca. 21	520 °C 15 h $K_N = 5$ $K_C^B = 0.1$	ca. 23	ca. 77
520 °C 15 h $K_N = 5$	ca. 89	ca. 11	550 °C 15 h $K_N = 5$ $K_C^B = 0.1$	ca. 68	ca. 32

Figure 4 shows the sum of the nitrogen and carbon concentration determined with GD-OES for the nitrocarburized samples as a function of depth. In the range of the first 5–6 μm the sum of nitrogen and carbon content was clearly higher than 7 mass.-%. For the two compound layers whose thickness fell within this range. Therefore a higher ε-(carbo-)nitride content than after nitriding could be assumed. This has also been confirmed by X-ray analysis (see Table 4). The thicker compound layer formed at the temperature of 550 °C, which is more typical for nitrocarburizing, shows the ε-(carbo-)nitride in the surface near area and a plateau at just over 7 wt.% under this phase. This indicates a two-phase compound layer structure with predominantly ε-(carbo-)nitride in the upper part of the compound layer and γ′-(carbo-)nitride below. Accordingly, the X-ray phase analysis revealed a high proportion of γ′-(carbo-)nitride.

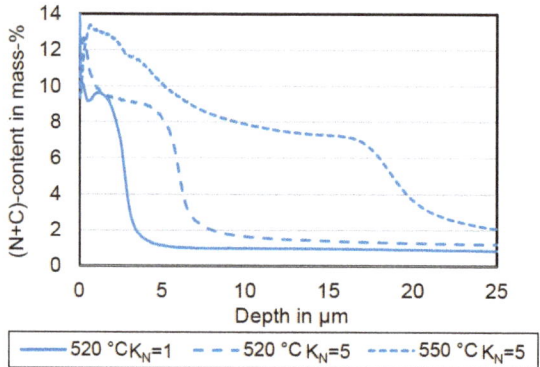

Figure 4. Sum of GD-OES nitrogen and carbon depth profiles.

3.2. One-Stage Deep Nitriding

In the one-stage treatments, a continuous compound layer build-up should take place while the nitriding hardness depth was reached. Table 5 shows the compound layer and thicknesses of the porous layer achieved in the single-stage treatments as well as the nitriding hardness depths. The targeted nitriding hardness depth of 0.8 mm was achieved almost exactly with all treatments. A comparison with the data for the compound layer and thickness of the porous layer of the reference (Table 1) makes it clear that only the treatment with a nitriding potential of $K_N = 3$ had led to a comparable compound layer thickness. However, the porous zone was significantly thicker than in the reference.

Table 5. Compound layer thickness (CLT), thickness of the porous zone (CLT_P) and nitriding hardness depth (NHD) after the single-stage nitriding treatments.

530 °C 170 h $K_N = 0.5$	530 °C 170 h $K_N = 1$	530 °C 170 h $K_N = 3$	550 °C 120 h $K_N = 0.5$
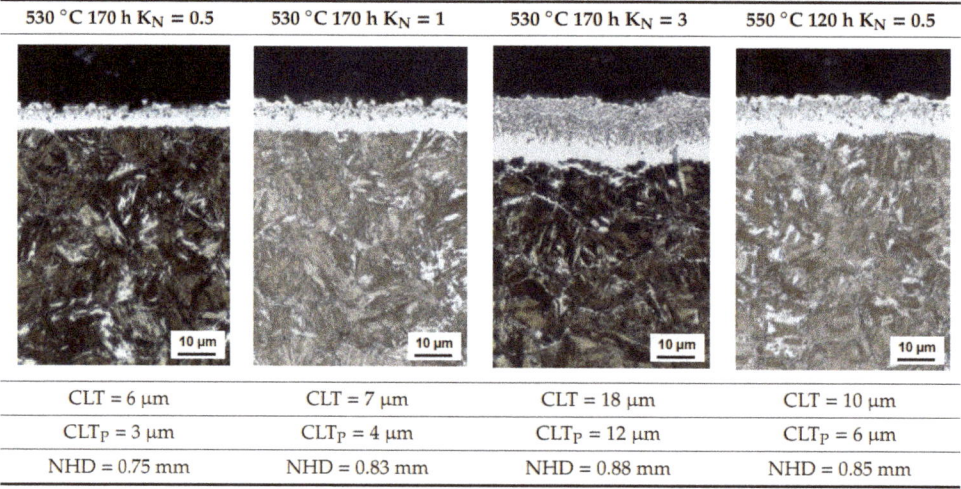			
CLT = 6 µm	CLT = 7 µm	CLT = 18 µm	CLT = 10 µm
CLT_P = 3 µm	CLT_P = 4 µm	CLT_P = 12 µm	CLT_P = 6 µm
NHD = 0.75 mm	NHD = 0.83 mm	NHD = 0.88 mm	NHD = 0.85 mm

Since deep nitriding at 530 °C requires a high nitriding potential to build up a thick compound layer and the long nitriding time required at the same time favors pore formation, single-stage deep nitriding did not appear to be suitable for producing thick and at the same time compact compound layers during deep nitriding.

3.3. Two-Stage Deep Nitriding

In the two-stage treatments, the process control should influence the compound layer structure in such a way that a compact compound layer with a similar thickness and composition to the reference is produced. In the first two-stage treatment, the compound layer was first built up in analogy to the procedure developed by Floe [10,11] where thin compound layers and high nitriding hardness depths can be achieved. Since in the current investigations a thick compound layer was aimed, in the first process stage a comparatively high nitriding potential of $K_N = 5$ was chosen to form a compound layer. In analogy to the short-term treatments, a compound layer with a thickness of approximately 12 µm should already be formed in this 15 h treatment stage at 520 °C. In the second process stage, a relatively low nitriding potential was selected to limit the further growth of the compound layer and the porous layer. Table 6 shows that the compound layer is in fact similar in thickness to that obtained after the short-term treatment corresponding to the first stage. One difference, however, is the thickness of the porous zone in the compound layers. In the long-term treatment, additional pores could form in the process stage to reach the nitriding hardness depth within the long treatment period, so that the proportion of the porous zone in the compound layer is greater than desired.

Table 6. Compound layer thickness (CLT), thickness of the porous zone (CLT_P) and nitriding hardness depth (NHD) after the two-stage nitriding treatments.

520 °C 15 h $K_N = 5$ 530 °C 170 h $K_N = 0.5$	530 °C 170 h $K_N = 0.5$ 520 °C 15 h $K_N = 5$	530 °C 170 h $K_N = 0.5$ 550 °C 15 h $K_N = 5$	530 °C 170 h $K_N = 0.5$ 550 °C 15 h $K_N = 5$ $K_C^B = 0.1$
CLT = 12 µm	CLT = 13 µm	CLT = 17 µm	CLT = 20 µm
CLT_P = 7 µm	CLT_P = 6 µm	CLT_P = 8 µm	CLT_P = 8 µm
NHD = 0.75 mm	NHD = 0.82 mm	NHD = 0.81 mm	NHD = 0.85 mm

Since the aim of the current investigations is a thick, compact compound layer at a high nitriding hardness depth, the sequence of the process stages was reversed in the further experiments, i.e., nitriding was first carried out at a comparatively low nitriding potential to achieve the desired nitriding hardness depth and then the compound layer was built up in a second process stage. In the first stage, it must be ensured that the nitriding potential is still sufficiently high to saturate the material with nitrogen, otherwise diffusion and thus the achievement of the nitriding hardness depth is slowed down.

Table 6 shows that reversing the sequence of the process stages led to the desired result. The layer thickness after the two treatments, which differ only in the sequence of the process stages, is similar, but as intended the proportion of porous zone has been significantly reduced by reversing the process stages.

Since the targeted compound layer thickness of 15–17 µm has not yet been achieved, the parameters of the second process stage were changed in two further experiments in such a way that compound layer growth was promoted. In the third two-stage experiment, the temperature in the compound layer formation stage was increased from 520 to 550 °C, and in the last two-stage experiment, a carbon donator was added to the nitriding atmosphere, so that this was a nitrocarburizing stage. With regard to the compound layer and thickness of the porous zone, a similar result to the reference was achieved with these treatments.

Table 7 shows the phase fractions of γ'- and ε-nitride of the compound layers formed in the two-stage experiments determined by X-ray analysis. It can be stated that the phase composition of the compound layers formed by pure nitriding was approximately the same. Like the compound layer of the reference (86% γ'-nitride and 14% ε-nitride), these consisted mainly of γ'-nitride with small amounts of ε-nitride. A comparison with the 15 h short-term treatment at 520 °C with a nitriding potential of $K_N = 5$ also shows that not only the thickness but also the phase composition of the compound layer was approximately the same after the short-term treatment and the corresponding long-term treatment.

Table 7. Average phase composition of the nitride layers after the two-stage long time treatments measured by X-ray diffraction.

Two-Stage Nitriding	γ'-Nitride in %	ε-Nitride in %
520 °C 15 h $K_N = 5$ 530 °C 170 h $K_N = 0.5$	ca. 88	ca. 12
530 °C 170 h $K_N = 0.5$ 520 °C 15 h $K_N = 5$	ca. 85	ca. 15
530 °C 170 h $K_N = 0.5$ 550 °C 15 h $K_N = 5$	ca. 88	ca. 12
530 °C 170 h $K_N = 0.5$ 550 °C 15 h $K_N = 5$ $K_C^B = 0.1$	ca. 60	ca. 40

In the two-stage treatment, in which the compound layer was formed during a nitrocarburizing stage at the end of the process, a higher proportion of ε-nitride was formed. This also corresponded approximately to the proportion of ε-nitride formed in the corresponding short-term treatment.

Therefore, from the results of the short-term treatments and the two-stage treatments it can be concluded that both the thickness of the compound layer and its phase composition in the two-stage processes were essentially determined by the process stage for compound layer formation. This opens up the possibility of influencing the compound layer formation during two-stage deep nitriding in a targeted manner.

4. Conclusions

The investigations have shown that during deep nitriding of the material 31CrMoV9 it is possible to control the nitriding layer structure through process control. In order to specifically influence the thickness and phase composition of the compound layer, a two-stage process control is suitable, in which the diffusion layer and thus the nitriding hardness depth is first adjusted in a long process stage with a comparatively low nitriding potential. Here it must be ensured that the nitriding potential is still sufficiently high enough to saturate the material with nitrogen and form a thin, closed compound layer, as otherwise diffusion is slowed down and uniformity would be negatively affected. In the second short process stage, the formation of the compound layer can be controlled by selecting the nitriding or nitrocarburizing parameters, since in such a process the thickness and phase composition of the compound layer are essentially dependent on the treatment parameters in the second process stage.

The two-stage nitriding treatments developed made it possible to produce compound layers comparable to those of the reference material 32CDV13. The question of whether a similar tooth root and flank load-bearing capacity can also be achieved with these nitrided layers is the subject of further investigations. Another subject of future investigations is the question, if the two-stage processing for a specific compound layer design is also suitable for other nitriding and tempering steels, since the current investigations are limited to the specific material 31CrMoV9.

Author Contributions: S.H.: project administration, conceptualization, investigation, methodology, writing—original draft preparation. M.S.: funding acquisition, supervision, writing—review and editing. H.-W.Z.: resources, writing—review and editing. All authors have read and agreed to the published version of the manuscript.

Funding: This IGF Project IGF 19594 N of the Forschungsvereinigung Antriebstechnike. V. is supported via AiF within the programme for promoting the Industrial Collective Research (IGF) of the German Ministry of Economic Affairs and Energy (BMWi), based on a resolution of the German Parliament. Supported by Federal Ministry of Economics and Energy by decision of the German Bundestag.

Conflicts of Interest: The authors declare no conflict of interest. The funders had no role in the design of the study, in the writing of the manuscript and in the decision to publish the results.

References

1. Conrado, E.; Gorla, C.; Davoli, P.; Boniardi, M. A comparison of bending fatigue strength of carburized and nitrided gears for industrial applications. *Eng. Fail. Anal.* **2017**, *78*, 41–54. [CrossRef]
2. König, J.; Hoja, S.; Tobie, T.; Hoffmann, F.; Stahl, K. In Increasing the Load Carrying Capacity of Highly Loaded Gears by Nitriding. In Proceedings of the MATEC Web of Conferences, Varna, Bulgaria, 19–22 June 2019. [CrossRef]
3. Niemann, G.; Winter, H. Band 2: Getriebe allgemein, Zahnradgetriebe-Grundlagen. In *Maschinenelemente*; Springer: Heidelberg/Berlin, Germany, 2003.
4. Hoja, S.; Hoffmann, F.; Zoch, H.W.; Schurer, S.; Tobie, T.; Stahl, K. Entwicklung von Prozessen zum Tiefnitrieren von Zahnrädern. *HTM J. Heat Treatm. Mat.* **2015**, *70*, 276–285. [CrossRef]
5. Hoja, S.; Hoffmann, F.; Steinbacher, M.; Zoch, H.W. Untersuchung des Anlasseffekts beim Nitrieren. *HTM J. Heat Treatm. Mat.* **2018**, *73*, 335–343. [CrossRef]
6. Bräutigam, F. Nitrieren im Ammoniakgasstrom: Ein Oberflächenhärte-Verfahren mit großer Zukunft. *Industrie Anzeiger* **1981**, *103*, 36–39.
7. Limodin, N.; Verreman, Y. Fatigue strength improvement of a 4140 steel by gas nitriding: Influence of notch severity. *Mater. Sci. Eng. A* **2006**, *435–436*, 460–467. [CrossRef]
8. Zlatanović, M.; Popović, N.; Mitrić, M. Plasma processing in carbon containing atmosphere for possible treatment of wind turbine components. *Thin Solid Films* **2007**, *516*, 228–232. [CrossRef]
9. Wu, K.; Liu, G.Q.; Wang, L.; Xu, B.F. Research on new rapid and deep plasma nitriding techniques of AISI 420 martensitic stainless steel. *Vacuum* **2010**, *84*, 870–875. [CrossRef]
10. Floe, C.F. A study of the nitriding process. Effect of ammonia dissociation on case depth and structure. *Trans. ASM* **1944**, *32*, 44–171.
11. Floe, C.F. Method of Nitriding. U.S. Patent No. 2,437,249, 17 April 1946.
12. Hassani-Gangaraj, S.M.; Moridi, A.; Guagliano, M.; Ghidini, A.; Boniardi, M. The effect of nitriding, severe shot peening and their combination on the fatigue behavior and micro-structure of a low-alloy steel. *Int. J. Fatigue* **2014**, *62*, 67–76. [CrossRef]
13. Bossy, E.; Noyel, J.P.; Kleber, X.; Ville, F.; Sidoroffd, C.; Thibault, S. Competition between surface and subsurface rolling contact fatigue failures of nitrided parts: A Dang Van approach. *Tribol. Int.* **2019**, *140*, 105888. [CrossRef]
14. Hoja, S.; Schurer, S.; Hoffmann, F.; Tobie, T. Tiefnitrieren von Zahnrädern. In *FVA-Nr. 615 II, FVA-Forschungsheft Nr. 1147*; Forschungsvereinigung Antriebstechnik e. V.: Frankfurt, Germany, 2015.
15. Ochoa, E.A.; Wisnivesky, D.; Minea, T.; Ganciu, M.; Tauziede, C.; Chapon, P.; Alvarez, F. Microstructure and properties of the compound layer obtained by pulsed plasma nitriding in steel gears. *Surf. Coat. Technol.* **2009**, *203*, 1457–1461. [CrossRef]
16. Zornek, B.; Hoja, S.; Tobie, T.; Hoffmann, F. Tribologische Tragfähigkeit nitrierter Innen- und Außenverzahnungen bei geringen Umfangsgeschwindigkeiten. In *FVA-Forschungsheft Nr. 1206*; Forschungsvereinigung Antriebstechnik e. V.: Frankfurt, Germany, 2017.
17. Hoffmann, F.; Bujak, I.; Mayr, P.; Löffelbein, B.; Gienau, M. Verschleißwiderstand nitrierter und nitrocarburierter Stähle. *HTM-Härterei-Techn. Mitt.* **1997**, *52*, 376–386.

18. Liedtke, D. Beitrag zum Technisch-Wirtschaftlichen Optimieren des Nitrocarburierens von Bauteilen. Ph.D. Thesis, TU Bergakademie Freiberg, Freiberg, Germany, 1986.
19. Binder, C.; Bendo, T.; Hammes, G.; Klein, A.N.; de Mello, J.D.B. Effect of nature of nitride phases on sliding wear of plasma nitrided sintered iron. *Wear* **2015**, *332–333*, 995–1005. [CrossRef]
20. Hollomon, J.H.; Jaffe, L.D. *Ferrous Metallurgical Design*; Wiley: Hoboken, NJ, USA, 1947.
21. Huchel, U.; Klümper-Westkamp, H.; Liedtke, D. Lichtmikroskopische Bestimmung der Dicke und Porigkeit der Verbindungsschichten nitrierter und nitrocarburierter Werkstücke. In *Prüfvorschrift des AWT-FA3*; Arbeitsgemeinschaft Wärmebehandlung + Werkstofftechnik e. V.: Bremen, Germany, 2008.
22. Hoffmann, R.; Mittemeijer, E.J.; Somers, M.A.J. Verbindungsschichtbildung beim Nitrieren und Nitrocarburieren. *HTM-Härterei-Techn. Mitt.* **1996**, *51*, 162–169.

© 2020 by the authors. Licensee MDPI, Basel, Switzerland. This article is an open access article distributed under the terms and conditions of the Creative Commons Attribution (CC BY) license (http://creativecommons.org/licenses/by/4.0/).

MDPI
St. Alban-Anlage 66
4052 Basel
Switzerland
Tel. +41 61 683 77 34
Fax +41 61 302 89 18
www.mdpi.com

Metals Editorial Office
E-mail: metals@mdpi.com
www.mdpi.com/journal/metals

www.ingramcontent.com/pod-product-compliance
Lightning Source LLC
LaVergne TN
LVHW070427100526
838202LV00014B/1542